STATE OF THE
WORLD
1990

Other Norton/Worldwatch Books

Lester R. Brown et al.

State of the World 1984

State of the World 1985

State of the World 1986

State of the World 1987

State of the World 1988

State of the World 1989

STATE OF THE WORLD

1990

A Worldwatch Institute Report on Progress Toward a Sustainable Society

PROJECT DIRECTOR
Lester R. Brown

ASSOCIATE PROJECT
DIRECTORS
Christopher Flavin
Sandra Postel

EDITOR
Linda Starke

SENIOR RESEARCHERS
Lester R. Brown
Alan B. Durning
Christopher Flavin
Jodi Jacobson
Sandra Postel
Michael Renner

RESEARCHERS
Hilary F. French
Marcia D. Lowe
John E. Young

W·W·NORTON & COMPANY

NEW YORK   LONDON

Printed in the United States of America

ISBN 0-393-20788-0

ISBN 0-393-30614-3 {PBK.}

W. W. Norton & Company, Inc., 500 Fifth Avenue, New York, N.Y. 10110
W. W. Norton & Company Ltd., 37 Great Russell Street, London WC1B 3NU

3 4 5 6 7 8 9 0

Acknowledgments

State of the World has become something of an institution. Each year it is used by governments and citizens throughout the world as a sort of touchstone of progress in the search for a sustainable course on the planet. But its growing stature makes its production none the easier.

Dozens of people labored in one way or another to make possible this year's edition. Our exceptional crew of research assistants undertakes the Sisyphean task of tracking developments in dozens of fields simultaneously. For their enthusiasm and fresh insight we thank Holly Brough (Chapters 8 and 9), Meri McCoy-Thompson (Chapter 5), and John Ryan (Chapters 3 and 10). For their exceptional endurance and seasoned instincts, we thank Nicholas Lenssen (Chapters 2 and 10) and John Young (Chapters 1 and 10).

The *State of the World* series has benefited from the unflagging support of a core group of funders from its inception: principally, The Rockefeller Brothers Fund and Winthrop Rockefeller Trust. In addition, the report draws on research funded by several foundations, including the Geraldine R. Dodge, George Gund, William and Flora Hewlett, W. Alton Jones, William D. and Catherine T. MacArthur, Andrew W. Mellon, Curtis and Edith Munson, Edward John Noble, Jessie Smith Noyes, Public Welfare, and Rockefeller foundations. Additional research funding was provided by the United Nations Population Fund and, through the Sasakawa Environment Prize, the United Nations Environment Programme.

Chapters of this year's edition were reviewed on a moment's notice by a host of scholars outside the institute. Among them were James Broadus, Chris Calwell, Robert Chambers, Andy Clarke, Mohamed El-Ashry, Malin Falkenmark, Jonathan Feldman, Ellen Fletcher, Frank Gable, Graham Giese, Peter Gleick, Ralph Hirsch, Ken Hughes, Charles Komanoff, Michael Lipton, Philip Micklin, John Milliman, Setty Pendakur, Paul Quigley, W. Robert Rangeley, Michael Replogle, Peter Sand, Mark Svendsen, and Fiona Weir. Our thanks go to them all.

While the research staff labored on preparing *State of the World 1990*, our capable administrative and financial staff, skillfully guided by Vice President Blondeen Gravely, kept sales of *State of the World 1989* flowing briskly and managed the massive daily tide of letters and telephone calls. Linda Doherty (in-house recycling coordinator) saw to it that tens of thousands of publications orders coming in were quickly filled. Guy Gorman distributed millions of facts among the researchers as he maintained our ever expanding library. And Reah Janise Kauffman deftly managed our growing computer network, all the while shouldering the heavy burden of assisting the Institute's President. These Worldwatch veterans were joined by gifted newcomers Joseph Gravely, Steve Kaufman, and Barbara Granzen along with part-timer Jennifer Hamilton.

Director of Communications Stephen Dujack, assisted ably by Denise Byers, edited the Worldwatch Paper series, oversaw a hectic mailing schedule, and facilitated relations with publishers, reporters, and citizens on every continent (except Antarctica). James Gorman, aided by Howard Youth (in-house ornithologist), carried the banner of our award-winning magazine *World Watch* during the onslaught of *State of the World* production.

Some long-standing members of our team severed ties with the institute this year to pursue other ambitions. Among them was Susan Norris, who retreated from urban life to the farm country of southern Virginia. Cynthia Pollock Shea, now in Calgary, Canada, divides her time between motherhood and Worldwatch, and Lori Heise is on a year's sabbatical doing health work with Indian women in Guatemala.

We owe a debt of gratitude to Iva Ashner and Andy Marasia at W.W. Norton & Company in New York, who accommodate our idiosyncracies without complaint, and to Bart Brown, indexer par excellence, for his speedy work with our page proofs.

The cycle of birth and death has been deeply felt in our small community this year. Joyously, we welcomed a daughter to Cynthia Pollock Shea and a son to Michael Renner. Our youngest research assistants have already been put to work proving that disposable diapers are no better than cloth.

Sadly, we bid farewell to Felix Gorrell, treasurer since Worldwatch's very inception in 1974. More than just our treasurer, he was initially a midwife, helping bring the Institute into the world, and then a godfather, always there with advice when we needed it. Over the years, he ceaselessly went beyond the call of duty to see that our resources were secure, well invested, and carefully monitored. This volume is dedicated to his memory; in it, we aspire to his unique blend of vision and pragmatism.

Lester R. Brown, Christopher Flavin,
and Sandra Postel

Contents

List of Tables and Figures

LIST OF TABLES

LIST OF FIGURES

Foreword

During 1989, politically astute national leaders sensed the mounting public concern over the future of the planet, and moved quickly to catch up. Some, such as U.K. Prime Minister Thatcher and French President Mitterand, hosted conferences. Those assembled at the Group of Seven economic summit in July issued a communiqué. All in all, these initiatives were long on rhetoric and short on action.

In the short run, political leaders can get by with speeches and professions of concern. But over the longer run, building an environmentally sustainable economy requires specific steps and tough choices. People everywhere are worried about the environmental degradation of the planet. They want action. This tide of concern may mark the beginning of an environmental decade, a time when we will be increasingly preoccupied with trying to restore the earth's health.

The trends of environmental degradation described in the previous six volumes in this annual series all continue unabated: Forests are shrinking, deserts expanding, and soils eroding. The depletion of the stratospheric ozone layer that protects us from harmful ultraviolet radiation appears to have escalated. The levels of carbon dioxide and other heat-trapping gases in the atmosphere continue to build in an all too predictable fashion.

Only a monumental effort can reverse the deterioration of the planet. As the East-West ideological conflict wanes, it will free the time and energy of political leaders to concentrate on environmental threats to security. It will also facilitate a reordering of priorities, providing resources to reforest the earth, stop population growth, and develop energy sources that will stabilize the earth's climate.

Within this environmental decade, 1990 promises to be the Year of the Environment. Earth Day 1990, to be celebrated on April 22nd, will mark the twentieth anniversary of the first Earth Day and provide an organizing focus in the spring for environmental education and policy initiatives of all kinds. The airing of a major series for public television based on the *State of the World* reports will provide a focal point in the fall. Entitled "Race to Save the Planet," this series, produced by the Nova unit at public television station WGBH in Boston, will provide a global overview of what it will take to reverse the degradation of the planet. Given the intense interest in these issues, we expect the series will be seen throughout the world.

Increasingly, efforts to deal with environmental issues are becoming international. Yet the United Nations does not produce an integrated analysis of the global interaction of energy, environment, food, population, and economic trends of the sort contained in the annual Worldwatch volume. Nor does anyone else. In the absence of such an analysis, *State of the World* has become a basic resource for policymakers and for the thousands of environmental action groups that have sprung up all over the world.

From the outset, we at the Institute have seen our constituency as a global one. With the addition of French and Portuguese, *State of the World* is now published in all the world's major languages: Spanish, Arabic, Chinese, Japanese, Indonesian, German, Italian, Polish, French, Portuguese, and Russian, in addition to English. It also is slated for publication in several less widely spoken languages, including Norwegian, Swedish, Dutch, Hungarian, and Korean.

The market for integrated or interdisciplinary research of the sort that characterizes Worldwatch publications is growing by leaps and bounds. This is evident in sales of the Worldwatch Paper series and in subscriptions for *World Watch* magazine, launched at the beginning of 1988. Income from publication sales and royalties reached $1.6 million and covered 60 percent of the Institute's budget for 1989.

Distribution of *State of the World* in all languages is climbing, with English the first to break the 100,000 mark. Although sales go largely to policymakers and concerned individuals, the latest volume was adopted for use in more than 800 courses in some 600 U.S. colleges and universities. Its use as a supplemental text has climbed rapidly—our contribution to the environmental literacy of the next generation.

The goal of the Institute is to help raise the level of public understanding of global environmental threats to the point where it will support policies needed to reverse these trends. Our global public education effort is making progress, but we have a long way to go.

As the new decade begins, the world has a fresh opportunity to reorder priorities, focusing on the real threats to our common future. Just as no one could imagine the rate at which the waves of reform swept across Eastern Europe in late 1989, so too may we be surprised by the pace of environmental reform in the early nineties as we mobilize to save the planet.

Lester R. Brown
Project Director

Christopher Flavin
Sandra Postel
Associate Project Directors

Worldwatch Institute
1776 Massachusetts Ave., N.W.
Washington, D.C. 20036

December 1989

STATE OF THE
WORLD
1990

1

The Illusion of Progress

Lester R. Brown

For most of the nearly four fifths of humanity born since World War II, life has seemed to be a period of virtually uninterrupted economic progress. Since mid-century, the global economic product has nearly quintupled. On average, the additional economic output in each of the last four decades has matched that added from the beginning of civilization until 1950.[1]

World food output during this period also grew at a record pace. Soaring demand fueled by population growth and rising affluence provided the incentive, and modern technology the means, to multiply the world's grain harvest 2.6 times since mid-century. No other generation has witnessed gains even remotely approaching this.[2]

Such gains would seem to be a cause for celebration, but instead there is a sense of illusion, a feeling that they overstate progress. The system of national accounting used to measure economic progress incorporates the depreciation of plant and equipment, but not the depletion of natural capital. Since mid-century, the world has lost nearly one fifth of the topsoil from its cropland, a fifth of its tropical rain forests,

Units of measurement are metric unless common usage dictates otherwise.

and tens of thousands of its plant and animal species.[3]

During this same period, atmospheric carbon dioxide (CO_2) levels have increased by 13 percent, setting the stage for hotter summers. The protective ozone layer in the stratosphere has been depleted by 2 percent worldwide and far more over Antarctica. Dead lakes and dying forests have become a natural accompaniment of industrialization. Historians in the twenty-first century may marvel at this economic performance—and sorrow over its environmental consequences.[4]

Throughout our lifetimes, economic trends have shaped environmental trends, often altering the earth's natural resources and systems in ways not obvious at the time. Now, as we enter the nineties, the reverse is also beginning to happen: environmental trends are beginning to shape economic trends.

The environmental degradation of the planet is starting to show up at harvest time. The cumulative effects of losing 24 billion tons of topsoil each year are being felt in some of the world's major food-producing regions. Recent evidence indicates that air pollution is damaging crops in both auto-centered economies of the West and coal-burning economies of the East. Meteorologists cannot yet be certain, but the hotter summers and drought-reduced harvests of the eighties may be early indications of the greenhouse effect.[5]

Environmental degradation undoubtedly contributed to the slower growth in world grain output during the eighties. The 2.6-fold gain in world grain output just mentioned occurred between 1950 and 1984; since then, there has been no appreciable increase. The 1989 estimated harvest (1.67 billion tons) was up only 1 percent from that of 1984, which means that grain output per person is down nearly 7 percent. Some two thirds of this fall in production has been offset by drawing down stocks, reducing them to a precariously low level; the remainder, by reducing consumption. Although five years is obviously not enough time to signify a long-term trend, it does show that the world's farmers are finding it more difficult to keep up with growth in population.[6]

The world's farmers are finding it more difficult to keep up with growth in population.

Nowhere is this more clear than in Africa, where the combination of record population growth and widespread land degradation is reducing grain production per person. A drop of 20 percent from the peak in 1967 has converted the continent into a grain importer, fueled the region's mounting external debt, and left millions of Africans hungry and physically weakened, drained of their vitality and productivity. In a 1989 report sketching out several scenarios for this beleaguered continent, World Bank analysts termed the simple extrapolation of recent trends the "nightmare scenario."[7]

In both Africa and Latin America,

food consumption per person is lower today than it was when the decade began. Infant mortality rates—a sensitive indicator of nutritional stress—appear to have turned upward in many countries in Africa and Latin America, reversing a long-term historical trend. Nations in which there are enough data to document this rise include Brazil, the Dominican Republic, El Salvador, Ghana, Madagascar, Mexico, Peru, Uruguay, and Zambia.[8]

Environmental degradation is affecting more than economic and social trends: 1989 was the year environmental issues moved into the political mainstream. In Western Europe, environmentalists won resounding gains in legislative races. And environmental issues moved to the forefront of political debate in Poland, the Soviet Union, Japan, and Australia. Unfortunately, rising political awareness has not yet translated into policies that will reverse the deteriorating situation.[9]

On the two most important fronts in the race to save the planet—stopping population growth and stabilizing climate—the world is losing ground. Some progress has been made in slowing the rate of population growth since 1970, but the decline has been so gradual that the annual increment grows larger each year. During the eighties, world population increased by 842 million, an average of 84 million a year. (See Table 1–1.) During the next 10 years it is projected to grow by 959 million, the largest increment ever for a single decade. As the annual excess of births over deaths continues to widen, the date of population stability is pushed ever further into the future.[10]

Progress in stabilizing climate is equally disappointing. (See Chapter 2.) Carbon emissions from fossil fuel use declined for several years as countries invested heavily in energy efficiency measures. But in the last few years they

Table 1-1. World Population Growth by Decade, 1950–90, With Projection to 2000

Year	Population	Increase by Decade	Average Annual Increase
	(billion)	(million)	
1950	2.515		
1960	3.019	504	50
1970	3.698	679	68
1980	4.450	752	75
1990	5.292	842	84
2000	6.251	959	96

SOURCE: United Nations, Department of International Economic and Social Affairs, *World Population Prospects 1988* (New York: 1989).

have started to rise again. Leading industrial economies, such as the United States and Japan, are the primary contributors to this unfortunate global upturn. In 1987, global carbon emissions from fossil fuels rose 1.5 percent and in 1988, 3.7 percent, reaching a record total of 5.7 billion tons.[11]

Reading the daily newspapers gives the impression that changes in economic indicators such as the gross national product (GNP), interest rates, or stock prices are the keys to the future. But it is changes in the biological product that are shaping civilization. It is changes in the size of the photosynthetic product that determine ultimately how many of us the earth can support and at what level of consumption.

THE EARTH'S DECLINING PRODUCTIVITY

Three biological systems—croplands, forests, and grasslands—support the world economy. Except for fossil fuels and minerals, they supply all the raw materials for industry; except for seafood, they provide all our food. Forests are the source of fuel, lumber, paper, and numerous other products. Grasslands provide meat, milk, leather, and wool. Croplands supply food, feed, and an endless array of raw materials for industry such as fiber and vegetable oils.

Common to all these biological systems is the process of photosynthesis, the ability of plants to use solar energy to combine water and carbon dioxide to produce carbohydrates. Although an estimated 41 percent of photosynthetic activity takes place in the oceans, it is the 59 percent occurring on land that underpins the world economy. And it is the loss of terrestrial photosynthesis as a result of environmental degradation that is undermining many national economies.[12]

The biological activity that supplies the bulk of our food and raw materials takes place on the nearly one third of the earth's surface that is land, some 13 billion hectares. According to a U.N. Food and Agriculture Organization tabulation for 1986, 11 percent of this—nearly 1.5 billion hectares—is used to produce crops. Roughly 25 percent is pasture or rangeland, providing grass or other forage for domesticated livestock and wild herbivores. A somewhat larger area (31 percent) is in forests, including open forests or savannahs only partly covered with trees. The remaining 33 percent of the world's land supports little biological activity. It is either wasteland, essentially desert, or has been paved over or built on.[13]

The share of land planted to crops increased from the time agriculture began until 1981, but since then the area of newly reclaimed land has been offset by that lost to degradation and converted to nonfarm uses. The grassland area has shrunk since the mid-seventies, as overgrazing slowly converts it to desert. The forested area has been shrinking for centuries, but the losses accelerated at mid-century and even more from 1980 onward. The combined area of these three biologically productive categories is shrinking while the remaining categories—wasteland and that covered by human settlements—are expanding.[14]

Not only is the biologically productive land area shrinking, but on part of it productivity is falling. In forests, for example, output is being lowered on some remaining stands, apparently by air pollution and acid rain. (See Chapter 6.) Evidence of this damage in industrial countries is now widespread. In the United States, it can be found throughout much of the country, and in Europe it stretches from the Atlantic coast in the west to the remote reaches of Siberia in the east.[15]

Even an experienced forester often cannot see any changes in the trees that would indicate slower growth; only careful measurements over time show how much pollutants are stressing trees. A Forest Inventory Analysis conducted regularly by the U.S. Forest Service reports that the annual growth of yellow pines, a major species covering some 42 million hectares in the Southeast, declined by 30–50 percent between 1955 and 1985. From 1975 to 1985, the dead pines increased from 9 percent of all trees to 15 percent. Soviet foresters report a decline in tree growth rates in central Siberia over the last few decades that is remarkably similar.[16]

In forests, output is being lowered, apparently by air pollution and acid rain.

While forest productivity is being diminished by chemical stress, that of grasslands is being reduced by the physical stress of overgrazing. Widespread grassland degradation can now be seen on every continent. Although the data for grassland degradation are even more sketchy than for forest clearing, the trends are no less real. This problem is highly visible throughout Africa, where livestock numbers have expanded nearly as fast as the human population. In 1950, 238 million Africans relied on 272 million livestock. By 1987, the human population had increased to 604 million, and the livestock to 543 million.[17]

In a continent where grain is scarce, 183 million cattle, 197 million sheep, and 163 million goats are supported almost entirely by grazing and browsing. Everywhere outside the tsetse-fly belt, livestock are vital to the economy, but in many countries their numbers exceed grassland carrying capacity by half or more. A study charting the mounting pressures on grasslands in nine southern African countries found that the capacity to sustain livestock is diminishing. As grasslands deteriorate, soil erosion accelerates, further reducing the carrying capacity and setting in motion a self-reinforcing cycle of ecological degradation and deepening human poverty.[18]

Fodder needs of livestock in nearly all developing countries now exceed the sustainable yield of grasslands and other forage resources. In India, the demand by the end of the decade is expected to reach 700 million tons, while the supply will total just 540 million. The National Land Use and Wastelands Development Council there reports that in states with the most serious land degradation, such as Rajasthan and Karnataka, fodder supplies satisfy only 50–80 percent of needs, leaving large numbers of emaciated cattle. When drought occurs, hundreds of thousands of these animals die. In recent years, local governments in India have established fodder relief camps for cattle threatened with starvation, much as food relief camps are set up for people similarly threatened.[19]

Overgrazing is not limited to the Third World. In the United States, where the Bureau of Land Management (BLM) is responsible for 66 million hectares of government-owned grazing land, overgrazing is commonplace. A 1987 survey found that only 33 percent of the BLM's rangeland was in good to excellent condition; 58 percent was fair to poor. (See Table 1–2.)

As the deterioration of grazing lands continues, some of it eventually becomes wasteland, converted to desert by the excessive demands of growing livestock populations. And as the forage available to support animals diminishes, pressure shifts to croplands to produce more grain to feed livestock, thus intensifying the competition between humans and animals for scarce food supplies.

The loss of productive woodland, grassland, and cropland to nonfarm uses is also progressing on every continent, though at varying rates. Each year, millions of hectares of biologically productive land are paved over or built upon. Growth in the world's automobile fleet, though it has slowed dramatically over the last decade, is nonetheless leading to the paving of more and more of the earth's surface with streets, roads, and parking spaces. Each car added to the world fleet competes with farmers.[20]

Stanford University biologist Peter M. Vitousek and his colleagues estimate that humans now appropriate close to 40 percent of the land's net primary biological product. In other words, nearly 40 percent of the earth's land-based photosynthetic activity is devoted to the satisfaction of human needs or has been lost as a result of human degradation of natural systems. As our own share continues to increase, it becomes more difficult for other species to survive. Eventually, life-supporting systems could begin to unravel.[21]

To summarize, at a time when demand for various biological products is rising rapidly, the earth's biological production is shrinking. The even greater annual additions to world population in prospect for the nineties will further reduce the earth's ability to supply our food and raw materials. These two trends cannot continue indefinitely. At some point, the continuing decline in the photosynthetic product will translate into a decline in the economic product.

RECALCULATING ECONOMIC PROGRESS

Looking at the basic biological systems just discussed, the world is not doing very well. Yet key economic indicators show the world is prospering. Despite a slow start at the beginning of the eighties, global economic output expanded by more than a fifth during the decade. The economy grew, trade increased, and millions of new jobs were created. How can basic biological indicators be so bearish and economic indicators so bullish at the same time?[22]

The answer is that the economic indicators are flawed in a fundamental way: they do not distinguish between resource uses that sustain progress and those that undermine it. The principal

Table 1-2. Condition of Bureau of Land Management Grazed Land, 1987

Condition	Percent[1]
Excellent	3
Good	30
Fair	39
Poor	19

[1]Column totals 91 because the condition of 9 percent of land was not reported.

SOURCE: U.S. Department of the Interior, Bureau of Land Management, *Public Land Statistics, 1987* (Washington, D.C.: 1988).

measure of economic progress is the gross national product. In simple terms, this totals the value of all goods and services produced and subtracts depreciation of capital assets. Developed a half-century ago, GNP accounts helped establish a common means among countries of measuring changes in economic output over time. For some time, this seemed to work reasonably well, but serious weaknesses are now surfacing. As noted earlier, GNP includes depreciation of plant and equipment, but it does not take into account the depreciation of natural capital, including nonrenewable resources such as oil or renewable resources such as forests.[23]

This shortcoming can produce a misleading sense of national economic health. According to the conventional approach, for example, countries that overcut forests actually do better in the short run than those that manage forests on a sustained-yield basis: the trees cut down are counted as income but no subtraction is made to account for depletion of the forest, a natural asset. The advantage is short-lived, however, as overcutting eventually destroys the resource base entirely, leading to a collapse of the forest products industry.[24]

To illustrate the flaws in current GNP accounting, economist Robert Repetto and his colleagues at the World Resources Institute recalculated the GNP of Indonesia, incorporating the depletion of natural capital. Considering only oil depletion, soil erosion, and deforestation, he showed that Indonesia's economic growth rate from 1971 to 1984, originally reported at 7 percent, was in reality only 4 percent. The conventional system not only sometimes overstates progress, it may indicate progress when there is actually decline. In Repetto's revised system of national economic accounting, natural capital depletion gets a line entry just as depreciation of plant and equipment does.[25]

Including changes in the stock of natural capital represents a major advance in national economic accounting, but if this system is to be a basis for policymaking in an era when environmental issues loom large, it will have to go one step further and incorporate the environmental effects of economic activity. For instance, the deforestation that led to a net loss in Indonesia's natural capital also contributed to the buildup of CO_2 around the world, thus hastening global warming. How much will it cost to cope with the share of climate change due to deforestation in Indonesia?

Or consider the oil produced in Indonesia, which Repetto incorporated as a net reduction in the country's natural capital. To what extent is it contributing to the serious air pollution problem in Jakarta and to respiratory illnesses among the residents? How much is the Indonesian oil burned in the Netherlands contributing to the air pollution and acid rain destroying lakes in Scandinavia and forests in West Germany? It is certainly true that data on the costs of lost forest productivity in Europe or of global warming are not very good. But is that a good reason to ignore them entirely rather than try to make some estimates, however crude they are, and incorporate them into the national economic accounts? The consequences are so profoundly important that it would be better to include even the roughest of estimates.

Another way to grasp the importance of natural capital depreciation would be to look at a particular sector of the world economy, such as food, and subtract from national accounts the value of output that is produced unsustainably. This would also help determine how much of our consumption is at the expense of future generations. Grain is currently produced, for example, by cultivating highly erodible land that will eventually become wasteland or by intensifying farming in ways that lead to excessive soil erosion and cropland loss.[26]

In the United States, an estimated 13 million hectares of cropland was losing topsoil so rapidly that Congress has provided for its conversion to grassland or woodland before it becomes wasteland. If this were all Great Plains wheatland, it would produce roughly 2.5 tons of grain per hectare, for a total of 33 million tons. Since a minor share is higher yielding midwestern cornland, this is a lower bound estimate for unsustainable grain output.[27]

In addition, one fifth of the 20 million hectares of U.S. irrigated land is being watered by drawing down water tables by 15–152 centimeters (6 inches to 5 feet) per year. If, for purposes of calculation, it is assumed that water tables are eventually stabilized under 2 million hectares of this 4 million hectare total by increasing water use efficiency and that the other 2 million hectares reverts to dryland farming, grain output on the latter would be reduced by perhaps 4 tons per hectare or 8 million tons. Thus the 41 million tons of grain produced with the unsustainable use of land and water in the United States alone would offset most of the excess production capacity in world agriculture.[28]

If adjustment were made for all grain produced with the unsustainable use of land and water worldwide, it would show a grain output well below consumption and provide a much bleaker sense of global food security. When the unsustainable use of land and water is eventually abandoned, it will dramatically tighten world food supplies, pushing prices upward.

With the existing economic accounting system, those who overplow and overpump appear to be doing very well in the short run, even while facing a disastrous collapse over the long run. Although the loss of topsoil does not show up in the national economic accounts or resource inventories of most countries, it is nonetheless serious. And it is largely unrecognized, since the intensification of cropping patterns and the plowing of marginal lands that lead to excessive erosion over the long run can lead to production gains in the short run, thus creating the illusion of progress and a false sense of food security.

The loss of topsoil does not show up in national economic accounts or resource inventories.

If all the environmental consequences of economic activity—from resource depletion to the numerous forms of environmental damage—were included, real economic progress would be much less than conventional economic measures indicate. The challenge to governments is to revise national accounting systems so as to reflect more precisely real changes in output. Just as we deflate reported economic growth with a price deflator, we must also apply an ecological deflator if we are to measure real progress. Without this, we will continue to delude ourselves into thinking we are making progress when we are not.

A few governments and international organizations are starting to move in this direction. Laws enacted by the U.S. Congress in 1989 require the federal government to calculate a "gross sustainable productivity" for the United States each year in conjunction with the annual GNP figures, and to work with international organizations and agencies to revise national accounting systems. Further, the United Nations is considering a revision of its own system.[29]

THE BOTTOM LINE

We know that we cannot continue to damage our life-support systems without eventually paying a price, but how will we be affected? What will the price be? Is it likely to be a buildup of carcinogens in the environment so severe it increases the incidence of cancer, dramatically raising death rates? Or will the rising concentration of greenhouse gases make some regions of the planet so hot that they become uninhabitable, forcing massive human migrations? Or will it be something we cannot even anticipate yet?

We also know we cannot keep adding ever more people to the planet each 12 months. Adding 88 million people a year—the equivalent of the population of the United Kingdom combined with those of Belgium, Denmark, Ireland, Norway, and Sweden—will eventually get us into trouble. What form will that trouble take? And is it imminent or in the distant future?[30]

Amidst the uncertainty, food scarcity in developing countries is emerging as the most profound and immediate consequence of global environmental degradation, one already affecting the welfare of hundreds of millions. All the principal changes in the earth's physical condition—eroding soils, shrinking forests, deteriorating rangelands, expanding deserts, acid rain, stratospheric ozone depletion, the buildup of greenhouse gases, air pollution, and the loss of biological diversity—are affecting food production negatively. Deteriorating diets in both Africa and Latin America during the eighties, a worldwide fall in per capita grain production since 1984, and the rise in world wheat and rice prices over the last two years may be early signs of the trouble that lies ahead.[31]

After a generation of record growth in world food output, it sometimes seemed that the rapid ascent in production of this essential commodity could continue indefinitely. The 2.6-fold increase in the world grain harvest between 1950 and 1984 (nearly 3 percent a year) raised per capita production by more than one third. But between 1984 and 1989, overall output rose by only 1 percent. (See Figure 1–1.) This five-year period is too short to show a trend because weather fluctuations could be partly responsible for the slowdown, but it is an unsettling interruption in food output growth.[32]

The downturn in grain production per capita that became firmly established in Africa during the seventies and spread to Latin America during the early eighties thus engulfed the entire world in the late eighties. In 1989, higher grain prices to farmers, the return to production of idled U.S. cropland, and near-normal weather were expected to restore production and permit a rebuilding of depleted grain stocks. But this did not happen. With carryover stocks now amounting to little more than pipeline supplies, there is little cushion for short harvests in the future. Falls in production will translate directly into falls in consumption.[33]

Why has production not increased during this five-year period, a time in

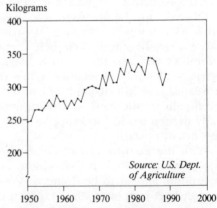

Figure 1-1. World Grain Production Per Capita, 1950–89

which the world's farmers invested billions of dollars to expand output, and in which fertilizer use increased by 18 million tons, a gain of 14 percent? There is no single explanation. Three historical trends are converging to make it more difficult to expand world food output. One is the growing scarcity of new cropland and fresh water that affects most of the world. The second is the lack of any new technologies, such as hybrid corn or chemical fertilizer, that can dramatically boost output. And the third is the negative effects of planetary environmental degradation on food production. Any one of these trends could slow the growth in food output. But the convergence of all three could alter the food prospect for the nineties in a way the world is not prepared for.[34]

The sad reality is that the contributions to increased output of greater fertilizer use, continuing modest expansion in irrigated area, and other technological advances are being partly offset by soil erosion, the waterlogging and salinity of irrigated cropland, air pollution, and several other forms of environmental degradation. (See Chapter 4 for a more detailed analysis of these environmental effects.)

The first concrete economic indication of broad-based environmental deterioration now seems likely to be rising grain prices. After reaching an all-time low of $224 per ton in 1986 (in 1988 dollars), rice prices rose to an estimated $309 for 1989, an increase of 38 percent. Wheat prices, which bottomed in 1987 at $117 per ton, rose 48 percent, to an estimated $173 per ton in 1989. (See Figure 1–2.) If world grain production should surge upward in the nineties, this price rise will be checked. If, on the other hand, the world's farmers still have difficulty expanding production at the 28 million tons a year required merely to maintain current per capita consump-

Figure 1-2. World Wheat and Rice Prices, 1950–89

tion levels, prices could rise dramatically—as they did in the mid-seventies.[35]

The combination of falling incomes in most of Africa and Latin America and the grain price rises of the last few years has forced consumption below the level of survival for millions. An estimated 40,000 infants die each day in the Third World as a result of severe nutritional stress and infectious disease. Tragically, higher grain prices also mean that food assistance, which is determined by budgetary allocations, is actually falling just as chronic food scarcity becomes a part of the landscape in much of the Third World.[36]

If before stocks are rebuilt the United States experiences a drought-reduced harvest, grain exports will slow to a trickle.

The world enters the nineties not only with a low level of grain in reserve, but with little confidence that the "carryover" stocks can be rebuilt quickly. Sketching out the consequences of a poor harvest while stocks are this low sounds like the outline of a science fiction catastrophe novel. If before stocks are rebuilt the United States experiences a drought-reduced harvest similar to that of 1988, which dropped grain production below domestic consumption, U.S. grain exports will slow to a trickle. Nearly everything produced would be needed to satisfy domestic consumption. By early fall, the more than 100 countries that import U.S. grain would be competing for meager exportable supplies from Argentina, Australia, and France. Fierce competition could double or triple grain prices, driving them far above any level previously experienced.[37]

By early winter, the extent of starvation, food riots, and political instability in the Third World would force governments in affluent industrial societies to consider tapping the only remaining food reserve of any size—the 600 million tons of grain fed to livestock. If they decided to restrict livestock feeding and use the grain saved for food relief, governments would have to devise a mechanism for doing so. Would they impose a meat tax to discourage consumption and free up grain? Would they ration livestock products, much as was done in many countries during World War II?[38]

Forcing millions of the world's poor to the brink of starvation would be the most immediate and tragic consequence of such a food emergency, but the international monetary system would also be in jeopardy. Third World governments desperately trying to import enough high-priced grain to avoid widespread starvation would have little foreign exchange for debt payments. Whether all the major international banks could withstand such a wholesale forfeiture of income is problematic. At this point, the effects of food scarcity would spread beyond the Third World hungry, driving interest rates upward and threatening the integrity of leading international fi-

nancial institutions. The linkage between the environmental degradation of the planet and the economic prospect would have become all too apparent.

The number of hungry people in the world has increased dramatically during the eighties. Reversing the spread of hunger will depend on a massive effort to cut the world rate of population growth and restore the health of the planet. We can no longer separate our health from that of our home. If the health of the planet continues to deteriorate, so will that of its inhabitants.[39]

In many ways, changes in the amount of grain that can be sustainably produced per person may be the best single indicator of success in the battle to save the planet. If an upward trend can be restored in the years immediately ahead—one that reduces hunger and malnutrition—it will be a clear sign of victory. If, on the other hand, per capita grain supplies continue to decline, leading the world into an era of food scarcity, rising food prices, and political instability, it will be a sure sign that the battle is being lost.

A POLITICAL AWAKENING

Scenes of people and nations competing for scarce grain supplies after another drought thankfully did not make the front pages last year. But the news of the planet's health was dire enough to keep the environment on magazine covers and in television broadcasts day after day. And it finally brought the issue into the political mainstream. Concern over the future of the planet pushed the environment toward center stage in political structures at all levels, from town councils to the U.N. General Assembly. International diplomacy, national political campaigns, and grassroots political activity are increasingly being shaped by environmental issues. In every corner of the world, environmentalism is on the rise.

Initially, international environmental issues were confined to local transboundary concerns, such as discussions on acid rain between the United States and Canada or the pollution of internationally shared rivers like the Rhine in Europe. More recently, genuinely global issues such as ozone depletion, climate change, and the preservation of biological diversity have attracted the attention of national political leaders and the international community. Environmental security issues now share the stage with more traditional economic and military concerns, inaugurating a new age of environmental diplomacy.

Political leaders organized several important meetings in 1989. In early March, Prime Minister Thatcher—impressed by new scientific evidence on depletion of the ozone layer—hosted a three-day international conference on the subject. Specifically, her aim was to convince more countries to support the Montreal Protocol goal of halving the use of chlorofluorocarbons (CFCs), the family of chemicals responsible for the damage, by the end of the nineties. By the close of the London conference, 20 more countries had agreed to sign the protocol, bringing the total to 66. And 11 more indicated they were seriously considering doing the same.[40]

A meeting in the Hague the following week, organized by the governments of France, the Netherlands, and Norway, considered the challenges to international governance posed by the twin problems of global warming and ozone depletion. Political representatives of 24 countries, including 17 heads of government, attended the conference. The three-page Hague Declaration they agreed to began: "The right to live is the right from which all other rights stem.

Guaranteeing this right is the paramount duty of those in charge of all States throughout the world. Today, the very conditions of life on our planet are threatened by the severe attacks to which the earth's atmosphere is subjected."[41]

Environmental security issues now share the stage with more traditional economic and military concerns.

This tone set the stage for a statement in which the international community went the furthest ever in recommending that the United Nations be given enforcement powers. It argued for a new or newly strengthened U.N. institutional authority to deal with global warming and ozone depletion that would "involve such decision-making procedures as may be effective even if, on occasion, unanimous agreement has not been achieved." And it discussed the need to "develop instruments and define standards to enhance or guarantee the protection of the atmosphere and monitor compliance," questions on which would be referred to the International Court of Justice.

Though it remains unclear when—if ever—the Hague Declaration's call will be heeded, it is becoming increasingly clear that some reform of the current international environmental machinery is on the horizon. It is widely expected that institutional reform will figure prominently on the agenda of the 1992 World Conference on Environment and Development, on the twentieth anniversary of the Stockholm meeting. The various proposals on the table at the moment include a British one to give the existing U.N. Security Council an environmental role, a Soviet suggestion to establish an ecological council separate from but equal in stature to the present Security Council, a private proposal to transform the moribund Trusteeship Council into an environmental body, and near-universal support for a strengthened U.N. Environment Programme.[42]

In late March 1989, 116 countries met in Basel to negotiate a treaty on the international transport and disposal of hazardous waste. The result was a compromise between industrial countries, which wanted to retain the right to export toxic materials under controlled conditions, and some developing countries that wanted to ban international movements of such materials altogether. Following by 18 months the adoption of the Montreal Protocol to reduce CFC use, this treaty also underlined the growing importance of environmental issues in international affairs.[43]

The Basel treaty was signed on the spot by 34 countries, and 71 more nations tentatively agreed to it, exceeding by a wide margin the 20 required for the accord to take effect. Mostafa Tolba, Executive Director of the U.N. Environment Programme, recognized that the treaty was a step in the right direction, albeit a small one, when he said, "Our agreement has not halted the commerce in poison, but it has signaled the international resolve to eliminate the menace that hazardous wastes pose to the welfare of our shared environment and to the health of the world's peoples."[44]

The single event of 1989 that best symbolized the new age of environmental diplomacy was the agenda at the Group of Seven's annual summit meeting hosted by President Mitterand in Paris during July. These meetings were initiated in 1976 as an economic summit when the world was in a turmoil following the doubling of grain prices and the fourfold rise in oil prices; by 1989, much of the meeting was devoted to environmental issues. Discussions on climate change, deforestation, and ozone depletion consumed a good part of the weekend conference and dominated the communiqué released at its end.[45]

In Europe, the results of elections for the European Parliament provided unmistakable evidence that the electorate cares about the environment. The Greens gained 19 seats in the European Community's governing body, for a total of 39 of the 518-seat assembly. Although still small compared with the Socialist Party, which led the multiparty assembly with 180 seats, they became nonetheless a force to be reckoned with. The results in local terms were even more impressive. In France's elections for the European Parliament, the Greens won 11 percent of the vote, up from 3 percent in the 1984 elections. And in the United Kingdom, the Green share of the vote increased from less than 1 percent in 1984 to 15 percent.[46]

In national parliaments, the Greens are also now represented. For example, they hold 20 seats of a 349-seat total in the Swedish Riksdag and 13 of 630 seats in the Italian House of Deputies. In the Australian state of Tasmania, Greens hold the balance of power, with 5 out of 35 seats in the parliament. They captured 1,800 city council seats in France last year, establishing a foothold in local governments there as well. In West Germany, where they have been represented for many years, gains in 1989 were less impressive, partly because the federal government has already adopted many of the Greens' issues, such as disarmament.[47]

Similar opportunities to deal with global issues developed in other key areas. In Poland, environmental issues figured prominently in the Round Table discussion between the Communist Party and Solidarity that paved the way for the power-sharing arrangement. Elections to the Supreme Soviet and the Congress of People's Deputies also recognized the need to broaden the base of participation if the Soviet Union's economic and environmental issues were to be resolved. And the success of Japan's Socialist Party in wresting control of the Upper House from the long-entrenched Liberal Democratic Party reflected a growing concern in the electorate over both social and environmental issues.[48]

Three governments—the Netherlands, Norway, and Australia—announced long-term environmental plans in 1989. In varying degrees, each recognized both domestic and global environmental issues.

The Dutch National Environmental Policy Plan developed by the federal government was presented to the Parliament for debate following the September 1989 election. Given the combination of high population density and uncontrolled use of automobiles, pollution problems in this small nation are severe. The government proposes to adopt various incentives, disincentives, and regulations to encourage greater use of bicycles for trips of 5–10 kilometers. For longer trips, it will encourage people to use trains, not cars or planes. It plans to install emission controls in power plants to reduce sulfur dioxide and nitrogen oxide discharges sharply. The plan also calls for building 2,500 megawatts of cogeneration capacity this decade to increase energy efficiency. The goal is to halt the growth in CO_2 emissions by the end of the nineties.[49]

The Norwegian plan, which like the Dutch one is designed as a national response to the report of the World Commission on Environment and Development, also sets the goal of stabilizing carbon dioxide emissions by the year 2000 at the latest. Combined with efforts to reduce CFCs and nitrous oxide, this is intended to reduce total greenhouse gas emissions by Norway from the end of the decade on.[50]

The Australian plan is much less comprehensive than the other two. The major specific action it calls for is the planting of 1 billion trees during the nineties. If successful, this would replace roughly half the tree cover removed since European settlement began two centuries ago. Recognizing that Australia has lost 18 species of mammals and 100 of flowering plants since European settlement, the plan also launched an endangered species program; this is designed to save the 40 species of mammals at risk of extinction and the 3,300 rare or endangered plants. Although the plan recognizes the near-desperate need to conserve Australia's topsoil, it includes no detailed program for doing so. Likewise, it talks about the need to reduce CO_2 emissions, but has no specific steps to lower fossil fuel use.[51]

The Australian plan calls for the planting of 1 billion trees during the nineties.

A number of national governments responded to specific environmental threats in 1989. Thailand, for example, concerned about increasingly destructive floods and landslides associated with deforestation, announced a ban on logging. Brazil, responding in part to international criticism of the burning of the Amazon rain forest, withdrew tax incentives from ranchers doing much of the burning.[52]

Many local governments throughout the world have taken their own specific steps on particular environmental issues, without waiting for a national strategy to materialize. In the United States, for instance, some states and numerous municipal governments have adopted mandatory recycling laws. Vermont has banned several major applications of

CFCs, including their use in car air conditioners by 1993. Even more ambitious, Governor Madeleine Kunin is developing a state-level program to combat global warming.[53]

Grassroots groups are also joining the battle. Their efforts range from those of the rubber tappers in the Amazon to protect the rain forest to those of Soviet groups organizing to block construction of nuclear plants in their communities. No one knows how many grassroots environmental groups now exist, but it must surely be in the tens of thousands. Issues they commonly organize around include the cleanup of toxic waste sites, recycling, and the protection of forests. Increasingly, people are realizing that protecting the environment requires organizing at the local level.[54]

Public opinion polls show rising concern with environmental degradation throughout the world. A U.N.-commissioned poll conducted in 14 disparate countries—rich and poor, East and West, North and South—found "in every major area of the world nothing less than alarm about the state of the environment."[55]

Is the recent political awakening too little and too late? Overall, the effort to protect the earth's life-support systems is lagging badly. Despite growing public concern, government expenditures to defend against military threats still dwarf those to protect us from environmental ones. For example, the United States plans to spend $303 billion in 1990 to protect the country from military threats but only $14 billion to protect it from environmental threats, a ratio of 22 to 1. Unfortunately, this national distribution of resources in response to these two threats is not atypical.[56]

All in all, 1989 was a year of much activity, a year of many conferences and declarations, but with little concrete action. Few major new laws were passed or new programs adopted. Some progress was made on reducing the threat to the ozone layer: as noted earlier, 66 countries have signed or agreed to sign the Montreal Protocol. A few countries, such as Sweden, have decided to phase out CFC use entirely. So, too, have some of the major manufacturers, such as U.S.-based Du Pont and Allied Chemical.[57]

With the major threats to food security, such as population growth, climate change, and soil erosion, little or no progress was made last year. The annual addition to world population, which reached a record high in 1989 of 88 million people, is likely to average 96 million during the nineties. As noted earlier, the buildup of carbon dioxide, the principal heat-trapping gas in the atmosphere, is accelerating; not a single national government has adopted a plan to reduce CO_2 emissions. And as far as soil goes, only the United States, which has reduced losses by one fifth since 1985, has succeeded in slowing the depletion of this essential resource.[58]

This past year brought the promise of change, but little real change. Proclamations and statements of concern were many, but actual steps to restore the planet's health were few. If the world does not seize the opportunities offered by the promise of change, the continuing environmental degradation of the planet will eventually lead to economic decline.

2

Slowing Global Warming

Christopher Flavin

Global warming promises to be one of the central environmental issues of the nineties. After a decade of scientific concern but popular neglect, the eighties ended with a growing political as well as scientific consensus that the world can no longer afford to procrastinate about this issue. Plans call for the drafting of an international climate treaty in 1990, and for its formal adoption at a global environmental summit in 1992.[1]

This action can hardly come too soon. Changes to the earth's atmosphere are by nature global and—for all practical purposes—irreversible, not only in our lifetimes but in those of our children and grandchildren as well. Lending urgency to the problem is the fact that the chemical composition of the earth's atmosphere is already quite different than it was just a century and a half ago. We have unknowingly committed ourselves to more climate change than many societies will be able to cope with.

Nitrogen and oxygen are still the main constituents of the atmosphere, but several more complex gases are building steadily: carbon dioxide (CO_2) is 25 per-

An expanded version of this chapter appeared as Worldwatch Paper 91, *Slowing Global Warming: A Worldwide Strategy.*

cent higher than preindustrial levels, nitrous oxide 19 percent, and methane 100 percent. Chlorofluorocarbons (CFCs), a class of synthetic chemicals not normally found in the atmosphere, have added further to this blanket of gases that allow sunlight in but trap the resulting heat.[2]

Global average temperatures are now about 0.6 degrees Celsius warmer than they were 100 years ago. No conclusive proof links this recent heating to the greenhouse effect, but circumstantial evidence has convinced many scientists that this is the cause. Of more concern, however, is the much faster warming that is predicted by a half-dozen computer models—reaching an increase of 2.5–5.5 degrees Celsius (4.5–9.9 degrees Fahrenheit) late in the next century. The difference between the warming of the past century and that expected in the decades ahead is like that between a mild day in April and a late-summer scorcher.[3]

Although climate change is a young science, many aspects of which are uncertain, this is no excuse for delay. Societies invest in many programs, such as defense, to protect against an uncertain but potentially disastrous threat. Similarly, investing in strategies to slow global warming is a sort of insurance policy—against catastrophes that have far greater odds of occurring than most events for which we buy insurance. And many of these programs are economical investments in their own right, cutting energy bills and air pollution as well as helping to restore the carbon balance.

Coping effectively with global warming will force societies to move rapidly into uncharted terrain, reversing powerful trends that have dominated the industrial age. This challenge cannot be met without a strong commitment on the part of both individual consumers and governments. In terms of the earth's carbon balance, the unprecedented policy changes that have now become urgent include a new commitment to greater energy efficiency and renewable energy sources, a carbon tax on fossil fuels, a reversal of deforestation in tropical countries, and the rapid elimination of CFCs.

THE GLOBAL CARBON BUDGET

The element carbon has become one of the largest waste products of modern industrial civilization. During 1988, some 5.66 billion tons were produced by the combustion of fossil fuels—more than a ton for each human being. Another 1–2 billion tons were released by the felling and burning of forests, mainly in tropical areas. Each ton of carbon emitted into the air results in 3.7 tons of carbon dioxide, the seemingly innocuous gas that is now one of the principal threats to humanity's future.[4]

Global carbon emissions from fossil fuel use have grown rapidly during the postwar period: it took 10 years for them to go from 2 billion to 3 billion tons, but just six more years to get up to 4 billion. This trend has of course been fueled by other exponential growth rates—namely of population and economic output, which translated into ever-greater use of fossil fuels. Increases in oil use have been particularly rapid, but during the eighties the use of coal and natural gas has also swelled.[5]

The past four decades of growth can be broken into three distinct periods. (See Figure 2–1.) From 1950 to 1973, the annual increase in carbon emissions was a remarkably steady 4.5 percent; from 1973 to 1983, emissions gyrated wildly but on average increased at a yearly rate of just 1.0 percent; since then, more rapid growth has resumed, at an average rate of 2.8 percent a year. In 1988, carbon emissions went up 3.7 percent, the largest annual increase in almost a decade.[6]

If expansion had continued at the pre-1973 rate, annual emissions today would be almost 3 billion tons higher. Of course, the slowdown was not the result of actions to protect the atmosphere. It stemmed from the effects of the two oil crises, energy policy changes in some countries, and global economic prob-

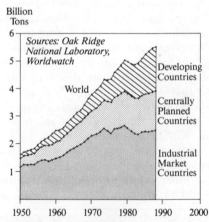

Figure 2-1. Carbon Emissions from Fossil Fuels, 1950–88

lems that have hit developing countries particularly severely.[7]

More than one third of the drop in the carbon emissions growth rate since 1973 —representing 1,100 million tons—can be traced to the declining energy intensity of many national economies, although a large disparity in per capita carbon emissions persists. (See Table 2–1.) The impact of improved energy effi-

ciency would have been even greater except that the efficiency revolution largely bypassed other parts of the world. Rising use of renewable energy and nuclear power played a significant but smaller role since 1973, offsetting about 500 million tons of annual emissions.[8]

In Eastern Europe and the Soviet Union, where most industries and individuals are shielded from market in-

Table 2-1. Carbon Emissions from Fossil Fuels, Selected Countries, 1960 and 1987

Country	Carbon		Carbon Per Dollar GNP		Carbon Per Capita	
	1960	1987	1960	1987	1960	1987
	(million tons)		(grams)		(tons)	
United States	791	1,224	420	276	4.38	5.03
Canada	52	110	373	247	2.89	4.24
Australia	24	65	334	320	2.33	4.00
Soviet Union	396	1,035	416	436	1.85	3.68
Saudi Arabia	1	45	41	565	0.18	3.60
Poland	55	128	470	492	1.86	3.38
West Germany	149	182	410	223	2.68	2.98
United Kingdom	161	156	430	224	3.05	2.73
Japan	64	251	219	156	0.69	2.12
Italy	30	102	118	147	0.60	1.78
France	75	95	290	133	1.64	1.70
South Korea	3	44	274	374	0.14	1.14
Mexico	15	80	446	609	0.39	0.96
China	215	594	n.a.	2,024	0.33	0.56
Egypt	4	21	688	801	0.17	0.41
Brazil	13	53	228	170	0.17	0.38
India	33	151	388	655	0.08	0.19
Indonesia	6	28	337	403	0.06	0.16
Nigeria	1	9	78	359	0.02	0.09
Zaire	1	1	n.a.	183	0.04	0.03
World	2,547	5,599	411	327	0.82	1.08

SOURCE: Worldwatch Institute, based on Gregg Marland et al., *Estimates of CO₂ Emissions from Fossil Fuel Burning and Cement Manufacturing, Based on the United Nations Energy Statistics and the U.S. Bureau of Mines Cement Manufacturing Data* (Oak Ridge, Tenn.: Oak Ridge National Laboratory (ORNL), 1989); Gregg Marland, ORNL, private communication, July 6, 1989; Economic Research Service, *World Population by Country and Region, 1950–86, and Projections to 2050* (Washington, D.C.: U.S. Department of Agriculture, 1988); U.S. Central Intelligence Agency, *Handbook of Economic Statistics, 1988* (Washington, D.C.: 1988); World Bank, *World Development Report 1989* (New York: Oxford University Press, 1989).

centives, energy intensities have remained essentially unchanged, and efficiency in most sectors is only about half that in Western Europe. Thus although Soviet per capita economic output (measured in purchasing power) is only two thirds that of Western Europe, per capita carbon emissions are almost twice as high. Planned economic restructuring may improve energy efficiency and thus reduce carbon emissions, but a Soviet consumer society with automobiles and larger houses could well push carbon emissions above their current level.[9]

The wealthy, energy-intensive quarter of the world's population has an obvious responsibility to lead in the search for solutions.

The Third World currently burns fossil fuels at far lower rates than in the industrial world. With many people continuing to live in poverty in the countryside, fossil fuels are often too expensive or simply unavailable. In many of these countries, a sizable portion of energy needs is met through the combustion of wood, straw, and other biomass fuels, which despite their importance are not generally counted in international energy statistics. Yet burning these also emits carbon, and when they are not replaced by new trees or other plants, the atmosphere receives an additional burden.[10]

During the past two decades, Third World cities have begun to fill with automobiles, oil-burning factories, and air-conditioned buildings. Rapidly expanding economies, such as South Korea's, have increased their use of fossil fuels commensurately. This slowed somewhat during the eighties, under the crush of debt burdens. For many nations, imported oil and capital-intensive power plants have become unaffordable. As a result, the world's carbon emissions are only slightly more evenly distributed today than they were in the sixties. If and when developing countries overcome their economic problems, however, the potential for growth in fossil fuel use is enormous.[11]

Many developing countries are adding far more carbon to the atmosphere through deforestation than through fossil fuel combustion. Brazil, for example, contributes some 336 million tons of carbon each year through deforestation, over six times as much as through burning fossil fuels. (See Table 2-2.) Combining these two sources makes Brazil the fourth largest carbon emitter in the world. Deforestation also pushes Indonesia and Colombia into the top 10 global emitters.

Although considerable uncertainty remains about the precise rate of deforestation, strong circumstantial evidence indicates it is accelerating. Tropical countries report ever more rapid losses of forests, and satellite reconnaissance confirms this. As with fossil fuel use, this suggests a challenge ahead. It is one thing to accomplish a goal from a standing start. It is quite another to turn around an ongoing and powerful trend.[12]

The U.S. Environmental Protection Agency has estimated that to stabilize atmospheric concentrations of CO_2 at the current level, carbon emissions must be cut by 50–80 percent, taking them back to the level of the fifties. Scientists and policymakers meeting in Toronto in June 1988 offered a short-term goal: cutting them by 20 percent by 2005.[13]

Even a 20-percent cut in carbon emissions would mark a dramatic shift in direction and require wholesale changes in energy policy and land use planning around the world. Such goals, though feasible, will force policymakers to consider familiar issues in a dramatically

Table 2-2. Estimated Carbon Emissions from Deforestation and Fossil Fuels in Selected Tropical Countries

Country	Deforestation[1]	Fossil Fuels[2]	Total
	(million tons)		
Brazil	336	53	389
Indonesia	192	28	220
Colombia	123	14	137
Thailand	95	16	111
Côte d'Ivoire	101	1	102
Laos	85	< 1	85
Nigeria	60	9	69
Philippines	57	10	67
Malaysia	50	11	61
Burma	51	2	53
Others[3]	509	181	690
Total	1,659	325	1,984

[1]1980. [2]1987. [3]65 countries.
SOURCES: R.A. Houghton et al., "The Flux of Carbon from Terrestrial Ecosystems to the Atmosphere in 1980 Due to Changes in Land Uses: Geographic Distribution of the Global Flux," *Tellus,* February–April 1987; Gregg Marland, Oak Ridge National Laboratory, private communication, July 6, 1989.

different context. In the past, international emissions of carbon were left to the marketplace. But markets by nature ignore environmental costs. Continuing with business as usual will result in a rapid undermining of the habitability of the planet during the next 20–30 years.

Any realistic strategy must start with the fact that one fourth of the world's population accounts for nearly 70 percent of the fossil-fuel-based carbon emissions. This wealthy, energy-intensive quarter has an obvious responsibility to lead in the search for solutions. But limiting carbon emissions in developing countries remains an extraordinarily difficult issue. It is a simple fact of atmospheric science that the planet will never be able to support a population of 8 billion people generating carbon emissions at the rate, say, of Western Europe today.[14]

The growth that would imply, to just more than 16 billion tons of carbon

emissions per year during the next few decades, would result in a concentration of CO_2 that is perhaps three times the preindustrial level—well above even the doomsday scenarios developed by climate modelers. At the recent annual growth rate of about 3 percent, carbon emissions in the year 2010 would reach nearly twice the current level, and by 2025 would be three times where we are today.[15]

A frighteningly large gap looms between projected growth rates in carbon emissions and the level that atmospheric scientists believe is necessary to maintain a climate that can meet human needs. World leaders meeting at the Hague environmental summit in March 1989 and the Paris economic summit in July 1989 agreed on a need to cut carbon emissions. But no one has begun to wrestle with the crucial equity issues this raises.[16]

How much of a burden should the

United States, as a wealthy and energy-intensive country, have to bear? And what about Japan, which is frugal in its consumption of fossil fuels but is one of the few nations with capital surpluses? Should the use of fossil fuels be constrained in developing countries or should most of their effort be devoted to reversing deforestation? And what of the Soviet Union, which is embarking on an economic program that could boost energy consumption even higher than it already is?

A growing world population and the associated demand for energy, land, and other resources will also have an impact on carbon emissions. In countries such as Kenya and the Philippines, a 3-percent annual growth rate of emissions is just enough to stay even with population, implying static per capita carbon emissions. Successful efforts to slow population increases would allow such nations to cut emissions more easily. Indeed, unless Third World population growth does slow drastically, it is hard to imagine any global program of carbon reductions that is both sufficient and equitable.[17]

ALTERNATIVES TO FOSSIL FUELS

Fossil fuels today provide 78 percent of our energy (oil provides 33 percent; coal, 27 percent; and natural gas, 18). But we have always known that the world must stop using them eventually since nonrenewable resources are by definition limited and will one day run out. Now the specter of global warming requires us to phase fossil fuels out during the early part of the twenty-first century, long before reserves are depleted. Not only do they contain carbon, but the ex-

traction and use of fossil fuels contribute a significant share of the emissions of two other greenhouse gases—methane and nitrous oxide.[18]

Although it is important that the transition away from fossil fuels begin in the near future, the process will extend over a period of decades. Many new technologies will have to be developed and, even then, installing replacement sources will take time.

Carbon emissions among the fossil fuels vary widely. Oil contains about 44 percent more carbon per unit of energy than natural gas does, while coal contains 75 percent more. Some analysts have argued that switching to natural gas should be a major element of a global warming strategy. But in many nations, including the United States, gas reserves are too limited to make this possible. And although natural gas is plentiful globally, it may have limited greenhouse benefits. A study by Dean Abrahamson of the University of Minnesota found that, compared with oil heating, the methane leaking from natural gas distribution systems has such a powerful greenhouse effect that it offsets any CO_2 reduction benefits of switching to gas heating.[19]

Another possibility is to "scrub" the carbon either before or during combustion, an idea with more hypothetical than practical appeal. While up to 90 percent of the carbon could in theory be removed, either by chemically treating the fuel before combustion or by introducing chemical reagents into the flue gas, these are unproven technologies. Further, carbon makes up 73 percent of the weight of the average kind of coal, and huge quantities of it would have to be extracted and disposed of in the deep oceans, something that would both be costly and require the construction of long pipelines to the sea.[20]

Such a program could nearly double the cost of coal-fired power generation, according to engineering studies. The equipment costs alone for a single 500-megawatt coal-fired power plant are estimated at between $500 million and $1.5 billion. While carbon scrubbing would be an enormous task even in large power plants, it would be a practical impossibility for the many small, decentralized consumers of fossil fuels such as home furnaces or automobiles. For the time being, at least, policymakers should assume that carbon scrubbing is not an economical strategy.[21]

Using energy more efficiently, on the other hand, has the combined advantage of low cost and enormous abundance. Although improved efficiency cannot replace fossil fuels entirely, neither can any alternative likely to be available in the next few decades. Energy efficiency has the immediate potential to cut fossil fuel use at least 3 percent annually in industrial countries, reducing CO_2 emissions commensurately. And in developing nations, efficiency can be used to limit fossil fuel growth while economic development continues.[22]

Worldwide, lighting accounts for about 17 percent of electricity use, or 250 million tons of annual carbon emissions. These emissions continue to grow rapidly, as electric lighting spreads in developing countries. If the world were to double the efficiency of these systems by the year 2010, it could cut the projected 450 million tons of carbon emissions from lighting in half. For household use, compact 18-watt fluorescent bulbs already exist that provide the same illumination as 75-watt incandescent ones.[23]

Energy efficiency's potential is also evident in transportation. Each year, the world's nearly 400 million cars spew about 550 million tons of carbon into the atmosphere, 10 percent of the total from fossil fuels. Projections show these emissions increasing 75 percent, along with vehicle numbers, by 2010. But if these cars got 50 miles to the gallon instead of the current average of 20, emissions would decline slightly. And if, in addition, improved mass transit, greater use of bicycles for short trips (see Chapter 7), and a carbon tax could keep the world fleet to 500 million cars, carbon emissions from autos would fall to half what they are today.[24]

Overall, energy efficiency improvements worldwide between 1990 and 2010 could make a 3-billion-ton difference in the amount of carbon being released to the atmosphere each year. This would imply an annual rate of energy efficiency improvement of 3 percent, keeping carbon emissions from fossil fuels to 6 billion tons in 2010, rather than the 9 billion that would result if efficiency improved at only 1 percent a year. No other approach offers as large an opportunity for limiting emissions in the next two decades.[25]

The other way to reduce emissions is by developing non-carbon-based energy sources. Both nuclear power and renewable energy sources have been used to displace fossil fuels in the recent past: increased use of nuclear power during the past 15 years is now displacing 298 million tons of carbon emissions annually—5 percent of the annual total—while increased use of renewables displaces 192 million tons. Each could be employed more extensively in the future.[26]

Nuclear power, however, is less ready for massive deployment today than it was a decade ago. The last 10 years have been marked by serious accidents at Three Mile Island and Chernobyl, enormous cost escalations, the failure to develop long-term disposal sites for nuclear waste, and a precipitous decline in public acceptance. Today, just 94 nuclear plants are under construction worldwide, the smallest number in 15 years. If all these are completed, they would displace an additional 110 million tons of carbon annually. Indeed, between 1995 and 2005, nuclear energy generation may actually decline as a growing number of plants reach the end of their operable lives.[27]

Some nuclear advocates argue that this history of mistakes provides justification for an attempted reinvention of nuclear power—with global warming as a key rationale. When he was U.K. Environment Secretary, Nicholas Ridley said in late 1988, "If we want to arrest the greenhouse effect we should concentrate on a massive increase in nuclear generating capacity." Although such a stance might be supportable if nuclear power's problems were the result of a few engineering mistakes, mismanagement, or a lack of government commitment, this is not the case. The problems stem from basic unresolved technological issues such as vulnerability to accidents and reliance on highly dangerous materials.[28]

Nuclear power is simply not a medium-term option for slowing global warming.

As far as new plant designs are concerned, scientists have yet to settle even on which to pursue. While the proposed redesigns have promising features such as modularity and "passive" safety systems, experts question whether they can ever be as economical or even as safe as their proponents hope. Indeed, the entire concept of inherent safety may be an engineering mirage. A study by the U.K. Atomic Energy Authority concludes that the proposed designs may be just as vulnerable to structural failures as conventional plants are. *Nucleonics Week* states that "experts are flatly unconvinced that safety has been achieved—or even sub-

stantially advanced—by the new designs."[29]

Even if safety problems could be overcome, it would be the end of the decade before a prototype reactor could be completed and tested, meaning that it would be close to the year 2010 before a significant number of new plants come on-line. And for nuclear power to offset even 5 percent of global carbon emissions would require that worldwide nuclear capacity be nearly double today's level. Nuclear power is simply not a medium-term option for slowing global warming. Policymakers must decide whether the hundreds of billions of dollars needed to achieve a new generation of reactors are worth the uncertain returns—particularly considering other, more attractive alternatives.[30]

Leading that list of alternatives are several renewable energy sources poised to develop rapidly. Wind, geothermal, solar thermal, photovoltaic, and various biomass technologies are among the energy sources with potential to move strongly into the market during the nineties. And renewable energy technologies as a group have a big advantage over nuclear power as a means of reducing CO_2 emissions: they can be used not only to produce electricity but also to displace many other uses of fossil fuels, such as running automobiles.

Wind power is one renewable energy source that has come of age during the eighties. Today, more than 20,000 electricity-producing wind machines are in use worldwide, with a capacity of about 1,600 megawatts. Most of these are in California and Denmark, though wind farms have also begun to appear in India, West Germany, and other nations. Wind power is now commercially competitive, with generating costs of 6–8¢ per kilowatt-hour. Analyst Robert Lynette has estimated that with increased government support, between 6,100 and

31,000 megawatts of additional wind capacity could be installed in the western United States by the end of this decade. Northern Europe, North Africa, India, and the Soviet Union have similar potential. By 2030, wind power could provide more than 10 percent of the world's electricity.[31]

Geothermal power can be another substantial component of an energy system run on renewables. The world now has over 5,000 megawatts of installed geothermal capacity, at costs of 4–8¢ per kilowatt-hour. Rapid growth continues, with the main focus on developing difficult-to-tap resources such as the pressurized hot water reserves found in abundance in many regions. In the United States, installed geothermal capacity by the end of this decade could range from 4,200 megawatts (double the current level) to 18,700 megawatts. Other areas likely to rely heavily on this renewable energy source include Central America, New Zealand, the Philippines, and the Soviet Union.[32]

Solar thermal power is the most recent renewable to break into the commercial market. Since 1984, some 194 megawatts' worth of generating capacity has been installed in southern California in the form of mirrored troughs that focus sunlight onto oil-filled tubes. These convey heat to a turbine and generator that produce power. Solar thermal systems convert up to 22 percent of the sunlight that hits them into electricity, at a cost that will soon reach 8¢ per kilowatt-hour. Natural gas serves as a backup source to run the turbines when the sun is not shining, which allows the system to provide reliable power at times of peak demand. Luz International, the company that builds the systems, is exploring even larger projects in Nevada and Brazil.[33]

Solar photovoltaics, which converts the sun's radiation directly into electric- ity using semiconductor materials, is the renewable energy technology likely to advance most rapidly in the years ahead. Capable of being mounted on rooftops, or installed in massive quantities as desert power plants, solar cells could become as ubiquitous as petroleum. Annual installations have grown from just 1 megawatt in 1978 to 35 megawatts in 1988. Third World use of solar cells is expanding particularly rapidly, with India reporting that 6,000 village systems have been installed. In many remote areas, photovoltaics is a practical alternative to expansion of the electricity grid—providing power sooner, more reliably, and at lower cost.[34]

Scientists at the U.S. Solar Energy Research Institute (SERI) estimate that new technologies may yield photovoltaic systems by the late nineties that are competitive with conventional power produced during peak periods. This could result in 1,000 megawatts of capacity in the United States alone by the end of this decade, with more rapid growth thereafter. SERI further estimates that photovoltaics are capable of supplying over half of U.S. electricity four to five decades from now; many nations are likely to find similar potential.[35]

Biomass sources such as wood, agricultural wastes, and garbage also have great potential to fuel a sustainable energy system. Already, biomass supplies about 12 percent of world energy, a figure that reaches as high as 50 percent or more in some developing countries. Much of this use is not sustainable, however, and could even exacerbate global warming. In the Brazilian Amazon, for example, smelters fueled by wood from virgin forests that are not being replanted emit more carbon dioxide than if they were fueled by coal. Yet with careful replanting, management, and efficient conversion, biomass fuel could play an important role.[36]

Wood- and waste-fired power plants are now being built in many nations at costs competitive with fossil fuel plants. They produce minimal air pollution and, as long as they consume waste materials or wood from forests that are being replanted, they do not contribute to CO_2 buildup. Power plants that burn methane building up in landfills are particularly effective at slowing global warming, since they consume a gas with 25 times the greenhouse strength of carbon dioxide. A California study found that a kilowatt-hour of electricity produced this way removes methane equivalent to the carbon released by 10 kilowatt-hours generated by a coal-fired plant.[37]

Among the transport fuels that yield substantially lower carbon emissions are alcohol fuels from biomass. Brazil has the world's largest alcohol fuels program, with about 72 million barrels of ethanol derived from sugarcane annually. In 1988, this provided 62 percent of the country's automotive fuel. Although this huge program has helped reduce Brazil's dependence on imported oil, it has done so at the price of enormous government subsidies. The United States is the second largest producer of alcohol fuels—20 million barrels per year, derived mainly from corn. But this program is also subsidized and is based on a crop grown on prime land. Because fossil fuels are used in the production, U.S. ethanol yields only a 63-percent reduction in carbon emissions on an energy-unit basis.[38]

To make biomass fuels a more important energy supply, the world will have to turn to crops grown on marginal lands, to conversion of waste materials, and to the development of integrated cropping systems that allow the same land to produce as much food as at present, and to produce fuel as well.

As societies adopt the slowing of climate change as a guiding principle in selecting energy strategies, it is impor-

tant that they seize alternatives that are both practical and economical. The relative cost of avoiding carbon emissions is one central consideration for policymakers to keep in mind. In many nations, coal-fired electricity is still the mainstay of the power system, so it makes sense to evaluate the potential of various alternatives to displace the carbon emitted by this form of energy. Since the fuel and operating costs of a coal-fired power plant are about 2¢ per kilowatt-hour, and environmental pollution costs about 1.5¢ per kilowatt-hour, anything above 3.5¢ can be attributed to "carbon avoidance."[39]

At the high end of the spectrum are nuclear power and photovoltaics, both expensive means of reducing carbon emissions. (See Table 2–3.) Opinion is divided about the potential to reduce nuclear costs, but the expense of solar cells is almost certain to fall precipitously. While newer, more efficient fossil fuel technologies such as combined-cycle coal plants or steam-injected gas turbines are relatively inexpensive, they only partially reduce carbon emissions, leaving the cost of carbon avoidance high. Alcohol fuels produced from corn are also uneconomical—$350 per ton of carbon saved. Costs have to be cut by at least a third to make this a practical means of reducing carbon emissions.[40]

Significantly more economical are a number of the emerging renewable options, including wind, wood, geothermal, and solar thermal, with carbon avoidance costs as low as $95 per ton. Moreover, all these costs are falling. The cheapest way to reduce carbon emissions, however, is energy efficiency— enormously abundant and costing $16 or less per ton of avoided carbon. Indeed, many efficiency investments are less expensive than operating existing coal-fired power plants, meaning that the cost of avoided carbon emissions is zero.[41]

Table 2-3. Costs of Avoiding Carbon Emissions Associated with Alternatives to Fossil Fuels, 1989

Fossil Fuel Alternative	Generating Cost[1]	Carbon Reduction	Estimated Pollution Cost	Carbon Avoidance Cost[2]
	(cents/kwh)	(percent)	(cents/kwh)	(dollars/ton)
Improving Energy Efficiency	2.0–4.0	100	0.0	<0–16[3]
Wind Power	6.4	100	0.0	95
Geothermal Energy	5.8	99	1.0	110
Wood Power	6.3	100	1.0	125
Steam-injected Gas Turbine	4.8–6.3	61	0.5	97–178
Solar Thermal (with gas)	7.9	84	0.2	180
Nuclear Power	12.5	86	5.0	535
Photovoltaics	28.4	100	0.0	819
Combined-cycle Coal	5.4	10	1.0	954

[1]Levelized cost over the life of the plant, assuming current construction costs, and a range of natural gas prices. [2]Compared with existing coal-fired power plant. [3]Some energy efficiency improvements cost less than operating a coal plant, so avoiding carbon emissions is actually free.
SOURCE: Worldwatch Institute estimates.

SUPPORT FOR THE TRANSITION

Accelerating the development of energy efficiency and renewable energy sources will require a comprehensive package of policies—some local, some national, and still others at the international level. Cost-effective strategies have already been identified in some countries: in Canada, a major government-supported report found that although a 20-percent reduction in carbon emissions could cost the government $108 billion, it would save the nation $192 billion.[42]

The first step is to reverse the policies that bias many countries toward increasing reliance on fossil fuels. Tax codes, research and development programs, and other measures often frustrate efforts to stabilize the climate. In the United States, for example, $2.3 billion is slated to go into "clean coal" research during the next five years—a program that would move the country toward even higher levels of carbon emissions.

And the 1989 proposal by the Bush administration to develop methanol fuel could commit the U.S. automobile fleet to a highly carbon-intensive fuel source well into the next century.[43]

There is a pressing need to end such counterproductive programs and to develop energy policies that will address various energy and environmental issues in a complementary way. One example is the program developed by the South Coast Air Quality Management District to clean up southern California's air over the next 15 years, in part by phasing out gasoline and revamping the region's transportation system. (See Chapter 6.) Although global warming played no role in the formulation of the plan, which is aimed at meeting air pollution goals, officials later calculated that a side benefit would be a 20-percent reduction in regional carbon emissions.[44]

A key priority is levying a "carbon tax" on fossil fuels. Such a tax would allow market economies to consider the global environmental damage of fossil fuel combustion. It would encourage individuals and companies to choose fuels based in part on their relative contribution to global warming. Coal would be taxed the highest, oil would be next, and natural gas would follow. Renewable energy sources that do not contribute to the buildup of carbon dioxide would not be taxed at all. A carbon tax has been proposed by the U.K. environment secretary and is now under discussion internationally. In countries that do not wish to raise total taxes, the carbon tax could be offset by reductions in other taxes.[45]

Picking the correct tax is a complicated exercise. One way of determining the appropriate level is to assess the environmental costs of using a particular fuel, then internalize those costs through taxes. (In addition to minimizing global warming, the benefits of reduced air pollution that would come from lowering fossil fuel consumption should also be considered in determining such a tax.) Yet no reliable estimates have been made of how much climate change will cost the international economy, and scientists doubt whether accurate projections will ever be possible.[46]

A "carbon tax" would encourage individuals and companies to choose fuels based in part on their relative contribution to global warming.

Suffice it to say that the potential cost is in the hundreds of billions of dollars. The costs of limiting emissions are almost certainly lower than the possible future costs of warming, and so these can be used as the basis for calculating such a tax. As indicated earlier, carbon avoidance costs vary widely, but a conservative estimate is that large-scale efforts will cost on average at least $50 per ton of carbon. This in effect represents

the cost of avoiding climate change and should be incorporated in the price of fuels. Ideally, such taxes will be agreed on internationally, so that additional expenses do not hit different economies disproportionately. The revenues could be used in part to develop permanent and stable funding for improving energy efficiency and developing renewable sources. This would, in the words of West Germany's Social Democratic Party, endorsing such a tax, "provide an ecological redistribution of taxes."[47]

In the United States alone, revenues from such a tax would equal about $60 billion ($240 per person); in India, they would come to $7.5 billion ($9 per person). A tax at this level would raise the price of gasoline in the United States by 17¢ a gallon and of electricity by 28 percent. It would spur a surge in efficiency investments similar to the one of the late seventies and early eighties. A gradually levied carbon tax would avoid the dislocations caused by oil price hikes while encouraging the world to get on a course of declining carbon emissions.[48]

Stepped-up research and development is also essential. Although governments have supported efficiency and renewable technologies for over a decade, their commitment has been erratic. This is particularly true in the United States, where R&D budgets for renewables and efficiency soared in the late seventies and then were cut by four fifths between 1981 and 1988. Despite these reductions, many new technologies, such as wind turbines, highly efficient fluorescent lighting systems, and new insulating materials, have entered the marketplace. The next step is to encourage more widespread use of these technologies.[49]

Most of the progress so far in slowing growth in fossil fuel consumption has been stimulated by market forces. For energy efficiency to realize its full potential to protect the environment, however, major institutional reforms will be needed. (See also Chapter 6.) While energy producers look to new supply options with payback periods of up to 20 years, energy consumers rarely invest in efficiency measures with paybacks longer than two. Legislation that encourages consumers to make longer-term investments—perhaps with the aid of utilities—would help close this gap.[50]

Electric utilities can help overcome these problems—supplying highly efficient bulbs instead of building more generators, for instance, as is already being done in some areas. Since the late seventies, a policy known as "least-cost planning" has been used by regulators and legislators to push utilities into adopting conservation programs. Using this approach, utilities invest in improving customers' efficiency as long as greater efficiency costs less than new power plants. The cheapest options sell first in a market, and they are chosen first in least-cost planning.[51]

Going a step further, a more competitive power industry would foster innovation and the development of technologies that are environmentally preferable. By opening the generation market to competitive bidding for all new power suppliers, policymakers can allow cogeneration and renewable energy projects to sell power to utilities at their "avoided cost" of new generation. Some regulators have also begun to include bids from companies that provide "saved energy" that can postpone or offset the construction of new facilities. A logical next step is to incorporate global warming costs in the bidding process and modify regulations. Utilities' rates of return can be increased when they lower the total economic and environmental costs to consumers by investing in improved efficiency—a "win-win" situation that would end the current bias toward ever greater electricity sales.[52]

Market incentives are generally more efficient than legislated standards in encouraging change, but mandatory standards do have a role to play. They can ensure that the least efficient or most wasteful practices are eliminated, so that efficiency improvement does not bypass industries and markets where institutional impediments persist. Efficiency standards for automobiles, buildings, and appliances offer enormous energy savings while cutting down on pollution. The 1986 U.S. appliance standards, for example, will have saved $28 billion worth of electricity and gas and have kept over 340 million tons of carbon out of the atmosphere by the end of this decade.[53]

In the Third World, the challenge of raising sufficient funds to invest in efficiency is particularly severe. Already deeply in debt, most developing countries are chronically short of capital and lack access to many of the highly efficient technologies developed in the eighties. One answer is to redirect a portion of the enormous flow of international lending currently devoted to building power plants and electric lines. This would free billions of dollars that could be invested in improving the efficiency of industry, transportation, and buildings.

OTHER GREENHOUSE STRATEGIES

As noted in the opening of this chapter, two other important components of efforts to slow global warming are the reversal of deforestation, especially in the Third World, and the elimination of chlorofluorocarbons. The first of these relies on using forests and agricultural lands as a carbon sink. Living plants and their soils constantly accumulate carbon; indeed, the approximately 120 billion tons in flux each year in the biosphere is about 20 times that released by fossil fuels. Net loss of forests is estimated to have resulted in the release of at least 1 billion tons of carbon annually in recent years.[54]

Several analysts have attempted to estimate the carbon-fixing potential of various kinds of forest, and to design appropriate tree planting programs. In *State of the World 1989* we considered the likely contribution of 130 million hectares of tropical forest—an area twice the size of France. Trees covering that much territory, planted around the world, could meet the fuelwood and timber needs of the Third World and restore degraded lands. At the same time, they would sequester about 5.5 tons of carbon per hectare. Such a "carbon bank" would absorb 660 million tons of carbon each year until the trees reach maturity in about three decades. This is more than a third of the amount currently thought to be emitted each year by deforestation, and a little under 10 percent of total net carbon emissions.[55]

Calculations like these have spurred efforts to develop new multiple-use forestry programs, part of whose justification would be to slow climate change. The most concrete project so far is one in Guatemala established to offset the 387,000 tons of carbon to be emitted annually by a planned 183-megawatt, coal-fired power plant in Connecticut. The plant's builder, Applied Energy Services, has joined with the Washington-based World Resources Institute and the private development group CARE to create 12,000 hectares of woodlots, plant 60,000 hectares of agroforest (combining trees and crops) that will be harvested on a sustainable basis, and protect additional forestland from fires. The 10-year project is expected to cost $16.3 million and involve 40,000 families in planting and maintaining roughly 52 million trees.[56]

Although such an approach cannot offset all or even most carbon emissions from power plants—in the United States alone it would have to be duplicated 1,200 times over—it is a comparatively inexpensive method of reducing emissions and slowing deforestation. The Guatemala project benefits from the fact that no land need be purchased and that many of the workers will not have to be paid. Another study has suggested that tree planting will cost $16 per ton of carbon avoided, a number that can only be bettered by the most attractive energy efficiency measures. In wider implementation of such a plan, costs would likely be far higher.[57]

Applied Energy Services is fully aware that a project in a country like Guatemala—with oppressive rural poverty, a population whose numbers will nearly double in 20 years, and serious political problems—may never achieve its full climate stabilization benefits. The project also illustrates the difficulties of designing a large-scale afforestation program. Finding sufficient and appropriate land, raising the needed financing, and mobilizing social institutions are three challenges that must be faced in devising a practical global plan. The rate of deforestation is accelerating, however, and without policy changes the process will contribute growing amounts of carbon. Slowing this trend, let alone making forests net absorbers of the chief greenhouse gas, will require the adoption of a long list of policy changes.[58]

Slowing deforestation is the first priority, and requires that tropical countries end financial incentives for land speculators and settlers to move into virgin forests and for loggers to export hardwoods. Governments and international aid agencies also need to work actively to support sustainable development projects such as agroforestry and woodlots that allow people to make a living from forests that are left standing rather than cutting them down.[59]

Remarkably, in a 1989 policy paper on global warming, the World Bank concluded that "the economics of vigorously pursuing [reforestation] are probably not favorable at this time." This statement ignores the impact of the current rate of deforestation and the potential leverage of the international community in easing the pace of destruction. Stopping deforestation within their own borders is by far the largest contribution that many developing countries can make to global climate stabilization, as well as to their own economic futures.[60]

Countries in temperate regions can also help restore the earth's carbon balance by planting trees. Surveys show that Europe and Japan are the only parts of the world currently increasing their total forested area. Even in Canada and the United States, forests are shrinking, largely due to the spread of land-intensive suburban and commercial development. And in both North America and the Soviet Union, the clear-cutting of virgin forests not only continues but is subsidized by governments. Relatively minor policy changes could convert North American, central Asian, and Australian forests into net carbon absorbers. The prime minister of Australia has taken the lead in such efforts with his recent announcement of a program to plant 1 billion trees by the end of this decade.[61]

A major step in this direction would be to convert large areas of marginal crop and grazing lands to trees, which stabilizes soils at the same time as it increases the rate of carbon fixing. Trees planted, for example, on the 13 million hectares of erodible cropland set aside in the United States since 1986 under the Conservation Reserve Program would absorb 65 million tons of carbon annually for the first few decades. This would lower net U.S. carbon emissions by about 5 percent, a major step. Once the new trees reach maturity, in 20–30 years, and cease to absorb carbon, they could be harvested on a sustainable basis for use as fuel to replace oil or coal, further lowering carbon emissions.[62]

A number of cities have already decided that the local benefits of tree planting are so great that they will not wait for national governments or the international community to adopt new measures. The American Forestry Association launched a program in late 1988 called Global ReLeaf, which aims to encourage U.S. communities to plant 100 million trees by 1992. This project was spurred by concern over global warming and the desire of individuals to make a contribution. (One hundred million growing trees would sequester an estimated 5 million tons of carbon annually.)[63]

Governments and international aid agencies need to support projects such as agroforestry and woodlots that allow people to make a living from forests left standing.

Global ReLeaf also recognizes that tree planting will improve the urban environment itself, moderating summer heat and improving aesthetics. The mayor of Los Angeles announced in February 1989 that the city hopes to plant between 2 million and 5 million trees by 1994, and in September the mayor of Houston, Texas, set forth a plan to plant 2 million trees by 2000. Other cities are expected to join in.[64]

The real challenge facing these programs is maintaining the trees once they are in the ground. In the past, trees put in during crash campaigns, such as in China, have been plagued by high mortality rates. This can only be prevented by careful nurturing of the saplings. If such difficulties can be overcome, tree planting will make a small but important contribution to climate stabilization. Newly planted trees have an even larger educational and symbolic value, enlisting individuals and their communities in the fight to slow global warming.

Global ReLeaf aims to encourage U.S. communities to plant 100 million trees by 1992.

A third essential greenhouse strategy—the elimination of chlorofluorocarbons—will in some ways be the easiest element of the effort to slow climate change, and a critical early test of worldwide commitment to do so. CFCs, an important class of modern industrial chemicals, were responsible for an estimated 25 percent of the added greenhouse effect during the eighties. In the United States, the leading CFC producer, these chemicals constitute 40 percent of the country's greenhouse emissions. In Japan, CFCs account for over half the country's contribution. Moreover, although growth in the use of CFCs has slowed since the seventies, their concentration is expanding more rapidly than any other greenhouse gas.[65]

Because CFCs are also responsible for depletion of the ozone layer in the stratosphere—which shields the earth from deadly ultraviolet radiation—efforts to limit their use are further along than for the other greenhouse gases. Beginning in the seventies, the United States banned the use of CFCs in aerosol sprays, and several nations followed suit. International efforts to restrict CFCs began in the early eighties, spurred by growing evidence of ozone depletion. After years of negotiations, a large group of countries meeting in Montreal in 1987 agreed to freeze all production of CFCs immediately and to halve output by 2000. Although this agreement is rightly considered a landmark in environmental diplomacy, evidence has mounted in the past two years that it is not sufficient.[66]

It is estimated that, even under the Montreal Protocol, atmospheric concentrations of CFC-11 and CFC-12 will increase by 77 percent and 66 percent respectively by 2040. This could raise the greenhouse burden on the atmosphere by almost one fifth, even if all the other greenhouse gases were stabilized. This large prospective increase is in part because the protocol makes exceptions for the countries that have lagged in CFC production, such as the Soviet Union and developing nations. It means that the actual emissions reductions by the end of this decade will be less than 40 percent. In addition, large quantities of CFCs already contained in refrigerators, air conditioners, and foam products will gradually find their way into the atmosphere in the years ahead.[67]

Immediate efforts to strengthen the Montreal Protocol and accelerate the elimination of CFCs are priorities in the global warming fight. Curbing CFC releases should be relatively painless because alternatives are already available for some of these chemicals, others are being developed rapidly, and technologies to keep them out of the atmosphere are well along.

During the past year numerous corporations and governments have announced plans to eliminate certain uses of the compounds. And, with the ratification ink not yet dry, the Montreal Protocol is already being reviewed with an eye to tightening it. In May 1989, some 81 nations signed the Helsinki Declaration, pledging in principle to phase out five CFCs implicated in ozone depletion by 2000, if additional substitute chemicals are developed. For the first time, developing countries were actively engaged in the process, enticed by the agreement of industrial nations to establish a fund to assist in obtaining substitutes. There is no time for delay. It is unconscionable to allow CFC concentrations to keep building up in the atmosphere when they are so clearly implicated in two of the world's most pressing environmental problems.[68]

A National and Global Policy Agenda

This past year has been marked by a flurry of proposals to deal with global warming. According to one U.S. count, 130 bills were introduced in 22 state legislatures during the first half of 1989. Major proposals are also being debated in national parliaments. (See Table 2–4.) And in some areas, city governments

Table 2-4. Climate Policies, Enacted and Proposed, November 1989

Nation or State	Policy	Status
The Netherlands	Proposal to freeze CO_2 emissions by 2000 and increase spending on efficiency	Parliament debating proposals; government considering four-year, 8-percent reduction in CO_2 emissions
Norway	Plan to stabilize CO_2 emissions by 2000, then reduce emissions	White paper approved by Parliament in June 1989
Sweden	Plan to freeze CO_2 emissions at current levels; tax CO_2 emissions	Parliament approved emissions freeze, 1988; tax planned by 1991
United Kingdom	Considering control of methane leakage, improved energy efficiency	House of Commons Energy Committee made recommendations
United States	Comprehensive legislation to cut carbon emissions 20 percent	Several bills pending in U.S. Congress
California	Comprehensive policy being developed	Government report to be submitted June 1990
Oregon	Law requiring 20 percent reduction in greenhouse gases by 2005	Enacted July 1989
Vermont	Order to decrease greenhouse gas emissions, re-evaluate state energy policy	Proposal announced by Governor September 1989
West Germany	Comprehensive policy under discussion	Parliament Commission formulating proposals; report due mid-1990

SOURCE: Worldwatch Institute, based on various sources.

and private groups are getting into the act as well. Rarely has a new policy issue taken hold so quickly.[69]

In the United States, two comprehensive global warming bills were first introduced in Congress in late 1988. They proposed a national goal of cutting carbon emissions 20 percent over the next 10 years and included programs to implement national least-cost planning, improve automobile fuel economy, develop renewable energy sources, plant trees, and assist developing countries in slowing population growth and deforestation. In the ensuing months, however, the bills ran into the opposition of entrenched industries, and key elements of the legislation failed to move forward.[70]

The picture in Europe is more encouraging. The public is strongly concerned about climate change, and is electing politicians who are ready to act. The Netherlands, Norway, and Sweden are considering plans to freeze or cut national CO_2 emissions. In Sweden, one oil-fired power project has already been put on hold pending evaluation of its greenhouse impact. The United Kingdom and West Germany report similar sentiments and are reviewing their energy policies. A special West German parliamentary commission is likely soon to suggest major policy initiatives, but already the opposition Social Democratic Party has called for higher energy taxes, new efficiency incentives, and a sweeping overhaul of utility laws. An inexorable environmental bandwagon may next push the European Parliament, with its dominant coalition of pro-environment parties, to take up related proposals.[71]

Other parts of the world are moving much more slowly on a greenhouse policy agenda. Brazil, China, Japan, and the Soviet Union are among the key countries that have done little beyond supporting more research and strong rhetoric. Canada, which has played a leading role in international discussions, failed to reach agreement on a 20-percent-reduction goal at an August 1989 meeting of provincial energy ministers. Some small countries, however, have actively supported new policy initiatives on global warming. These include nations like the Maldives and Malta that believe they may be big losers as climate changes.[72]

National goals to limit emissions of the four main greenhouse gases—carbon dioxide, CFCs, methane, and nitrous oxide—are among the most important features of any meaningful global warming strategy. But beyond establishing goals, it is important that credible policies be put in place to achieve them, including giving relevant government agencies responsibility to implement appropriate measures.

It is encouraging that so many governments have begun to mobilize to slow global warming, but an international agreement to stabilize the climate is still needed. Indeed, global warming presents an unprecedented challenge to the global community, forcing everyone from prime ministers to the general public to understand that we inhabit a single planet and share responsibility for its health. National differences and old rivalries will have to be cast aside, so that countries can act in concert as never before.

This seemed to be recognized by leaders of the seven largest industrial democracies, meeting in Paris for their annual summit in July 1989: Environmental proposals occupied one third of the resulting communiqué. On global warming, the leaders said "we strongly advocate common efforts to limit emissions of carbon dioxide and other greenhouse gases, which threaten to induce climate change."[73]

After the bold rhetoric, however, come complex and difficult negotiations. That process began in November 1988, when representatives of 30 countries met in Geneva under the auspices of the United Nations Environment Programme (UNEP) and the World Meteorological Organization. Following a pattern established in arms talks, these nations have formed an Intergovernmental Panel on Climate Change that meets periodically to forge an agreement, now scheduled to be in draft by late 1990.[74]

As a start, the panel is spearheading efforts to understand the issues of climate change more thoroughly and to consider possible responses. The group includes Brazil, China, India, Japan, the Soviet Union, the United States, and West Germany, as well as smaller countries such as Senegal and Sweden. International treaties on global warming will be far more complex than the agreements on depletion of the ozone layer. While that is caused mainly by a particular class of industrial chemicals (CFCs), many of which can be replaced, global warming is caused by gases that are central to the activities of modern societies.[75]

The first step is the establishment of ambitious but practical goals for the reduction of carbon emissions, particularly in those countries that currently use fossil fuels most heavily. Any plan that reduces carbon emissions is also likely to help cut back on two other greenhouse gases, nitrous oxide and methane. Only by reducing carbon emissions at least 10 percent in the next decade can the world get on course to at least halve emissions by mid-century.

In an effort to balance practicality and equity against the urgency of the problem, we have formulated a set of reduction targets based on today's per capita carbon emissions levels. (See Table 2-5.) Countries such as the United States and the Soviet Union that currently produce carbon dioxide at a high rate would be required to reduce emissions by about 35 percent in the next 10 years, while nations such as India or Kenya could continue to increase emissions. Those with carbon emissions between these two extremes, such as Italy and Japan, would have to reduce them less rapidly.[76]

These targets can be met by pursuing the technologies and policies described

Table 2-5. Proposed Carbon Emission Goals, Sample Countries

Current Carbon Emissions Level	Suggested Emissions Targets	Sample Countries
(tons/person)	(percent/year)	
<0.5	+3.0	Kenya, India, Niger
0.5–1.0	+1.5	China, Nigeria, Philippines
1.0–1.5	0	Indonesia, Mexico, South Korea
1.5–2.0	−0.5	Italy, France, New Zealand
2.0–2.5	−1.0	Japan, Thailand, Peru
2.5–3.0	−2.0	United Kingdom, West Germany, Brazil
>3.0	−3.0	Australia, United States, Soviet Union, Colombia, Côte D'Ivoire

SOURCE: Worldwatch Institute, based on Gregg Marland et al., *Estimates of CO$_2$ Emissions from Fossil Fuel Burning and Cement Manufacturing* (Oak Ridge, Tenn.: Oak Ridge National Laboratory, 1989); R.A. Houghton et al., "The Flux of Carbon from Terrestrial Ecosystems to the Atmosphere in 1980 Due to Changes in Land Use: Geographic Distribution of the Global Flux," *Tellus,* February–April 1987.

earlier in this chapter. The goals are designed to gradually narrow the disparities that now exist among national emissions levels. They are also practical, calling for a realizable 12-percent cut in global emissions by the year 2000. (See Table 2-6.) Under them, projected emissions of 6.4 billion tons are 38 percent below what they will be if the world continues on its current path. Although a 12-percent reduction would not by itself stabilize the climate, it would put the world on a course toward stabilization of global CO_2 concentrations by mid-century. (See Figure 2-2.)[77]

These goals put the onus on the industrial world to take the lead. Unless rapid reductions are achieved in wealthier countries, a stable climate could become unachievable for many decades. Indeed, if industrial countries only keep emissions at today's levels, Third World increases could raise global emissions 20–30 percent by the end of this decade and 50–70 percent by 2010. Unfortunately, no industrial countries have yet proposed lowering carbon emissions by the needed 20–35 percent in the next 10 years.

Since it is mainly deforestation that causes nations such as Brazil or Côte d'I-voire to have such high per-capita carbon emissions, these targets imply a major effort to slow forest clearing in these countries. A carbon reduction strategy can also begin to slow growth in Third World fossil fuel consumption, however, something that is essential if the planet is to avoid vast increases in carbon emissions in the decades ahead. China, for example, would need to cut its growth rate in coal use from 3.5 percent per year to 1.5. This change alone would reduce emissions estimated for this decade's end by more than 200 million tons annually.[78]

The world needs to end the production of CFCs and to cut global carbon emissions by 10–20 percent over the next decade.

What would a climate change agreement for the nineties look like? The first element would be a commitment to stabilize atmospheric concentrations of greenhouse gases by the middle of the twenty-first century, reducing net carbon emissions to a maximum of 2 billion tons per year. With a projected world population of 8 billion, this would imply a per capita rate of carbon emissions close to India's today, or one eighth the West European level. To get to this point, the world needs to end the production of CFCs and to cut global carbon emissions by 10–20 percent over the next decade, adopting country-specific targets based on the per capita figures described earlier. Within a year of being signed, each country would submit a plan to achieve the goals, and then issue progress reports every two years. Negotiators, meanwhile, would consider the adoption of stricter goals to begin in 2000.[79]

To accomplish these goals, UNEP

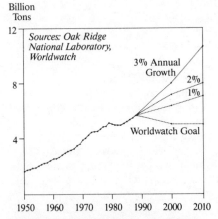

Billion
Tons

Figure 2-2. Carbon Emissions from Fossil Fuels, 1950–88, With Alternative Projections to 2010

Table 2-6. Global Carbon Emissions, 1988, and Goals for 2000 and 2010

Area	1988 Carbon (million tons)	1988 Per Capita (tons)	2000 Carbon (million tons)	2000 Per Capita (tons)	2010 Carbon (million tons)	2010 Per Capita (tons)
North America	1,379	5.07	897	3.03	662	2.13
Soviet Union and Eastern Europe	1,428	3.55	964	2.23	872	1.91
Oceania	336	2.27	284	1.79	270	1.65
Latin America	910	2.09	803	1.46	764	1.18
Western Europe	774	2.03	699	1.79	664	1.67
Middle East	187	1.14	187	0.83	217	0.74
Africa	534	0.86	646	0.73	749	0.64
Centrally Planned Asia	774	0.66	932	0.69	1,082	0.73
Far East Asia	833	0.55	998	0.52	1,158	0.52
World[1]	7,319	1.42	6,435	1.03	6,438	0.93

[1]The sums of the columns by region do not equal the world total due to U.N. reporting irregularities.
SOURCE: Worldwatch Institute estimates, based on Gregg Marland et al., *Estimates of CO_2 Emissions from Fossil Fuel Burning and Cement Manufacturing* (Oak Ridge, Tenn: Oak Ridge National Laboratory, 1989); Economic Research Service, U.S. Department of Agriculture, *World Population by Country and Region* (Washington, D.C.: 1988); R.A. Houghton et al., "The Flux of Carbon from Terrestrial Ecosystems to the Atmosphere in 1980 Due to Changes in Land Use: Geographic Distribution of the Global Flux," *Tellus*, February–April 1987; and British Petroleum, *BP Statistical Review of World Energy* (London: 1989).

would need to become a more powerful agency, charged with the tasks of coordinating research and of reviewing and assisting with national climate strategies. As with arms control, verification is essential to a credible agreement. But today UNEP has a minimal budget and is mainly chartered to coordinate the work of larger U.N. organizations. Recognizing this weakness, the leaders of 17 countries, including France, Japan, and West Germany, met in the Netherlands in March 1989. The resulting Hague Declaration called for the development within the United Nations of a new or newly strengthened institutional authority with powers to carry out the provisions of a global warming agreement.[80]

One issue that must be addressed is how to pay for massive investments in energy efficiency and reforestation in developing countries that are now strapped with huge debts. Many efficiency and forestry investments have broad economic and environmental value and can be justified on those grounds alone, but without sufficient capital they may be neglected. One important mechanism might be the carbon tax described earlier. Pledging 10 percent of the revenues from such a tax to Third World programs would create an annual fund of $28 billion. Such a fund could be managed by the United Nations, which would make dispersal contingent on the implementation of credible climate stabilization plans. Programs to improve energy efficiency, manage forests, plant trees, slow population growth, develop renewable energy sources, and design CFC substitutes would be eligible for support.

The World Bank and institutions such as the U.N. Development Programme can also play a role by increasing their lending for such projects and by including climate protection as a consideration when reviewing loans. In recent years the Bank has been increasingly receptive to the idea of lending for efficiency improvements, including a proposed $1-billion program in China, but few loans have been approved. In a September 1989 speech, World Bank President Barber Conable was still talking about the uncertainties of global warming and announced no significant new initiatives. The world community will need to broaden the mandate for these lending institutions, forcibly enlisting them in the effort to sustain a livable climate.[81]

As the scientific evidence mounts, the time has arrived for a global warming agreement—comprehensive, detailed, and prescriptive. Only a rapid turnaround in carbon emissions trends can begin to get the world on the path to a stable climate. Wholesale changes in energy, land use, and population policies are implied. Unless such action is taken in the next few years, however, the nineties will become a lost decade for the world's atmosphere, relegating the next generation to a world less able to meet growing human needs. No other environmental problem has such an exponential and cumulative dimension to it, a fact that argues persuasively for the immediate adoption of strong policies.

The benefits of such an effort extend well beyond stabilization of the climate. Economies would be strengthened, new industries created, air pollution reduced, and forests preserved for their economic and recreational benefits. For humanity as a whole, it would be another step in the evolution of society, demonstrating the ability to work cooperatively as a world community. It would be an auspicious beginning to the new millennium.

3

Saving Water for Agriculture

Sandra Postel

Irrigation, among the most ancient of technological activities, has turned many of the earth's sunniest, warmest, and most fertile lands into important crop-producing regions. Egypt could grow virtually no food without water drawn from the Nile or from underground aquifers. California's Central Valley and the Aral Sea basin—the fruit and vegetable baskets of the United States and the Soviet Union—could barely be cultivated without supplemental water supplies. And absent irrigation, output from the critical grain-growing areas of northern China, northwestern India, and the western U.S. Great Plains would drop by one third to one half.

As world population grew from 1.6 billion to more than 5 billion over the last 90 years, irrigation became a cornerstone of global food security. The higher yields farmers could get with a controllable water supply proved vital to feeding the millions added to our numbers an-

An expanded version of this chapter appeared as Worldwatch Paper 93, *Water for Agriculture: Facing the Limits.*

nually, especially as opportunities to cultivate new land dwindled.

It now looks, however, as though irrigation will not boost food supplies as much during the nineties as it has in recent decades. A confluence of forces has begun to slow irrigation's expansion and to chip away at the existing irrigated area. Rising costs for new projects, a host of ecological problems, growing concern about the social and environmental effects of large dams, and worsening regional water scarcities have all placed severe constraints on agricultural water use. Moreover, climate change from the buildup of atmospheric greenhouse gases casts ominous shadows over future supplies, and at the very least will require massive investments to adjust the global irrigation base.

Moving rapidly from profligacy to greater efficiency and equity in agriculture's use of water is the surest way to avert shortages and lessen irrigation's ecological toll. Farming accounts for some 70 percent of global water use. Much of the vast quantity diverted by and for farmers never benefits a crop: worldwide, the efficiency of irrigation systems averages less than 40 percent. The technologies and know-how exist to boost that figure substantially; what is needed are policies and incentives that foster efficiency instead of discouraging it.[1]

Greater attention to small-scale irrigation and to improving rainfed farming can also help prevent water scarcity from undercutting food production. Hope for balancing the water budgets of many countries with rapidly growing population, however, hinges as much on slowing birth rates as it does on raising water productivity. The struggle for a secure water future will not end until societies begin to bring human numbers and demands into line with water's natural limits.

THE END OF AN ERA?

Resource historians sizing up the twentieth century will almost certainly characterize it as an age of irrigation. Between 1900 and 1950, irrigated area worldwide nearly doubled, rising to 94 million hectares. An even faster surge since then has brought the total to about 250 million hectares. Today, one third of the global harvest comes from the 17 percent of the world's cropland that is irrigated. (See Table 3–1.) Many coun-

Table 3-1. Gross Irrigated Area, Top 15 Countries and the World, 1986

Country	Gross Irrigated Area[1]	Share of Cropland That is Irrigated
	(thousand hectares)	(percent)
India	55,000	33
China	46,600	48
Soviet Union	21,000	9
United States	19,000	10
Pakistan	16,000	77
Indonesia	7,300	34
Iran	5,800	39
Mexico	5,300	21
Spain	3,300	16
Turkey	3,300	12
Egypt	3,200	100
Thailand	3,200	16
Italy	3,000	25
Japan	3,000	63
Romania	3,000	28
Other	52,200	9
World	250,200	17

[1]As used here, gross irrigated area is the area equipped with irrigation facilities; it implies nothing about cropping intensity.
SOURCES: Irrigation estimates from W. Robert Rangeley, Washington D.C., private communication, June 1989; cropland area from U.N. Food and Agriculture Organization, *Production Yearbook 1987* (Rome: 1988).

tries—such as China, Egypt, India, Indonesia, Israel, Japan, North and South Korea, Pakistan, and Peru—rely on such land for more than half their domestic food production.[2]

Since the late seventies, though, expansion has slowed markedly. Whereas net irrigated area grew 2–4 percent per year during the sixties and seventies, since 1979 the annual rate has been on average about 1 percent—far lower than the growth in world population. (Net irrigated area is the land estimated to have actually been irrigated that particular year; it is less than the gross area, that estimated to be equipped with irrigation facilities.)[3]

Per capita irrigated area peaked at 47.9 hectares per thousand people in 1978, and then fell to 45.2 hectares in 1987, a drop of 5.6 percent. (See Figure 3–1.) Estimates compiled by the U.N. Food and Agriculture Organization (FAO) show that only 16 million hectares were added to the global irrigation base between 1980 and 1987, an average of 2.3 million hectares per year, roughly half the rate during the seventies.[4]

This slowdown occurred for several reasons. Low commodity prices, comparatively high energy costs, and general economic conditions during much of the eighties discouraged agricultural investments. The logical presumption is that when crop prices rise, irrigation will pick up speed again. Unfortunately, this optimistic outlook ignores several important problems.

The costs of adding irrigation capacity have been rising rapidly in many countries.

Lending for irrigation by the major international donors has declined sharply over the last decade. Total lending to 23 countries in Asia, North Africa, and the Middle East by the World Bank, the Asian Development Bank, the U.S. Agency for International Development, and the Japanese Overseas Economic Cooperation Fund fell more than 60 percent in real terms between 1977 and 1987. Given that large projects often take a decade or more to complete, this funding drop suggests that in a good deal of the Third World, growth in irrigated area will not quicken much in the nineties.[5]

A related cause of the slowdown is that the costs of adding irrigation capacity—building new dams, reservoirs, canals, and distribution channels—have been rising rapidly in many countries, making it harder to justify such investments on economic grounds. Engineers naturally exploited the best sites for water development first. Recent projects have been technically more complex, leading to greater expenditures.

In India, for example, the cost in real terms of large canal schemes more than doubled between 1950 and 1980. Today, capital costs for new irrigation capacity run about $1,500 per hectare in China, and between $1,500 and $4,000 per hectare for large projects in India,

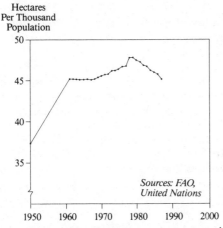

Hectares Per Thousand Population

Sources: FAO, United Nations

Figure 3-1. World Net Irrigated Area Per Capita, 1950–87

Indonesia, Pakistan, the Philippines, and Thailand. They climb toward $6,000 per hectare for public projects in Brazil, and toward $10,000 per hectare in Mexico.[6]

Developing new capacity costs the most in parts of Africa. Lack of roads and other infrastructure, the relatively small parcels of irrigable land, and the seasonal nature of river flows have driven per-hectare costs in many projects to between $10,000 and $20,000, or even higher. Not even double-cropping of higher-valued crops can make irrigation systems at the top end of this spectrum economical. Not surprisingly, Africa's irrigated area has expanded minimally since 1980, even though just 5 percent of the continent's cropland (excluding South Africa) is irrigated.[7]

The debt burden hanging over many developing countries makes the situation even bleaker. Irrigated area in Mexico, for example, has actually declined since 1985, in large part because its faltering economy has led to cuts in capital improvements, including expanding and rehabilitating these systems. Officials in Brazil view accelerated irrigation as critical to their goal of achieving greater food self-sufficiency. But their ambitious targets for the early nineties seem unlikely to be met, given the required investments.[8]

As costs rise and public investments decline, a larger share of whatever new irrigation does come on-line seems likely to support production of higher-valued cash crops rather than rice, wheat, and other food staples, especially if governments cannot afford to subsidize this farming heavily. Should grain prices rise because of diminished global stocks—a possibility in this decade (see Chapter 4)—investments will almost certainly pick up again. But the poor who suffer most from food shortages are unlikely to be able to afford crops at the high prices needed to inspire such investments. Neither scenario is comforting if irrigation's ultimate objective is to enhance food security.[9]

For the foreseeable future, therefore, irrigation's contribution will have to come more from improving existing systems than from expanding them to new land. This requires a fundamental shift in thinking: engineers are trained to design and build physical infrastructure, not to operate it effectively from the farmers' point of view. And many government irrigation agencies, intent on maintaining their large budgets, consider expansion their ultimate mission, rather than a tool to increase food output and raise rural incomes.

Behind the raw numbers lie vast discrepancies in irrigation's performance and quality. In many large surface-water systems, less than half the water diverted from the reservoir actually benefits crops. Much seeps through unlined canals before it gets to fields. An additional amount runs off the land or percolates unused through the soil because farmers apply water unevenly, excessively, or at the wrong times. Typical efficiencies for large projects in Asia run about 30 percent. Fortunately, not all this water is wasted: much of it returns to a nearby stream or joins underground supplies, and thus can be used again. But as it picks up salts, fertilizers, and pesticides along the way, it is often degraded in quality and contaminates water sources downstream.[10]

Lack of maintenance has caused many systems to fall into disrepair, further inhibiting performance. Over time, distribution channels fill with silt, increasing the likelihood of breaching, outlets break or are bypassed, and salts build up in the soil. In China, for example, more than 930,000 hectares of irrigated farmland have come out of production since 1980. Similarly, in the Soviet Union—the world's third largest irrigator—poor performance caused irrigation to cease on 2.9 million hectares between 1971 and 1985, equal to a quarter of the new area brought under irrigation.[11]

Worldwide, an estimated 150 million hectares—60 percent of the world's total irrigated area—need some form of upgrading to remain in good working order. An increasing share of available capital will thus go to rehabilitating existing systems. Although this is wise and necessary, it will further diminish funding for new projects.[12]

Poor management and inequities in the distribution of water are also to blame for irrigation's disappointing performance. Many systems deliver water on a fixed schedule that does not match crops' needs. In other cases, the supply is so unpredictable and unreliable that farmers cannot risk investing in other yield-raising inputs and activities, such as better seeds, fertilizers, or leveling their land to ensure efficient watering. As a result, output remains far below its potential.[13]

A study in Pakistan, for example, found that the average yields from irrigated fields of wheat, rice, and sugarcane were 60–70 percent below the upper range of yields some farmers achieved. This "yield gap" can largely be explained by ineffective management of the water supply and its influence on farmers' other management decisions.[14]

The plight of "tailenders"—farmers last in line to receive supplies from large canals—underscores the consequences of inefficient and unfair distribution. Typically, farmers at the head of a project area begin irrigating before the canals and distribution channels reach the end. By planting water-intensive crops, such as sugarcane or rice, they stake a claim to more water than was planned for their plots. Often wealthier and wielding considerable political influence, they then lobby effectively to maintain their heavy withdrawals even when the project is completed. Poorer farmers at the end of the line end up without enough water for their crops.

One project in the Indian state of Maharashtra, for example, had sugarcane occupying 12 percent of the irrigated area when the design called for only 4 percent. Since the water needed for a hectare of sugarcane could irrigate about 10 hectares of wheat, this practice causes much less area to be irrigated than was initially planned. Food production falls short of expectations, and the gap between wealthy and poor farmers widens.[15]

All these factors add up to much of the world's existing irrigation being underused, yielding far below its potential, and, in some cases, failing to enhance food and income security for those who most need it—the rural poor. To be sure, irrigation has greatly expanded world food output, alleviating hunger and improving diets in many areas. But as future gains get harder to come by, an important question becomes: irrigation for whom and for what?

THE ENVIRONMENTAL PRICE TAG

Each year, some 3,300 cubic kilometers of water—six times the annual flow of the Mississippi—are removed from the earth's rivers, streams, and underground aquifers to water crops. Practiced on such a scale, irrigation has had a profound impact on global water bodies and on the cropland receiving it. Waterlogged and salted lands, declining and contaminated aquifers, shrinking lakes and inland seas, and the destruction of aquatic habitats combine to hang on irrigation a high environmental price.[16]

Mounting concern about this ecological toll is making new water projects increasingly unacceptable. And ironically, the degradation of irrigated land from poor water management is forcing some land to be retired completely, offsetting a portion of the gains costly new projects are intended to yield.

By far the most pervasive damage stems from waterlogging and salinization of the soil. Without adequate drainage, seepage from unlined canals and overwatering of fields raises the underlying groundwater. Eventually, the root zone becomes waterlogged, starving plants of oxygen and inhibiting their growth. In dry climates, evaporation of water near the soil surface leaves behind a layer of salt that also reduces crop yields and, if the buildup becomes excessive, kills the crops.

The degradation of irrigated land from poor water management is forcing some land to be retired completely.

Even the best water supplies typically have concentrations of 200–500 parts per million (ppm). (For comparison, ocean water has a salinity of about 35,-000 ppm; water with less than 1,000 ppm is considered fresh; and the recommended limit for drinking water in the United States is 500 ppm.) Applying 10,-000 cubic meters of water to a hectare per year, a fairly typical irrigation rate, thus adds between 2 and 5 tons of salt to the soil annually. If it is not flushed out, this can build up to enormous quantities in a couple of decades, greatly damaging the land. Aerial views of abandoned irrigated areas in the world's dry regions reveal vast expanses of glistening white salt, land so destroyed it is essentially useless.[17]

No one knows for sure how large an area suffers from salinization. Estimates for the world's top five irrigators clearly suggest, however, that the problem is sufficiently widespread to be reducing world crop output. In India, salinity reduces yields on some 20 million hectares, and an additional 7 million have been abandoned as salty wasteland. China has about 7 million hectares of saline and alkaline agricultural land, and Pakistan 3.2 million. In the United States, salinity expert James Rhoades estimates that salt buildup lowers crop yields on 25–30 percent of the nation's irrigated land. Some 2.5 million hectares are salinized in the Soviet Union, most of them in the irrigated deserts of central Asia. Assuming the same share of world irrigated area is damaged as the collective share in these top five, salinization is reducing yields on 24 percent of irrigated land worldwide. (See Table 3–2.)

Other areas severely affected include Afghanistan, the Tigris and Euphrates river basins in Syria and Iraq, and Turkey. In Egypt, where virtually all cropland is irrigated, half is salinized enough to register diminished yields. And in Mexico, salinization is estimated to be reducing crop output by the equivalent of 1 million tons of grain per year—enough to feed 5 million people, roughly a quarter of Mexico City.[18]

In the western United States, excessive irrigation and poor drainage have spawned another set of problems. There, scientists have linked alarming discoveries of death, grotesque deformities, and reproductive failure in fish, birds, and other wildlife to agricultural drainage water laced with toxic chemicals. Since 1985, intensive investigations throughout the area have found lethal or potentially hazardous selenium concentrations at 22 different wildlife sites, including Kesterson National Wildlife Refuge, which has earned the unenviable reputation of being the Three Mile Island of irrigated agriculture.[19]

Table 3-2. Irrigated Land Damaged by Salinization, Top Five Irrigators and World Estimate, Mid-Eighties

Country	Area Damaged	Share of Irrigated Land Damaged
	(million hectares)	(percent)
India	20.0	36
China	7.0	15
United States	5.2	27
Pakistan	3.2	20
Soviet Union	2.5	12
Total	37.9	24
World[1]	60.2	24

[1]Estimated by assuming the share of remaining world irrigated area that is affected is the same as the share in the five leading irrigators.
SOURCES: D.R. Bhumbla and Arvind Khare, "Estimate of Wastelands in India," Society for Promotion of Wastelands Development, New Delhi, undated; Ministry of Water Resources and Electric Power, *Irrigation and Drainage in China* (Beijing: 1987); U.S. estimate from James Rhoades, Director, U.S. Salinity Laboratory, Riverside, Calif., private communication, September 1, 1989; Pakistan estimate from Ministry of Food and Agriculture, *Report of the National Commission on Agriculture* (Islamabad: 1988); Soviet figure from Philip Micklin, Western Michigan University, Kalamazoo, Mich., private communication, October 13, 1989.

Selenium, a natural element, is needed by humans and some other animals in very small amounts, but is highly poisonous at greater concentrations. Irrigation washed more selenium and other dangerous chemicals out of the soil in several decades than natural rainfall would have done in centuries. Either stored in ponds to evaporate or channeled away from farmland by rivers or canals, this toxic drainage has become a serious hazard to western wildlife—and possibly to humans, as harmful substances move up the food chain.

Potential solutions to the drainage dilemma, which include detoxification by soil microbes, mechanical filtration, and chemical treatment, range in cost from $30 million to more than $100 million per year for some indefinite period in California alone. With such price tags and the high ecological stakes of continued irrigation, it seems likely that some irrigated U.S. land will come out of production over the next decade. How much, and where, remains to be seen.[20]

Contamination of land and water by salts and toxic chemicals is only one indication that some irrigation is unsustainable. In much of the world, falling water tables signal that groundwater withdrawals exceed the rate of replenishment. As water levels decline, overpumping eventually can make irrigation too costly to continue, and can even drain some aquifers dry. In either case, land is forced out of irrigation temporarily or permanently.

In the United States, more than 4 million hectares—roughly a fifth of the nation's irrigated area—are watered by pumping in excess of recharge. By the early eighties, the depletion was particularly severe in Texas, California, Kansas, and Nebraska, four important food-producing states. In Texas, water tables were falling at least 15 centimeters (6 inches) per year beneath 1.54 million hectares, 72 percent of the state's total irrigated area. During the late seventies and early eighties, falling groundwater levels combined with high energy costs and low commodity prices to pull much irrigated land out of production. The total in Texas fell 30 percent between 1974, its peak, and 1987.[21]

In much of the world, falling water tables signal that groundwater withdrawals exceed the rate of replenishment.

Current year-to-year fluctuations in U.S. irrigated area reflect crop prices and government farm policies more than water availability and cost. Indeed, the 1989 total was likely the highest since 1982, with much of the recent growth occurring in the Southeast. Nonetheless, overpumping cannot continue indefinitely. The 4 million hectares currently watered unsustainably will eventually come out of irrigated production unless farmers reduce pumping to no more than the rate of aquifer charge.[22]

No other country has systematically assessed the extent of excessive groundwater pumping. But the situation is serious elsewhere as well, including in China and India, two of the three other major food producers. Groundwater levels are falling up to a meter per year in parts of the North China Plain, an important wheat-growing region. Heavy pumping in portions of the southern Indian state of Tamil Nadu reportedly dropped water levels as much as 25–30 meters in a decade; in the western state of Gujarat, overpumping by irrigators in the coastal districts has caused saltwater to invade the aquifer, contaminating village drinking supplies.[23]

Signs of irrigation overstepping ecological limits are evident among the world's rivers, lakes, and streams, as well. By far the most dramatic is the shrinking Aral Sea in the central Asian republics of the Soviet Union. Fully 95 percent of the Soviet cotton harvest is grown in this region, as well as a third of the country's fruits, a quarter of its vegetables, and 40 percent of its rice.[24]

Owing to the dry climate, 90 percent of the cropland in these Soviet republics is irrigated. As production expanded during the sixties and seventies, increasing amounts of water were diverted from the region's two major rivers, the Amu Dar'ya and Syr Dar'ya—the only sources of replenishment for the Aral Sea other than the meager rainfall. By 1980, flows in these rivers' lower stretches had been reduced to a trickle, and the sea had contracted markedly. (See Figure 3–2.)[25]

According to Philip Micklin of Western Michigan University, an expert on Soviet water issues, the Aral's surface area has shrunk by more than 40 percent since 1960, volume has dropped by two thirds, and salinity levels have tripled. All native fish species have disappeared, decimating commercial fisheries. Winds pick up salt from the dry seabed and annually dump 43 million tons of it on surrounding cropland, damaging harvests. Micklin warns that if no corrective measures are taken, by the end of this decade

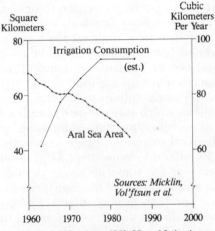

Figure 3-2. Sea Area, 1960–85, and Irrigation Consumption, 1963–87, in Aral Sea Basin

the Aral—once the world's fourth largest freshwater lake—"could consist of a main body in the south with a salinity well above the ocean and several small brine lakes in the north."[26]

Saving the Aral, albeit in some diminished form, is now a high priority in the Soviet Union. A long-standing plan to divert water from north-flowing Siberian rivers southwest into central Asia has been shelved, at least for the time being, because of concerns over the scheme's environmental effects and its exorbitant cost—estimated to range between 45 billion and 100 billion rubles ($70 billion to $156 billion at the mid-October 1989 official exchange rate).[27]

Planners hope that curtailing agriculture's water use in the Aral Sea basin will be sufficient to preserve the sea. In a September 1988 decree, the Communist Party Central Committee laid out a strategy aimed at raising average flows into the sea threefold over average levels during the eighties by the year 2005, largely by making irrigation systems more efficient. The plan could easily cost 30 billion rubles ($47 billion); whether the needed investments materialize remains to be seen. It may well turn out that preventing further ecological havoc will require taking some of Soviet central Asia's prized farmland out of production.[28]

All this visible ecological damage from large-scale irrigation has spawned strong opposition to new dams and diversion projects, even in the Third World, where water development remains a high priority. In February 1989, 10,000 people demonstrated against the Sardar Sarovar Dam, a centerpiece of India's Narmada Valley Development Project, which encompasses 30 large dams, several hundred medium-sized ones, and 3,000 small ones. Sardar Sarovar alone would flood vast areas of forests and farmlands and displace up to 100,000 people. Government officials

are pushing ahead, in part with a World Bank loan, even though the dam's impact on the river's ecology has not been adequately investigated. Nor, apparently, have studies been completed on how to manage the watershed to prevent erosion from rapidly silting up the planned reservoir and canals.[29]

In Africa, as elsewhere, planners often overestimate the benefits of large projects and underestimate social and environmental costs. Development anthropologist Thayer Scudder points out that large irrigation schemes have tended to benefit a small minority there while often destroying local production systems, such as floodwater farming and fisheries, that are vital to the poorer majority. Following completion of the Kainji Dam in northern Nigeria, fish catches and dry-season harvests from traditional floodplain agriculture dropped by more than half. A similar pattern was noted when the Bakolori Dam came on-line. And after the nearby Taiga Dam was completed, fish and floodplain harvests reportedly dropped even more than during the disastrous drought in the early seventies.[30]

In the United States, where the era of dam-building has largely ended and where water in many river basins is already overallocated, concerns center on the loss of free-flowing rivers, the destruction of fisheries from streamflow depletion, and damage to riverine and wildlife habitat. Efforts to restore and protect natural ecosystems could involve shifting water away from agriculture, by far the biggest consumer in the West.

Heightened awareness about these environmental effects may also upset the renewal of federal irrigation contracts, hundreds of which come due in the next couple of decades. The U.S. Environmental Protection Agency, backed by the Council on Environmental Quality, contends that extending at least some of these contracts requires the preparation of an Environmental Impact Statement under the National Environmental Policy Act. Should thorough impact studies be required, farmers would almost certainly be forced to reduce harm to wildlife, damage to the land, and toxic contamination generally—steps that, if too costly, could pull more land out of irrigation.[31]

SCARCITY, COMPETITION, AND CONFLICT

Regional shortages cropping up around the world form another growing threat to agriculture's heavy claim on water resources. (See Table 3–3.) Scarcities loom in a number of areas where irrigation is critical to farming, often accounting for 80 percent or more of water consumption. Many of these regions also are experiencing rapid population growth and urbanization—setting the stage for heated competition between cities and farms, and imposing constraints on agricultural development.

In much of northern and eastern Africa, increasing human numbers are on a collision course with scarce water resources. Swedish hydrologist Malin Falkenmark has found that societies typically experience "water stress" when annual renewable supplies per capita approach 2,000 cubic meters. By the end of the nineties, six out of seven East African countries and all five north African ones bordering the Mediterranean will exceed this measure of water scarcity. Six of these countries will have less than 1,000 cubic meters to tap per person. All of North Africa except Morocco already imports half or more of its grain. Under conditions of such severe water scarcity, boosting food self-sufficiency and meeting the domestic needs of populations growing on average at about 3 percent a year could well be impossible.[32]

Egypt, where rain is sparse, represents the extreme of the region's dilemma. Its 55 million people depend almost entirely on the waters of the Nile, none of which originates within the desert nation's borders. Virtually all its cropland requires irrigation. With its population leaping by 1 million every eight months, drinking-water and food needs are rising rapidly: water demands will likely exceed reliable supplies within a decade or so.[33]

To make matters worse, the Nile can diminish markedly during periods of low rainfall. During the summer of 1988, after several years of drought in the watershed, the Nile dropped to its lowest level in a century. Fortunately, good rains since then have restored supplies. But the reprieve may not last long. The Nile's runoff pattern exhibits periods of low flow early in each century. Diminished supplies superimposed on escalating demands could initiate a painful period of chronic water shortages.[34]

Just across the Suez Canal, the situation looks equally grim. Israel, Jordan, and Syria get most of their water from the Jordan River basin. Israel already uses 95 percent of the renewable supplies available to it, largely because it expanded the irrigated area sixfold since 1948. The World Bank predicts that if current consumption patterns continue, water demands in Israel, Jordan, and the West Bank will exceed all renewable supplies within six years. Both Israel and Jordan may soon face the politically difficult choice of shifting water away from farmers to supply growing domestic and industrial needs.[35]

Competition over water is heightening in China, as well, where dozens of cities already face acute shortages. The predicament is especially dire in and around Beijing, in the important industrial city of Tianjin, and in other portions of the North China Plain, a vast expanse

Table 3-3. Water Scarcity, Selected Countries and Regions

Country/Region	Observation
Africa, North and East	Ten countries likely to experience severe water stress by 2000; Egypt, already near its limits, could lose vital supplies from the Nile as upper-basin countries develop the river's headwaters.
China	Fifty cities face acute shortages; water tables beneath Beijing are dropping 1–2 meters per year; farmers in Beijing region could lose 30–40 percent of their supplies to domestic and industrial uses.
India	Tens of thousands of villages throughout India now face shortages; plans to divert water from Brahmaputra River have heightened Bangladesh's fear of shortages; large portions of New Delhi have water only a few hours a day.
Mexico	Groundwater pumping in parts of the valley containing Mexico City exceed recharge by 40 percent, causing land to subside; few options exist to import more fresh water.
Middle East	With Israel, Jordan, and the West Bank expected to be using all renewable sources by 1995, shortages are imminent; Syria could lose vital supplies when Turkey's massive Ataturk Dam comes on-line in 1992.
Soviet Union	Depletion of river flows has caused volume of Aral Sea to drop by two thirds since 1960; irrigation plans have been scaled back; high unemployment and deteriorating conditions have caused tens of thousands to leave the area.
United States	One fifth of total irrigated area is watered by excessive pumping of groundwater; roughly half of western rivers are overappropriated; to augment supplies, cities are buying farmers' water rights.

SOURCE: Worldwatch Institute, based on various sources.

of flat, fertile farmland that yields a quarter of the nation's grain. Water tables beneath the capital have been dropping 1–2 meters per year; already, a third of its wells have reportedly gone dry.[36]

Most supplies on the North China Plain have already been tapped, yet demands continue to grow rapidly. Shifting water from farm to city use may be the only way to balance the region's water budget. A management study for Beijing suggests that farmers in the vicinity could lose 30–40 percent of their current supply within the next 10 years. In dry years, some already lose out: when the levels of Beijing's two major reservoirs fell to record lows in 1985, supplies to farmers not growing vegetables were cut off.[37]

In pockets of the western United

States, a region plagued by numerous water deficits, supplies are being siphoned away from agriculture by thirsty cities willing to pay a premium to ensure water for their future growth. Where systems of water law and allocation establish clear property rights to water, markets can operate to transfer supplies between willing buyers and sellers for an agreed-upon price.

Agriculture will likely lose supplies wherever demands are nearing water's natural limits.

Clearly, if a farmer can earn more selling water to a nearby city than spraying it on cotton, alfalfa, or wheat, shifting that water from farm to city use is economically beneficial. If it prevents the city from damming another river to increase supplies, the transfer can also benefit the environment. Sale prices have varied greatly, with perpetual water rights having been sold for less than $200 per acre-foot in the Salt Lake City area, but for as high as $3,000 to $6,000 per acre-foot in the rapidly urbanizing Colorado Front Range. (An acre-foot equals 325,850 gallons or 1,233 cubic meters, enough to supply a typical four-person U.S. household for about two years.)[38]

Many large trades of this kind obviously could threaten crop production. But not all farm-to-city transfers require shifting water rights permanently out of agriculture. For example, the Metropolitan Water District of Southern California (MWD)—water wholesaler for roughly half the state's 28 million residents—has agreed to finance conservation projects in the neighboring Imperial Irrigation District in exchange for the estimated 100,000 acre-feet of water per year the investments will save. The annual cost per acre-foot conserved is estimated at $128, far lower than MWD's best new-supply option. Enough water is being transferred this way to meet the needs of 800,000 Californians, yet no cropland is being taken out of production and no irrigation water rights are actually being sold.[39]

At the other extreme are transactions in Arizona, where Tucson, Phoenix, and other burgeoning cities have taken to "water ranching." State law makes it difficult to buy rights to water independent of the land. As of April 1988, some 232,800 hectares of agricultural land had been purchased as water farms. In Pima County, where Tucson is located, irrigated agriculture is expected to disappear by 2020.[40]

The extent to which marketing leads to reductions in irrigated area will depend on how desperate for water cities become, and on whether farmers respond more by retiring land or improving efficiency. But whether marketing is encouraged or not, agriculture will likely lose supplies wherever demands are nearing water's natural limits, since almost everywhere this resource's value in crop production is far less than in other activities. Planners in China calculate, for example, that a given amount of water used in industry generates more than 60 times the economic value of that same amount used in agriculture.[41]

Growing concern about excessive stream-flow depletion could also heighten competition for farmers' water. The doctrine of "prior appropriation" governing western water allocations in the United States traditionally assigned rights only to water withdrawn from a river or stream for some "beneficial use." Historically, diversions for agricultural, industrial, and domestic purposes met this criterion, but no rights were granted for water left instream to preserve fish and wildlife habitat, recreational whitewater, and other ecological and aesthetic values. In recent years, all the western states except New Mexico have taken steps to protect instream water values. Most legally recognize these functions as a beneficial use and, in some cases, have authorized state agencies to buy water rights from private holders in order to boost streamflow levels.[42]

Another powerful U.S. lever is the "public trust doctrine," which asserts that governments hold certain rights in trust for the people and can take steps to protect those rights from private interests. Its application could have sweeping effects, since even existing water rights could be revoked to prevent violation of the public trust.[43]

WATER IN THE GREENHOUSE WORLD

Despite its many uncertainties, global warming guarantees that the future will not be a simple extrapolation of the past. (See Chapters 2 and 5.) Changes in water availability, among the profound transformations in store, add an entirely new dimension to the constraints irrigated agriculture faces during the next several decades.

Like a one-way filter, greenhouse gases allow the sun's energy to pass through the atmosphere but trap the longer-wave radiation emitted back toward space. As a result of their buildup from human activities, the earth's lower atmosphere will warm, which in turn will cause its hydrological cycle—the transfer of water between the sea, air, and land—to pick up speed. Both precipitation and evaporation, which balance each other in the planetary water budget, are expected to increase by 7–15 percent with the equivalent of a doubling of heat-trapping gases over preindustrial levels. Global averages, however, say little about what particular regions will experience. Temperatures will rise almost everywhere, though by more in the temperate and polar regions than in the tropics; rainfall will increase in some areas, decrease in others.[44]

Runoff, the primary measure of a region's renewable water supply, is a function of the amounts of water gained through precipitation and lost through evaporation or via transpiration by plants. Besides temperature and precipitation, global warming will alter winds, humidity, and cloud cover—all influences on evapotranspiration and, thus, runoff. Scientists have also found that plants grown in carbon-dioxide-enriched air partially close the leaf openings through which water vapor and other gases are exchanged with the atmosphere, enabling them to use water more efficiently. Studies have shown that doubling carbon dioxide levels can reduce transpiration by a third to half; across an entire watershed, this could augment runoff substantially. Scientists do not know, however, whether water efficiency would actually increase like this in natural environments, given the many other factors at work.[45]

Much of the limited research done to explore global warming's potential impact on regional water supplies has focused on the western United States, a food-producing area of great importance that relies heavily on irrigation. A large share of the West's runoff derives from the melting of high-mountain snowpacks, which are in effect vast reservoirs that release their supplies gradually through the spring and summer. A good deal of western agriculture depends on this meltwater becoming available during the growing season.

Moreover, engineers designed the numerous reservoirs in the West to meet hydropower, flood protection, and water supply needs assuming that historic runoff patterns would continue. Changes in those patterns would greatly complicate water management, and cost billions—either in greater flood damages, reduced hydropower, and more severe summer shortages, or in new reservoirs and other measures to avoid those consequences.[46]

Peter Gleick of the Pacific Institute for Studies in Development, Environment, and Security and John Schaake, a hydrologist with the National Weather Service, have each shown that the anticipated higher temperatures will dramatically alter the timing of runoff in key parts of the West. Schaake analyzed changes in runoff for the Animas River at Durango, Colorado. He found that an increase of 2 degrees Celsius, with no change in precipitation, would affect total annual runoff very little. Seasonal runoff patterns, however, would change markedly as a result of reduced winter snowpack and faster melting of the snow that does fall.[47]

Schaake's model showed that average runoff for January through March would increase by 85 percent, while in the critical months of July through September it would drop by 40 percent. Such drastic shifts would raise the risk of flooding, require costly trade-offs between hydropower and water supply needs, and lead to scarcities during the hot, dry summer, when water demands for crops and urban areas are greatest.[48]

To assess possible impacts for the entire western region, Charles Stockton and William Boggess examined the effects on seven river basins of a 2-degree-Celsius increase and a 10-percent decrease in precipitation—within the range of possible futures according to several leading climate models. They found that supplies in each of the seven basins dropped by 40–76 percent. In four of them—the Upper and Lower Colorado, the Missouri, and the Rio Grande—severe water deficits would likely arise. Since increasing scarcity has already shifted water out of agriculture, large reductions in available supplies would undoubtedly hasten that trend. Indeed, balancing the climate-altered water budgets of these seven regions could require taking as many as 4.6 million hectares out of irrigation, nearly a quarter of the U.S. total.[49]

Even as forces act to siphon water away from farms, a greater share of cropland may need the protection from soil-moisture deficits that irrigation offers. The late Professor Dean Peterson and consultant Andrew Keller estimated that a rise of 3 degrees Celsius would pull irrigation water needs up by 15 percent. If in addition rainfall increased by 10 percent, irrigation needs would rise only 7 percent. But the combination of the same temperature increase and a 10-percent drop in rainfall would mean the need for irrigation water would go up by 26 percent.[50]

With too little water available to expand irrigation in the West, cropland needing additional water for profitable production but unable to get it would likely come out of cultivation altogether. Peterson and Keller estimated that cultivated area in the western states could shrink by 24 percent, 15 percent, and 31 percent under the three different scenarios.[51]

Global warming's impacts on water resources and irrigation have not been studied as closely for other locations in the world. Yet for several reasons, climate change over the next several decades seems likely to act as an additional drag on expansion of the irrigation base. To be sure, some land will become suitable for irrigation. But even more will likely have to be retired.

It is probably conservative to assume that water scarcity induced by climate change will force irrigation to cease on 5 percent of the world's existing irrigated cropland. That would remove 13 million hectares from the global base. At today's rate of expansion, it would take half a decade to recoup those losses. Moreover, building the dams, canals, and other infrastructure to regain that loss would require supplemental investments on the order of $26–52 billion. (See Table 3–4.)

In addition, some cropland now watered only by rain will need irrigation to replace soil moisture lost to the atmosphere from the increased evapotranspiration that higher temperatures will bring. Peterson and Keller looked at the eastern United States, a region with little permanent irrigation now, and concluded that this increased need for new irrigation could be substantial. If 5 percent of the world's rainfed cropland requires irrigation to stay profitably productive, even after farmers switch crops and make other adjustments, some 60 million more hectares of irrigated area would be necessary. That would raise investments to adapt to new climatic regimes by $120–240 billion, bringing the total to roughly $150–300 billion.[52]

Whether such investments would materialize is far from certain. As noted earlier, lending for irrigation has been declining over the last decade. Perhaps more important, new investments will be hostage to uncertainties about just how the climate is changing in a given locale. Regional projections of soil moisture, evapotranspiration, runoff, and the other factors affecting water availability are not accurate enough to be the basis of important investment decisions. Worse, it could be decades before scientists' models acquire that degree of precision.

On a brighter note, some existing irrigated areas could benefit from increased rainfall that boosts regional water supplies. Models suggest such a possibility for India, for example, where irrigated crop production is currently constrained by water scarcity. Additional runoff could allow some irrigated lands to get optimum water amounts instead of current limited quantities, helping raise crop yields. But if the greater availability of water comes in the form of intensified monsoons, as models also suggest could happen, it could be as much a curse as a blessing. Much of it would run off in catastrophic floods rather than augmenting soil moisture and stable supplies.[53]

On balance, it appears that as climate changes over the next 20–30 years irri-

Table 3-4. Illustrative Irrigation Adjustments from Global Climate Change

Adjustment	Area Affected	Estimated Investment Cost[1]
	(million hectares)	(billion dollars)
Shifting 5 percent of world's existing irrigated area because of changes in water availability	13	26–52
Expanding irrigation to 5 percent of existing rainfed cropland to compensate for increased evapotranspiration	60	120–240
Total	73	146–292

[1]Assumes range of $2,000–4,000 per hectare; costs in some areas could be substantially higher. Costs were not discounted.
SOURCE: Worldwatch Institute.

gation is unlikely to be able to provide the degree of crop protection and yield enhancement the world now counts on. For some time, a mismatch between the cropland needing more water and that getting it could dampen irrigation's contribution to global food supplies.

STRATEGIES FOR THE NINETIES

Securing water to meet the world's growing food needs sustainably will not be easy. The slowdown in irrigation growth, mounting environmental damage from irrigation and from water projects generally, worsening regional scarcities, and the prospects of climate change all combine to severely constrain water use for crop production. Together these trends point to the need for a more diverse approach to watering crops—one that focuses on raising efficiency, on integrating irrigation more fully with basic development goals, and on improving water's productivity in rainfed farming.

Much of the pervasive overuse and mismanagement of water in agriculture stems from the near-universal failure to price it properly. Irrigation systems are often built, operated, and maintained by public agencies that charge next to nothing for these services. Farmers' fees in Pakistan, for example, cover only about 13 percent of the government's costs. Indeed, in most of the Third World, government revenues from irrigation average no more than 10–20 percent of the full cost of delivering water. Such undercharging deprives agencies of the funds needed to keep canals and other infrastructure in good working order.[54]

Pricing reforms are equally needed in the United States. The federal Bureau of Reclamation supplies water to a quarter of the West's irrigated land—more than 4 million hectares—under long-term contracts (typically 40 years) at greatly subsidized prices. Farmers benefiting from the huge Central Valley Project in California, for example, have repaid only 5 percent of the project's costs over the last 40 years: $50 million out of $931 million. Largely because of this underpricing, one third of the Bureau's water irrigates hay, pasture, and other low-value forage crops that are fed to livestock, even though so many other higher-valued activities need additional water.[55]

As climate changes, irrigation is unlikely to be able to provide the degree of crop protection and yield enhancement the world now counts on.

This free ride also explains for the most part why so few western farmers invest in efficiency improvements. Hundreds of federal irrigation contracts will be coming up for renewal during the nineties, as noted earlier. If the Bureau seizes this opportunity to establish contract terms that foster efficiency rather than discourage it, water stresses in the American West could ease measurably.

When prices reflect water's scarcity, when the costs of obtaining water go up, or when governments regulate its use, farmers begin to use the resource more prudently. Irrigators in Israel, parts of the Soviet Union, and Texas have shown that they can boost their efficiencies dramatically by adopting modern technologies and better management practices.

Large sprinklers can be made more efficient, for instance, by vertical drop tubes attached to the sprinkler arm. Called low-energy precision application (LEPA), these systems deliver water closer to the ground and in large droplets, cutting evaporation losses. Farmers can also conserve by using lasers to help level fields so that water gets distributed evenly, recycling used irrigation water, installing water-thrifty drip systems, and monitoring soil moisture so they irrigate only when necessary. The average amount of water applied per hectare by Texan farmers dropped 28 percent between 1974 and 1987 because they adopted LEPA and other efficiency measures. Similarly, upgrading of irrigation systems has reduced per-hectare withdrawals in the Soviet Union by 30 percent since 1970.[56]

In many developing countries, improving the performance of canal systems is critical to boosting irrigated crop yields closer to their potential. Involving farmers in the layout, operation, and management of the systems and providing flexibility in the scheduling of water deliveries are effective steps toward this end. Studies in the Philippines have shown, for example, that when farmers actively participate in the planning and management of projects, canals and other infrastructure function better, a greater proportion of the project area gets irrigated, and rice yields increase more.[57]

Variously called "irrigation associations" or "water user associations," organized groups of farmers who share a common water source can be instrumental in their effective involvement and in collecting fees. The Philippines National Irrigation Agency now receives no federal funding for operating and maintaining irrigation systems: it depends on revenues from the irrigation associations. This has made the agency more responsive to farmers' needs and more apt to solicit their views. Similar approaches are followed in China and South Korea, where canal systems tend to perform quite well.[58]

Greater attention to small-scale irriga-tion and techniques to make better use of rainwater can do much to increase food production at the village level—and to eliminate the need for expensive large dams with their attendant social and environmental costs. Smaller projects are decentralized, and focus on improving the productivity of water where it falls, rather than transporting it great distances. In contrast to the top-down character of large projects, they help build local self-reliance and, because of their greater flexibility, tend to be more responsive to farmers' needs as they perceive them.[59]

Simple techniques aimed at increasing soil moisture in the root zones of crops can markedly boost yields and make production more reliable. For instance, farmers can build check dams of earth and stone to capture runoff from hillsides, and then channel this water to their fields. Similar forms of "water harvesting" allowed some ancient farming cultures to thrive where annual rainfall averaged only 100 millimeters (4 inches). Vetiver grass, known in its native India as *khus*, can increase sustainable cropping on sloping land. When planted along the contours of a hillside, vetiver grass forms a vegetative barrier that slows runoff, allowing rainwater to seep into the soil. Sediment trapped behind it forms a terrace that further conserves soil and moisture for crops. Yields have increased by half over those in similar areas not using this technique.[60]

Although such down-to-earth practices rarely get the visibility and fanfare afforded a large, new dam, they nonetheless can produce the needed results. In Karnataka, India, for example, a watershed management effort involving some 600,000 hectares—nearly a third of the cultivated area—is based on low-cost techniques that farmers can use to increase soil moisture in their fields. Cropping intensity reportedly doubled in the project areas, to 220 percent.[61]

An analysis by the Environmental Defense Fund of Washington, D.C., shows that a variety of proven small-scale techniques collectively constitute a viable alternative to the irrigation component of the huge Sardar Sarovar Dam in western India, discussed earlier. Even the most expensive small-scale methods—which include small reservoirs to store rainfall, percolation tanks to replenish groundwater, and check dams and microcatchments to increase rainwater productivity—are estimated to cost less than half as much per hectare as irrigation with water from Sardar Sarovar would. Besides being cheaper, these alternatives would allow water development benefits to be distributed more equitably, reduce construction time, give farmers more control over their supply, and promote local employment. They clearly deserve a careful and objective examination by Indian officials and the World Bank, which is helping fund Sardar Sarovar.[62]

Small-scale irrigation and techniques to make better use of rainwater can do much to increase food production at the village level.

Besides directly funding more small-scale projects, development organizations can help foster private initiatives by giving small landholders access to credit and by bolstering local industries that supply needed tools and equipment. In Bangladesh, the progressive Grameen Bank is acquiring control of 100 tubewells and then offering loans to groups of five farmers who wish to buy one. The International Fund for Agricultural Development, a 12-year-old U.N. agency with an impressive track record, is among the global organizations now giving greater emphasis to small, farmer-managed systems. Its experience has been that they are "more economical, easier to manage, and better targeted at the poor."[63]

Tackling head-on the question of "irrigation for whom and for what?" is particularly critical in Africa. For much of the continent, irrigation will never work wonders the way it did in Asia. Water and irrigable land are too scarce, and the costs of developing water too high. Thus far, irrigation has tended to expand production of higher-valued crops grown for export or urban consumption, not basic food crops.[64]

Projects that supplement rather than replace traditional agricultural practices may be the most promising way to boost water productivity in Africa over the next couple of decades. African farmers' cultivation strategies tend to aim more toward minimizing risks of crop failure in bad years than maximizing yields in good ones. For many, complete dependence on irrigation would increase risk, since spare parts and power to run the systems may not always be available or reliable. Moreover, outside of Egypt and the Sudan, large-scale irrigation is foreign to most African cultures and has the potential to alienate farmers rather than benefit them.[65]

An important test of integrating modern water management with traditional practices is now under way in the Senegal River valley. Farmers there practice "flood-recession agriculture," an indigenous cultivation method used by millions that involves planting crops on river floodplains after seasonal floodwaters recede. By engineering controlled floods, the reliability and productivity of this method can be improved— indeed, some say yields can be doubled or tripled—without incurring much of the ecological damage, exorbitant cost, and disruption of local traditions that full-scale irrigation often brings. At least until irrigation facilities are in place, the Manantali Dam is being operated for optimum recession cropping of sorghum. The first controlled flood was released in September 1988, and plans are to continue them through 1993. If successful and sufficiently high-yielding, this modified version of traditional floodplain agriculture could become a permanent part of the dam's operations.[66]

Restoring deforested watersheds urgently needs greater international assistance. Especially in tropical regions, a large share of potentially useful water runs off in damaging floods, causing far more harm than good. Reforesting mountainous catchments will help slow runoff, enhance the infiltration of rainwater into the soil, and thereby increase downstream groundwater supplies for the dry season. Revegetating watersheds is especially important if, as some climate models predict, tropical monsoons begin striking with even greater ferocity. An important step forward would be for the Asian Development Bank, the World Bank, or another international body to work with the countries sharing the Himalayan watershed on a restoration plan for this badly degraded region.[67]

As fresh water becomes increasingly scarce, and as cities bid more supplies away from farmers, the use of treated urban wastewater for irrigation is likely to become commonplace. This returns valuable nutrients to the land, where they belong, and helps keep them out of rivers and streams, where they become troublesome pollutants. With proper treatment, and with care in how and where reclaimed wastewater is applied, the practice can be very beneficial.

Israel, with virtually no new fresh sources to tap, is now reusing 35 percent of its municipal wastewater, generally for irrigation. Some 15,000 hectares, most of them cultivating cotton, are now irrigated with the reclaimed water. Current plans are to reuse 80 percent of the

nation's total wastewater flow by the end of the decade.[68]

A relatively inexpensive way governments can buy insurance against future water constraints, especially in light of climatic change, is to increase funding of international agricultural research centers working to develop new strains of crops. Plants that are more salt-tolerant, drought-resistant, and water-efficient could play a vital part in securing adequate food supplies. (See Chapter 4.) Research suggests that wheat, for one, is a good candidate for breeding in greater tolerance to salt. This could allow this important grain to remain productive on salinized land where strains common today would not thrive.[69]

Salicornia, a succulent, can be irrigated with seawater and shows promise as a substitute for thirsty fodder crops in water-short regions. Its yield of oil seed compares well with soybeans, and it can contribute up to 10 percent of a fodder mix for cattle, sheep, and other livestock. Developing new strains that are sufficiently high-yielding and profitable to grow, however, takes time. Greater support for such efforts now could pay back handsomely in the decades ahead.[70]

No quick fix is going to solve agriculture's water problems any time soon. Transforming crop production into a water-thrifty but still highly productive enterprise is a monumental task. Added up, the varied ways of using irrigation water more efficiently and of increasing water productivity on rainfed lands can go a long way toward preventing water scarcity from undermining food security. But these diverse technologies and measures will only spread widely if pricing policies, markets, regulations, and international development agencies promote them.

Ultimately, emerging scarcities and other water constraints call upon societies to slow the growth in human numbers and adjust their activities to water's natural limits. Populations continue to grow fastest in some of the world's most water-short regions, notably northern Africa, the Middle East, and parts of the Indian subcontinent. When something so fundamental as water becomes scarce, only fundamental responses will suffice. Delaying action simply increases the risks that farmers—and crops—will be left high and dry.

4

Feeding the World in the Nineties

Lester R. Brown and John E. Young

As we enter the nineties, the world has little to celebrate on the food front. Between 1950 and 1984, the world's farmers raised world grain output 2.6-fold, an increase that dwarfed the efforts of all previous generations combined. Since then, unfortunately, little progress has been made. The share of hungry and malnourished people in Africa, Latin America, and parts of Asia has increased.[1]

Drought-damaged harvests in key producing countries in 1987 and 1988 brought world grain stocks to one of their lowest levels in decades—little more than is needed to fill the "pipeline" from field to table. They also brought world wheat prices in 1989 up by 48 percent from their all-time low of two years before. Rice prices, which in 1986 reached their lowest point, were up 38 percent.[2]

With higher prices and better weather in 1989, it was widely assumed that production would surge upward and stocks would be rebuilt. But this did not happen. The 1989 world grain harvest shortfall—18 million tons below the projected consumption of 1,685 million—reduced depleted stocks even further. If stocks cannot be replenished in years of near-normal weather, when can they be?[3]

Rebuilding stocks depends on pushing production above consumption, a task that is becoming more difficult. Growth in world food output is being slowed by environmental degradation, a worldwide scarcity of cropland and irrigation water, and a diminishing response to the use of additional chemical fertilizer. Meanwhile, the annual addition to our numbers is moving above 90 million people.[4]

DEGRADATION AFFECTING HARVESTS

Nearly all forms of global environmental degradation are adversely affecting food production: Soil erosion is slowly undermining the productivity of an estimated one third of the world's cropland. Deforestation is leading to increased rainfall runoff and crop-destroying floods. Damage to crops from air pollution and acid rain can be seen in industrial and developing countries alike. Data from experimental plots indicate that yields of some crops, such as soybeans, are reduced by the increased ultraviolet radiation associated with stratospheric ozone depletion. Waterlogging and salinity are lowering the productivity of a fourth of the world's irrigated cropland. (See Chapter 3.) Global climate change in the form of hotter summers may also be taking a toll.[5]

In this section, we attempt to estimate how much productive capacity the world has lost through these many forms of environmental degradation. We cannot present a detailed set of precise data on such crop losses, because such data do not exist. But we hope this presentation will help focus attention on how damaging these environmental influences have become, and that it will encourage the principal institutions involved in international agricultural development, such as the World Bank and the U.N. Food and Agriculture Organization (FAO), to research the effect of environmental degradation on grain output.

Reports of land degradation have come from every corner of the planet. In July 1989, Australian Prime Minister Robert Hawke said that "none of Australia's environmental problems is more serious than the soil degradation . . . over nearly two-thirds of our continent's arable land." *Pravda* reports that the Soviet Union is suffering from a catastrophic decline in soil fertility. Prime Minister Rajiv Gandhi has outlined the ecological and economic crisis India faces as the result of continuing deforestation and the associated degradation of land.[6]

Each year, the world's farmers lose an estimated 24 billion tons of topsoil from their cropland in excess of new soil formation. During the eighties, this translated into a loss of 240 billion tons, an amount more than half that on U.S. cropland.[7]

Agronomist Harold Dregne has classified agricultural land in each major geographic region into one of three categories according to the degree of degradation. Slight degradation includes all land where the yield potential has been reduced by less than 10 percent; moderate erosion, land where it has been reduced between 10 and 50 percent; and severe erosion, a reduced potential of more than 50 percent. All regions have some severely degraded land, but Africa, Asia, and South America have much more in this category than North America and Europe do. (See Table 4–1.)

According to FAO, soil erosion could reduce agricultural production in Africa by one fourth between 1975 and 2000 if conservation measures are not adopted. The shifting cultivation traditionally practiced in Africa to maintain soil fertility has begun to deteriorate under high population densities as farmers return to the same plot every 5–10 years instead of waiting 20–25 years, as they used to.[8]

Table 4-1. Estimated Land Degradation in the Late Seventies

Continent	Slight	Moderate	Severe	Total
	(percent of land surface)			
Africa	60	23	17	100
Asia	56	28	16	100
Australia	38	55	7	100
Europe	69	25	6	100
N. America	70	23	7	100
S. America	73	17	10	100

SOURCE: U.S. Department of Agriculture, Economic Research Service, *World Agriculture Situation and Outlook Report*, Washington, D.C., June 1989, based on data compiled by Harold E. Dregne.

As the fallowing cycle shortens and the land's vegetative cover diminishes, soil erosion and land degradation accelerate. Gbdebo Jonathan Osemeobo of Nigeria's Federal Department of Forestry notes that "shifting cultivation has a 'catalytic' effect on the encroachment of desert in the northern parts of Nigeria—namely in Sokoto, Kano, Kaduna, Bauchi, and Borno states." He believes that this land is now only marginal for either forestry or farming.[9]

The effect of topsoil losses on U.S. land productivity has been well established by 14 studies on corn and 12 on wheat. The former, done mostly in the Corn Belt states, showed that a loss of one inch of topsoil reduced corn yields by roughly 6 percent. Results for wheat, taking an average for the 12 studies, also showed a drop of 6 percent for each inch of topsoil lost.[10]

Applying this to the world as a whole, the loss of 24 billion tons of topsoil annually (equal to one inch on 61 million hectares) would reduce the grain harvest of 146 million tons from this area by 6 percent (assuming the world average yield of 2.4 tons per hectare), or 9 million tons per year. Alternatively, for purposes of calculation, we could assume that all the topsoil (typically seven inches deep) was lost from 9 million hectares, leading to the land's abandonment and the loss of more than 21 million tons of grain production per year. Choosing to be conservative, we have included the lower estimate in our tabulation of environmental losses of productivity.[11]

Land is also being degraded by changing soil chemistry, as rising concentrations of salt in irrigated soils reduce harvests on some of the world's most intensively farmed land. (These issues are discussed in more detail in Chapter 3.) In the haste to complete irrigation projects from 1950 to 1980, the drainage facilities essential to the long-term success of these systems were frequently omitted. When river water diverted onto farmland percolates downward and there is no drainage, it gradually raises underground water tables. When these get close to the surface, water evaporates into the atmosphere, concentrating salts in the upper few inches of soil.

In industrial and developing countries alike, waterlogging and salinity are lowering yields on an estimated 24 percent of the world's irrigated cropland. (See Chapter 3.) As with soil erosion, these processes gradually lower land productivity until eventually farming the land is no longer profitable and it is abandoned. Assuming an estimated 1-percent annual decline in output on 24 percent of the 180 million hectares of irrigated grain, mostly rice and wheat, with an average yield of 2.5 tons per hectare, an additional 1.1 million tons of grain output are lost each year. Alternatively, using the estimate by Soviet agronomist Victor Kovda of worldwide abandonment of 1–1.5 million hectares of irrigated land each year would yield an annual loss in grain output of 2.4–3.6 million tons. For our tabulation, we use the lower figure of 1.1 million tons.[12]

Deforestation affects food production in several ways. It directly alters the local hydrological cycles by increasing runoff and, perhaps less obviously, by affecting the recycling of rainfall inland. The former is now strikingly evident in the Indian subcontinent, where deforestation of the Himalayan watersheds is raising rainfall runoff, leading to increasingly severe flooding. The area subject to annual flooding in India has more than tripled since 1960. (See Table 4–2.)

This process was also tragically demonstrated in early September 1988, when two thirds of Bangladesh was under water for several days. This flood—the worst on record—damaged the country's fall rice crop, increasing its need to import grain.[13]

As long as deforestation continues,

**Table 4-2. India: Area Subject
to Flooding**

Year	Area
	(million hectares)
1960	19
1970	23
1980	49
1984	59

SOURCE: Centre for Science and Environment, *The Wrath of Nature: The Impact of Environmental Destruction on Floods and Droughts* (New Delhi: 1987).

firewood will become ever more scarce, forcing villagers to burn more cow dung and crop residues. This deprives the soil not only of nutrients, but also of the organic matter that helps maintain a healthy soil structure. One region for which data are available, the state of Madhya Pradesh in central India, is already far down this road: of the three principal cooking fuels used by its 62 million people, cow dung accounts for 9.6 million tons, firewood 9.5 million tons, and crop residues 6.9 million tons. Similar pressures are draining the land of productivity in other Indian states, in Bangladesh and Pakistan, in scores of African countries, and in China.[14]

Damage to crops from air pollution can now be measured in automobile-centered societies and coal-burning economies.

Researchers have only recently identified two other threats to future food security: cars and coal-fired power plants. Damage to crops from air pollution can now be measured in the automobile-centered societies of Western Europe and the United States and in the coal-burning economies of Eastern Europe and China. Much of what we know about this subject comes from a seven-year study jointly sponsored by the U.S. Environmental Protection Agency and the U.S. Department of Agriculture (USDA). The report concluded that ozone, sulfur dioxide, and nitrous oxides are the most damaging air pollutants in the United States.[15]

Of these pollutants, ozone does by far the most damage. Crops are incredibly sensitive to minute amounts of ground-level ozone, amounts measured in parts per million (ppm). As ground-level ozone rises above the naturally occurring 0.025 ppm, damage begins to show at 0.05 ppm if this exposure level lasts for more than 16 days of a growing season, at 0.10 ppm if it lasts for 10 days or more, and at 0.30 if it lasts 6 days or more. The potential for damage is evident in the readings from some 70 widely scattered sites across the United States where average ozone concentrations for the season ranged from 0.038 ppm to 0.065. Surprisingly, damage from ozone occurred in sparsely populated rural areas as well as in densely populated urban ones.[16]

Experimentally increasing ozone concentrations from 0.04 ppm to 0.09 ppm as a seasonal average reduced yields of crops rather dramatically. For corn, the loss went from 1 percent at 0.04 ppm to 13 percent at 0.09 ppm. For soybeans, from 7 percent to 31 percent. And for winter wheat, from 4 percent to 27 percent. Ground-level ozone at levels that prevailed during the eighties led to an estimated U.S. crop loss of at least 5 percent, and possibly as much as 10 percent.[17]

Using the lower figure of 5 percent for ozone damage and ignoring the harm caused by other air pollutants, the 1987 grain harvest for the most severely affected regions—North America, Europe (excluding the Soviet Union), and China—of 913 million tons was reduced by 48 million metric tons. Since most of the growth in fossil fuel burning, the source of the pollutants, has occurred over the last 40 years, we assume growth in the annual loss is roughly 1 million tons.[18]

While ground-level ozone reduces harvests, depletion of the ozone layer in the upper stratosphere and the resulting rise in ultraviolet radiation reaching the earth's surface also may be taking a toll, particularly of the more sensitive crops, such as soybeans. A U.S. government panel measured the worldwide loss of stratospheric ozone at roughly 2 percent from 1969 to 1986. For each 1-percent loss of stratospheric ozone, the ultraviolet radiation reaching the earth increases by 2 percent. Data from experimental plots indicate that each 1-percent increase in ultraviolet radiation lowers soybean yields by a like amount. This suggests that depletion of the stratospheric ozone layer may now be reducing measurably the output of the world's leading protein crop; unfortunately, no one is monitoring this loss to determine its precise dimensions.[19]

Of all the global changes we have set in motion, climate change is potentially the most disruptive. Already suffering from slower growth in food output, the world is now confronted with the prospect of hotter summers. The drought- and heat-damaged U.S. grain harvest in 1988, which fell below consumption for the first time in history, dramatically illustrated how hotter summers may affect agriculture over the longer term.[20]

Climate change will not affect all countries in the same way. The projected rises of 2.5–5.5 degrees Celsius (4.5–9.9 degrees Fahrenheit) by late in the next century are global averages. But temperatures are expected to increase much more in the middle and higher latitudes and more over land than over the oceans. They are projected to change little near the equator, while in the higher latitudes rises could easily be double

that projected for the earth as a whole. This uneven distribution will affect world agriculture disproportionately, since most food is produced on the land masses in the middle and higher latitudes of the northern hemisphere.[21]

Agriculture is affected by climate change in innumerable ways. Hotter summers lead to more evaporation during the growing season, increasing the likelihood of drought. The effects of higher temperatures on agriculture have so far focused mostly on on soil moisture and drought-related conditions, but extreme heat can interfere with the pollination of staple food crops. Corn pollination can easily be impaired by high temperatures during the 10-day period when fertilization occurs. Alan Teramura of the University of Maryland reports that even a rise of 2 degrees Celsius in summertime temperatures in some key Asian rice-growing regions can sharply reduce harvests.[22]

In 1988, the hottest year in the last century, each of the world's three leading food producers suffered a reduction in harvest as a result of heat and drought: the U.S. grain harvest was down from its 1987 level by 27 percent (74 million tons), the Soviet harvest by 8 percent (16 million tons), and the Chinese harvest by 2 percent (7 million tons). No one can say for sure whether any of this 97-million-ton drop was due to the buildup of greenhouse gases, but the summer of 1988 was the kind of season that global meteorological models project will occur with increasing frequency as heat-trapping gases continue to accumulate in the atmosphere. Crop losses from the hotter summers that are in prospect seem likely to far offset the gains.

In summary, we have a lot more to learn about the effects on world agriculture of these various forms of environmental degradation. Soil erosion, the waterlogging and salinization of irrigated land, flooding, air pollution, acid rain, stratospheric ozone depletion, and hotter summers may all be taking a growing toll on food production. We estimate that the world could be losing 14 million additional tons of grain output each year because of this damage to land and crops. (See Table 4–3.) And this does not include any allowance for the effects of hotter summers, such as those experienced during the eighties and projected for the nineties. Stated otherwise, if there were no additional investments in agriculture or technological progress, the world grain harvest would fall 14 million tons annually—nearly 1 percent a year—as a result of environmental degradation.

The key question is, How does this loss compare with the gains from greater investment in irrigation, fertilizer, and other inputs? If we assume that the newly irrigated land of 2.3 million hectares per year during the eighties boosts output an average 1.5 tons per hectare above its previous level, then more than 3 million tons of grain is added to the world harvest each year. The 3 million more tons of fertilizer used each year (assuming that each ton of fertilizer raises output by 7 tons) adds another 21 million tons. If we generously allow another 5 million tons of added output per year from other sources, the total gain is 29 million tons.[23]

The world could be losing 14 million additional tons of grain output each year because of damage to land and crops.

Subtracting the 14 million tons we estimate are lost each year from environmental degradation therefore leaves a net gain in output of 15 million, well

Table 4-3. Crude Estimate of Additional Loss of World Grain Output Each Year as a Result of Environmental Degradation

Form of Degradation	Grain Output Loss
	(million tons)
Land Degradation	
Soil erosion	9
Waterlogging and salting of irrigated land	1
Loss of soil organic matter from burning cow dung and crop residue	}
Shortening of shifting cultivation cycle	} 2[1]
Compaction of soil from heavy equipment	}
Crop Damage	
Air pollution	1
Flooding	}
Acid rain	} 1[1]
Increased ultraviolet radiation	}
Total	14

[1]Because the lack of data makes it difficult to quantify the crop losses from these three sources, their effect is estimated collectively.
SOURCE: Worldwatch Institute estimates.

below the 28 million tons required to merely keep pace with population growth. In effect, this shows a net gain in grain output of just under 1 percent per year while population growth is close to 2 percent. This does not allow for any rises in demand from rising affluence. This offsetting effect of pressures on the environment may partly explain why the growth in world grain output has slowed so dramatically recently.

LAND, WATER, AND FERTILIZER

From the beginning of agriculture until mid-century, growth in the world's cultivated area more or less kept pace with that of population, but starting around 1950 it slowed to a crawl. After actually falling in the mid-eighties, growth in cropland has recovered somewhat as the United States returned to production some land previously idled under agricultural supply management programs.[24]

Each year millions of hectares of cropland are lost either because the land is so severely eroded it is not worth plowing anymore or because new homes, factories, and highways are built on it. Losses are most pronounced in the densely populated, rapidly industrializing countries of East Asia, including China, Japan, South Korea, and Taiwan, where nonfarm uses claim roughly a half-million hectares of cropland each year.[25]

Other large, densely populated countries suffering losses include India and Mexico. Both the Soviet Union and the United States are pulling back from rapidly eroding land. Abandonment of such cropland plus an increase in alternate-year fallow to maintain yields has reduced the Soviet grain area some 13 per-

cent since 1977. Fearing similar losses, the United States adopted a five-year plan beginning in 1986 to plant 40 million acres (16 million hectares) of rapidly eroding cropland to grass or trees before it became wasteland. By mid-1989, 32 million acres were committed, but budget-cutting congressional committees—looking primarily at short-term costs—eliminated funding for further additions to the program.[26]

Worldwide, the potential for profitably expanding cultivated area is limited. A few countries, such as Brazil, will be able to add new cropland. But on balance, gains and losses in the nineties are likely to largely offset each other, as they did during the eighties. As a result, the global decline in grain area per person from 0.16 hectares in 1980 to an estimated 0.14 hectares in 1990 seems certain to continue. (See Table 4–4.)[27]

The prospect for expanding the world's gross irrigated area is only slightly more promising. After growing slowly during the first half of this century, the irrigated area increased from 94 million hectares in 1950 to 236 million in 1980, as investment in irrigation soared in both large-scale river-diversion projects and farmer-owned irrigation pumps using groundwater. This expanded the gross irrigated area per person by 43 percent. (See Table 4–5.)

By 1980, however, most of the economically attractive sites for large irrigation reservoirs had been developed, and the number of newly completed projects dwindled. The expansion in world irrigated area slowed dramatically, falling behind population growth. As it continues to slow, the gains from new capacity may be largely offset by losses from waterlogging and salinity, falling water tables, and the silting of existing reservoirs discussed earlier.

With the net gain in world irrigated land estimated at only 23 million hectares during the eighties, the irrigated

Table 4-4. World Grainland, Total and Per Capita, 1950–80, With Projections to 2000

Year	Total Grainland	Per Capita Grainland	Per Capita Change By Decade
	(million hectares)	(hectares)	(percent)
1950	593	0.23	
1960	651	0.21	− 8
1970	673	0.18	−15
1980	724	0.16	−11
1990	720	0.14	−16
2000	720	0.12	−15

SOURCES: Grainland 1950–80 from U.S. Department of Agriculture (USDA), Economic Research Service (ERS), *World Grain Harvested Area, Production, and Yield, 1950–88* (unpublished printout) (Washington, D.C.: 1989); 1990 and 2000 figures are authors' estimates; per capita figures derived from population data (based on data from U.S. Bureau of the Census) in Francis Urban and Philip Rose, *World Population by Country and Region, 1950–86, and Projections to 2000* (Washington, D.C.: USDA, ERS, 1988); figures are rounded, with calculations based on original unrounded data.

area per person shrunk by close to 8 percent over the last decade. Although the cropland area per person has been falling steadily since mid-century, this trend of shrinking irrigated area per person is new, making this the first decade in which both have declined. This, too, may help explain the markedly slower growth of food production during the late eighties.

The growth pattern in total irrigated area (see Figure 4–1) conforms to what is known as the S-shaped growth curve, which describes well any growth process

in a finite physical environment. It is also a trend that describes the historical growth of cropland area, with the trend leveling off during the eighties. In this case, the trend has made the final turn on the S and has flattened out.

This S-shaped curve pattern is particularly familiar to biologists who chart the growth of various organisms in finite environments. As a laboratory exercise, biology students learn about the concept by watching the increase of a population of fruit flies in a jar or of algae in a petri dish when they are given all the food they need. At first growth is slow, as a few organisms reproduce. Then it accelerates until some constraint such as the physiological stresses associated with high density or the accumulation of waste begins to slow it, eventually bringing it to a halt.

We know that the rise in grain yield per hectare, like any biological growth process in a finite environment, will also eventually conform to the S-shaped growth curve. What we do not know is how close grain yields in agriculturally advanced countries now are to the final turn on the S. Interestingly, world growth in the use of inputs that are rais-

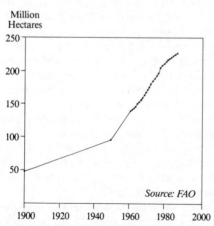

Million
Hectares

Source: FAO

Figure 4-1. World Net Irrigated Area, 1900–87

Table 4-5. World Gross Irrigated Area, Total and Per Capita, 1950–80, With Projections to 2000

Year	Total Irrigated Cropland	Per Capita Irrigated Cropland	Per Capita Change By Decade
	(million hectares)	(hectares)	(percent)
1950	94	0.037	
1960	136	0.045	+22
1970	188	0.051	+13
1980	236	0.053	+ 4
1990	259	0.049	− 8
2000	279	0.045	− 8

SOURCES: Data for 1950–80 adapted or derived from W. Robert Rangeley, Washington, D.C., private communication, June 1989, and from U.N. Food and Agriculture Organization, *FAO Production Yearbook 1988* (unpublished printout) (Rome: 1989); projections for 1990 and 2000 are Worldwatch Institute estimates; per capita figures derived from population data (based on data from U.S. Bureau of the Census) in Francis Urban and Philip Rose, *World Population by Country and Region, 1950–86, and Projections to 2000* (Washington, D.C.: U.S. Department of Agriculture, Economic Research Service, 1988).

ing yields—fertilizers and high-yielding varieties, as well as irrigation—are all conforming to the S-shaped growth curve as the nineties begin.

From mid-century on, the increasing use of chemical fertilizer has been the engine powering the growth in world food output. Between 1950 and 1989, world fertilizer use climbed from a meager 14 million tons to an estimated 143 million tons. If for some reason fertilizer use were abruptly discontinued, world food output would probably plummet some 40 percent or more.[28]

The contribution of the other two yield-raising inputs—irrigation and high-yielding varieties—derives heavily from their ability to boost the effectiveness of fertilizer. Hybrid corn, which was initially developed in the United States and has since spread throughout the world, was widely adopted because it responded readily to heavy applications of nitrogen fertilizer. Third World farmers switched from traditional varieties to the high-yielding dwarf wheats and rices that defined the Green Revolution precisely because the new varieties were so responsive to fertilizer.

Rapid growth in fertilizer use depends therefore on the continued spread of high-yielding seeds. But the adoption of these newly introduced crop varieties also follows an S-shaped trend. Their use by farmers increases slowly at first, as a few innovative growers plant them, and then more rapidly as large numbers of farmers see their advantage. Eventually, adoption slows and levels off as the new strains are planted on all suitable land.

The adoption curve for high-yielding wheats in India, which helped to double the wheat harvest between 1965 and 1972, illustrates this well. (See Figure 4–2.) These new strains are unlikely to reach all wheat-growing areas because much of what is left is semiarid land where the new varieties do not fare well. Graphs on the use of high-yielding corn in the United States, of high-yielding rices in Indonesia, and of high-yielding wheats in Mexico all show the same S-shaped curve.[29]

Once the new fertilizer-responsive varieties are planted on all suitable land, growth in fertilizer use also slows. Such a trend is now apparent in some major food-producing countries. After quintupling between 1950 and 1981, fertilizer use in the United States has fallen somewhat, and was slightly lower in 1989 than when the decade began. Within the Soviet Union, where heavy subsidies of fertilizers have led to their overuse, the move toward world market prices as part of the recent agricultural reforms is likely to actually reduce fertilizer use in the years ahead.[30]

Many developing countries are now also experiencing diminishing returns in fertilizer use. Both China and India have grain-fertilizer response ratios very similar to that of the United States, suggesting that future growth in fertilizer use will be much slower. And given the dependence of fertilizer use on water availability, the reduced growth in irrigation documented earlier is almost certain to affect fertilizer use as well. Fertilizer use will continue to expand as varieties with an even greater yield potential are developed, but where yields are already high by international standards, the output gains are likely to be modest compared with the quantum jump that came from the initial adoption of high-yielding varieties.[31]

The ultimate natural constraint on the rise of crop yields will be imposed by the upper limit of photosynthetic efficiency, a limit set by the basic laws of physics and chemistry. As the genetic potential of high-yielding varieties approaches this level, the response of crops to the use of additional fertilizer diminishes. Advances in plant breeding, including those using biotechnology, can help yields rise toward the photosynthetic limit, but there is little hope of altering the basic mechanics of photosynthesis.

BIOTECHNOLOGY: A LIMITED CONTRIBUTION

Grandiose claims about biotechnology and food production have been common since the first successful attempts at genetic engineering in the early seventies. As recently as 1984, one writer predicted that "in 5 to 10 years, Saudi Arabia may look like the wheat fields of Kansas." The unfortunate reality in 1989—when Kansas lost over a third of its wheat crop to drought—was that the wheat fields of Kansas came to resemble the still-fallow Saudi Arabian desert.[32]

Biotechnology has proved more difficult to apply to agriculture than its champions expected. It will not revolutionize agriculture overnight—but it could be an important new weapon in the fight against hunger.

To realistically gauge biotechnology's potential to help close the widening gap between the growth in world food supplies and that in human numbers, two questions must be answered. First, is the new biology technically capable of sharply boosting agricultural yields—particularly those of the cereal crops that feed most of the world's people—to a degree comparable to the increases of

Percent

Source: Agency for International Development

Figure 4-2. Share of Wheatland Planted to High-Yielding Varieties in India, 1965–83

the past four decades? And second, are the institutions now engaged in biotechnology research likely to achieve the breakthroughs needed to help feed the world's hungry?

It is difficult to overstate the revolutionary nature of the biological discoveries of the last two decades. The assortment of techniques referred to as biotechnology is opening new frontiers in areas from manufacturing to health care to pollution cleanup. Biotechnology has moved rapidly from universities into the laboratories and factories of industry. Like most revolutionary technologies, it has its strengths and weaknesses, and presents an array of potential difficulties as diverse as its promises.

Genetic engineering—the direct transfer of genes, which encode all life processes, from one organism to another—is the best-known and most fundamental of the new biological techniques. Tissue culture, an older technology that has advanced tremendously in recent years, is particularly important in plant biotechnology because it allows whole plants to be generated from single cells or small samples of tissue.[33]

In livestock production, a number of new techniques may allow biotechnology to live up to its technical promise. Hormones that speed and increase livestock weight gain per unit of feed and that increase the milk production of dairy cows can now be mass-produced by genetically altered bacteria. Similarly, veterinary vaccines and drugs can be made by microbes, and genetic techniques are allowing the quick and inexpensive diagnosis of many animal diseases. Livestock breeding may be speeded up and made more consistent by a variety of techniques involving manipulation of sperm and ova.[34]

Animal biotechnology has moved forward quickly because much more is known about the basic cell processes of animals than of plants, and because human health products are usually tested first in animals. Although the social impacts, safety, and long-term performance of some of the new animal biotechnologies are quite controversial, they are advancing rapidly, with some already commercialized.[35]

But plants—the ultimate source of all food—have been far less responsive to biotechnological manipulation. From the outset, it was molecular biologists and other scientists unfamiliar with agriculture and plant breeding who were most bullish on crop biotechnology. Most agricultural scientists have rightly remained skeptical.[36]

Plants—the ultimate source of all food—have been far less responsive to biotechnological manipulation.

Dr. Norman Borlaug, the plant breeder whose high-yielding wheats won him a Nobel Prize, recently pointed out that "the most productive applications of biotechnology and molecular genetics, in the near term, appear to be in medicine, animal sciences, and microbiology. . . . It will likely take considerably longer to develop biotechnological research techniques that will dramatically improve the production of our major crop species."[37]

Difficulties with identifying the gene or genes that code for particular crop characteristics, the unpredictable genetic variation that often occurs with gene transfer, and problems with tissue culture have all slowed the application of biotechnology to crops. Cereals—like all grasses—have been particularly resistant to genetic manipulation: they are not compatible with the bacterium most commonly used to introduce foreign genes into plant cells, and they are difficult to regenerate from altered cells. Only in 1988 did scientists succeed in transferring a foreign gene into a cereal crop, corn.[38]

Furthermore, once genes are transferred, conventional breeding techniques must still be used to evaluate the altered plants and make them suitable for distribution to farmers or for further breeding. The entire process—from gene transfer to dissemination of a new variety—can take as long as conventional breeding (typically 5–15 years). The new biotechnologies will allow scientists to specify the genetic structure of plants with increasing precision, but they will complement—not replace—plant breeding.[39]

Nonetheless, conventional breeding is limited to reproductively compatible species. Genetic engineering can bypass the reproductive barriers between species, giving plant breeders access to a new universe of genetic material. As the science advances, desirable qualities of unrelated plants will become available for transfer into crop species. Drought tolerance, nitrogen fixation, salt tolerance, and resistance to pests and disease are but a few often-mentioned possibilities.[40]

But it takes money to move genes. Economics and national and international policies will determine whether biotechnology can achieve its technical potential to provide more food for a hungry world.

Most of the money now going into biotechnology research comes not from public or philanthropic organizations but from multinational corporations. In the United States, industry spent an estimated $300 million on agricultural biotechnology research in 1987. The U.S. Department of Agriculture spent $85 million on such work in the same year— and an increasing share of such govern-

ment funds are spent on basic research, while applied research is now often the province of industry or industry-supported university laboratories.[41]

At the international level, the Consultative Group on International Agricultural Research (CGIAR)—the principal international crop research organization funded by governments—devoted only an estimated $11.6 million (5 percent of its 1989 operating budget of $229 million) to biotechnology research. Of this, more than two thirds was devoted to animal research and less than a third to plants.[42]

U.S. biotechnology companies are fond of describing miracle crops that will feed hungry millions, and often use the promise of such immense benefits to attract investors and justify minimal regulation. But the application of biotechnology in the rich nations may not yield the rather prosaic miracles needed by small farmers in developing countries. Corporations will use biotechnology to develop crops and agricultural products they expect to be profitable and, if possible, which they can patent. Much of their effort, for example, is now focused on fruits and vegetables and on expensive flavoring and sweetener crops.[43]

Several of the larger agrichemical firms see the most potential profit in selling farmers integrated packages of seeds, fertilizer, and pesticides. In conjunction with subsidiary or affiliated seed companies, they are combining research on chemicals and plants, developing crop varieties that will be compatible with their own products. Although pest and disease resistance are commonly touted as major goals in corporate crop development programs, resistance to herbicides—which will in fact increase the use of these chemicals—is receiving R&D priority.[44]

Chemical herbicides substitute for tillage, the traditional method of controlling weeds; they increase crop yields only if previous weed control methods were inadequate. No-till farming—which means fewer tractor trips through a farmer's field—can substantially reduce fuel use and soil erosion, but its reliance on herbicides poses greater risk of contaminated groundwater and adverse effects on wildlife. Some herbicides are toxic or carcinogenic, and questions remain about the safety of even the "new-generation" herbicides.[45]

Moreover, herbicides offer little to farmers in developing countries, where most of the world's hungry people live. Hand weeding is cheaper, because labor is both abundant and inexpensive, and it is safer than using chemicals. Herbicides make economic sense in high-capital agriculture and when labor costs are high; in most developing-country farming areas, the opposite situation exists.[46]

Similarly, bovine growth hormone (BGH)—one of the first animal biotechnology products to be commercialized, and one that can increase cows' milk production by as much as 10–25 percent—is not a practical technology for most livestock owners in the Third World. At roughly $100 per cow, BGH will cost as much per year as many people in developing countries spend on food. In addition, treated animals will require large, high-quality, consistent rations of feed, and frequent injections—hardly a practical technique in countries where vaccinations of humans for infectious diseases are still far from routine.[47]

Corporate biotechnology research is likely to lead to useful products for developing countries only if governments, private foundations, and international agencies have the opportunity—and the funding—to license corporate technologies. Ironically, such spinoffs are only likely to occur in crops for which corporations perceive little economic market. For example, corporate-funded research

on resistance to viral diseases in tobacco, tomato, and other species has yielded techniques useful for developing similar resistance in Third World crops such as cassava, yields of which are widely reduced by viruses.[48]

Allowing corporations to gain exclusive rights to the discoveries of university scientists has a chilling effect on the exchange of scientific information.

Thus, public funding for applied biotechnology research is extremely important at both the national and the international level. Unfortunately, public budgets for applied research are inadequate and shrinking even in the countries where biotechnology was pioneered.[49]

The danger of relinquishing this technology to the private sector is that the private sector will set the research agenda. Public money has directly and indirectly supported the development of biotechnology in public institutions. Allowing corporations to gain exclusive rights to the discoveries of university scientists in return for marginal contributions to the total research budget has a chilling effect on the exchange of scientific information and genetic materials, and skews research priorities.[50]

The international agricultural research centers—the institutions most likely to address the needs of farmers in developing countries—also deserve more support. CGIAR centers already use biotechnology techniques such as genetic probes, cell fusion, and tissue culture to detect plant diseases, introduce disease resistance into existing varieties, and create disease-free plants for distribution to farmers. But such research is only a small part of their overall international research, education, and extension effort.[51]

Most important in international agricultural research is a clear vision of priorities. Through their experience with the Green Revolution's successes and failures, the CGIAR centers are far better equipped than the private sector to develop crops and agricultural products that help feed the hungriest people rather than exacerbating existing nutritional inequities.

Ideally, the CGIAR system could develop new crop varieties, work with scientists in non-CGIAR international research centers, and train scientists from developing countries in biotechnology, thus improving national and regional crop development programs. This is already happening to some extent, but additional support for agricultural research is needed. The weakest links in the chain are between the international research centers, developing-country governments, and the farmers they ultimately serve.[52]

Controversy over genetic resources and their control could impede the development of biotechnology for developing countries. Realizing that the genes from native plants may be used to develop substitutes for their exports or be sold back to them in new products at high prices, developing countries have become understandably reluctant to allow unrestricted access to their genetic heritage. Though direct compensation schemes may be impractical, it is only just that industrial nations support development of biotechnology to feed the Third World—in full partnership with those countries.[53]

Furthermore, all private and public efforts in biotechnology will be undermined if the current wave of plant and animal extinctions goes unchecked. If the diversity of both crop and noncrop species is not safeguarded, much of the raw material now available for genetic manipulation will be lost. Biotechnology can move genes, but its ability to create them is virtually nonexistent.

THE SOVIET AGRICULTURAL PROSPECT

Efforts to identify possible dramatic gains in world food output in the near future often focus on the Soviet Union because of its recently announced program to reform agriculture by shifting to a market-oriented farm economy, with prices keyed to those of the rest of the world. Expectations are high for two reasons: the frequently made comparison between the productivity of private plots and of state farms, and the record-setting gains in China's agricultural output after reforms were adopted there a decade ago. (See Figure 4–3.)

Comparisons of the productivity on private plots with that of state farms are often more ideological than analytical, and hence misleading in several respects. First, private plots often grow specialty crops, rather than staples. Measuring the productivity of a household

Figure 4-3. Grain Output in the Soviet Union and China, 1950–89

garden plot on a Montana ranch against that of its wheat fields would yield comparative data similar to those frequently cited for the Soviet Union. The garden plot, carefully tended and watered with a garden hose and producing strawberries, tomatoes, and lettuce, would get a return per hectare many times that of the surrounding wheat fields.

Regarding the second source of hope, it is true that the agricultural reforms launched in China more than a decade ago and those recently set in motion in the Soviet Union have a common purpose—to raise output by strengthening the link between effort and reward. But there are several key differences. Many Chinese working on the land today managed family farms before the collectivization that followed the revolution in 1949. Since the Soviet revolution occurred much earlier, few people in that country remember how private farms operate in a market economy.

Given the enormous size of Soviet agricultural operations today, those who work on state and collective farms occupy highly specialized roles. Some drive tractors. Others milk cows. Some are bookkeepers. Others are mechanics responsible for equipment maintenance. A few people at the senior management level have some understanding of all phases of agricultural production and marketing, but they are in the minority. Farming requires an understanding of many things: land management, plant pathology, animal nutrition, agricultural mechanics, and marketing, to cite a few. Most Soviet farm workers lack this range of skills.

In contrast to agriculture in China, farming in the Soviet Union is highly mechanized, using heavy equipment not suited for family-sized production units. Time will be required to "tool up" manufacturing facilities to produce farm equipment more suited to smaller family-farm units.

Another key difference is the rate of fertilizer use. In both countries, reforms involve moving toward world market prices for agricultural inputs and farm products. Once these prices are established, the use of fertilizer will be adjusted to maximize profits. Farmers throughout the world facing the same crop prices and fertilizer costs will use similar amounts of fertilizer under similar conditions. Because fertilizer use in the mid-seventies in China was low by international standards, the reforms resulted in a sharp jump, accompanied by an impressive increase in grain output of nearly half between 1976 and 1984.[54]

In contrast, fertilizer use in the Soviet Union is already heavy due to high subsidies. The extent of overuse can be seen in the fertilizer-grain response ratios there compared with other grain-producing countries. In the United States, China, and India, where prices are keyed to the world market, farmers get 16 to 18 tons of grain for every ton of fertilizer used. (See Table 4–6.) In the Soviet Union, farmers get 8 tons of grain for each ton of fertilizer.

As Soviet agriculture moves toward the adoption of world market prices, the amount of fertilizer that can be profitably applied is likely to decline. Indeed, partial withdrawal of fertilizer subsidies actually reduced fertilizer use in some Soviet republics in 1988. A further reduction of subsidies in 1989 appears to have lowered fertilizer use nationwide. Although this did not measurably lower production, it does mean that the principal source of the growth in China's grain output following the reforms will not be operating at all in the Soviet Union.[55]

Nor is there likely to be much growth in net irrigated area in the Soviet Union in the years ahead. Barring the unlikely prospect of reviving the plan to divert southward the rivers that empty into the Arctic Sea, there is little hope of much irrigation expansion. (See Chapter 3.)

Table 4-6. Grain Production and Fertilizer Use in World's
Four Leading Grain-Producing Countries, 1986

Country	Grain Production	Fertilizer Use	Grain Produced Per Ton of Fertilizer
	(million tons)		(tons)
China	300	16.9	18
India	137	8.5	16
Soviet Union	202	25.4	8
United States	314	17.8	18

SOURCES: Grain production from U.S. Department of Agriculture, Economic Research Service, *World Grain Harvested Area, Production, and Yield, 1950–88* (unpublished printout) (Washington, D.C.: 1989); fertilizer use from U.N. Food and Agriculture Organization, *FAO Quarterly Bulletin of Statistics*, Vol. 1, No. 3, 1988.

Soviet agriculture is also plagued by its extreme northerly location. Moscow lies at roughly the same latitude as Canada's Hudson Bay, and Leningrad is actually in line with southern Alaska. The nation's climatic regime in terms of soils, rainfall, and solar intensity is thus much more similar to Canada's than to that in the United States.[56]

Crop production potential, like tree growth rates, is much lower in extreme northerly regimes than in the middle latitudes. It is more appropriate, therefore, to compare Soviet agricultural productivity with that of Canada than with that of the United States or China. Soviet wheat yields during the 1985–88 period averaged 1.7 metric tons per hectare, only slightly less than the 1.8 metric tons of Canada, where croplands are newer and less eroded.[57]

References in Soviet agricultural circles to an unfolding ecological catastrophe are becoming commonplace.

As information on agricultural conditions in the Soviet Union becomes more readily available, evidence of widespread environmental degradation is mounting.

Difficulties with the irrigation complex around the Aral Sea, which accounts for nearly 40 percent of Soviet irrigated land, may eventually cause the irrigated area in this region to contract.[58]

But the most serious threat to agricultural productivity may well be soil erosion and land degradation. With the largest unbroken expanses of cropland in the world and with little in the way of hedges, tree breaks, or other natural vegetation to interfere, wind and water erosion are taking a heavy toll. One Soviet academician believes this has reduced the inherent productivity of Soviet cropland an average of 30 percent.[59]

The extent of topsoil loss is staggering, and references in Soviet agricultural circles to an unfolding ecological catastrophe are becoming commonplace. An estimated 100,000 hectares of cropland are now claimed annually by an expanding network of gullies cutting through the land. Yuri Markish, a USDA analyst of Soviet agriculture, uses official Soviet sources to conclude that 152 million hectares, roughly two thirds of the arable land, has lost fertility as a result of water and wind erosion. This helps explain why Soviet grain output has increased little since 1980, even though fertilizer use has increased by more than half.[60]

Although the Soviet Union invests far

more in agriculture than any other country does, the returns on its investment have been disappointing. Indeed, the intensive agriculture program, patterned after the agricultural successes in Western Europe and designed to provide a complete package of inputs for use on the more productive land, has yielded only modest gains in output. Unfortunately, these are all too often offset by production losses resulting from soil erosion, soil compaction, and the waterlogging and salting of irrigation systems.[61]

The entire world has a stake in the success of Soviet agricultural reforms, but consumption gains will probably be limited, and are more likely to come from reducing waste than from expanding output. Soviet farmers face the difficult task of feeding a population of 289 million, who aspire to a much better diet and whose numbers are expanding by 3 million per year, with a steadily deteriorating resource base. Expectations that agricultural reforms will quickly end Soviet dependence on imported grain, perhaps even making it an exporter, are not realistic.[62]

A Tough Decade Ahead

Projecting future food production was once a simple matter of extrapolating historical trends. But as yields in many countries approach the upper bend on the S-shaped growth curve, this approach becomes irrelevant. The grain outputs of several countries, including China, Indonesia, Mexico, and the Soviet Union, have shown little or no growth since 1984. In addition, land degradation and hotter summers—the former difficult to measure and the latter impossible to project with precision—will shape future production trends.[63]

In one respect, projections are simpler now than in the past. Since the cultivated area is not likely to change appreciably during the nineties, assessing the production prospect becomes solely a matter of estimating how fast land productivity will rise. Historically, the largest gain in world cropland productivity, as measured by grain yield per hectare, came during the sixties, when it climbed 26 percent over the decade. It rose only 21 percent during the seventies, and an estimated 20 percent in the eighties.[64]

Given the adverse effects of environmental degradation discussed in this chapter, it seems likely that the rise in world cropland productivity will slow further during the nineties. Perhaps as a harbinger of the future slowdown, it increased very little between 1984 and 1989. (See Figure 4–4.) At this point, it is difficult to know whether this recent plateau is merely a short-term interruption in a long-term trend that will soon resume its vigorous climb, or an indication of how difficult it has become to continue raising land productivity rapidly in countries where yields are already high and in a world where environmental degradation is reducing production potential.[65]

For example, since 1950 U.S. corn yields and West European wheat yields

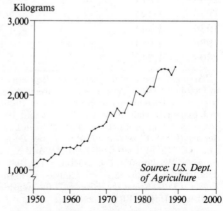

Kilograms

Source: U.S. Dept. of Agriculture

Figure 4-4. World Grain Yield Per Hectare, 1950–89

State of the World 1990

have nearly tripled. In the Third World, wheat yields in India have more than doubled since 1965, and rice yields in China have more than doubled since 1960. Obviously countries cannot continue doubling and tripling grain yield per hectare every few decades. But what kind of further increases can reasonably be expected? And how quickly will they come?[66]

Trends in land productivity in Japan may help answer these questions. Grain yields there started their long upward climb around 1880, several decades before those in other countries. The world grain yield today, taking into account the wide range of growing conditions, appears to be roughly where Japan's yields were in 1970. For instance, the 1989 rice yield in China, the world's largest rice producer, was an estimated 3.8 tons of milled rice per hectare, exactly the same as Japan's rice yield

in 1970. Today, yield per hectare of corn in the United States is nearly double that of rice in Japan in 1970. The yield of wheat in Western Europe is currently a fourth again as high. On the other end of the scale, 1989 rice yields in India were less than half those of Japan in 1970.[67]

Since 1970, Japan's rice yield per acre has risen an average of 0.9 percent per year, scarcely half the 1.7 percent projected annual growth in world population during the nineties. If the world can raise yields during the nineties at this rate, then grain output will increase by 158 million tons, an overall gain of 9 percent. But with world population expected to increase by more than 959 million (18 percent), per capita grain production would fall 7 percent during the decade. (See Table 4–7.) If the world cannot do any better over the next decade than Japan has over the last two, in other words, a steady deterioration of

Table 4-7. World Grain Production, Total and Per Capita, 1950–80, With Projections to 2000

Year	World Grain Production			Per Capita		
	Total	Change Per Decade		Total	Change Per Decade	
	(million tons)	(million tons)	(percent)	(kilograms)	(kilograms)	(percent)
1950	631			246		
1960	847	+216	+34	278	+32	+13
1970	1,103	+256	+30	296	+18	+ 6
1980	1,441	+338	+31	322	+26	+ 9
1990	1,684[1]	+243	+17	316	− 6	− 2
2000	1,842[2]	+158	+ 9	295	−21	− 7

[1]Assumes 1 percent increase over 1989 world harvest of 1,667 million metric tons as estimated by USDA in October 1989. [2]Assumes no appreciable gains or losses in world grain area and a rate of yield-per-hectare increase for world grain between 1990 and 2000 that will equal the 0.9 percent per year increase in Japan's rice yield between 1969–71 and 1986–88.
SOURCES: Grain production 1950–80 from U.S. Department of Agriculture (USDA), Economic Research Service (ERS), *World Grain Harvested Area, Production, and Yield, 1950–88* (unpublished printout) (Washington, D.C.: 1989); 1989 from USDA, World Agricultural Outlook Board, *World Agriculture Supply and Demand Estimates*, Washington, D.C., October 12, 1989; per capita figures derived from population data (based on data from U.S. Bureau of the Census) in Francis Urban and Philip Rose, *World Population by Country and Region, 1950–86, and Projections to 2000* (Washington, D.C.: USDA, ERS, 1988); increase in Japanese rice yield from USDA, ERS, *World Grain*.

diets for much of humanity and increasing hunger for many of us seems inevitable.[68]

The key question for the nineties is whether the world will even be able to match the Japanese record. Despite the powerful incentive of a domestic price support for their rice pegged at four times the world market, Japanese farmers have run out of agronomic options to achieve major additional gains in productivity. Farmers in the rest of the world—who are not as literate or as scientifically oriented as those in Japan—will find it difficult to do any better.[69]

A deterioration in diet and an increase in hunger for part of humanity is no longer a matter of conjecture. In Africa, both the absolute number of people and the share of population that is hungry are increasing. In Latin America, increasing poverty, declining food production per person, and rising food prices indicate a similar trend. Progress in reducing infant mortality, the most sensitive indicator of a society's nutritional state, has been slowed, stopped, or reversed in dozens of countries.[70]

If the world continues with business-as-usual policies in agriculture and family planning, a food emergency within a matter of years may be inevitable. It would extend beyond low-income people in the Third World, with its repercussions affecting the entire world. Soaring grain prices and ensuing food riots could both destabilize national governments and threaten the integrity of the international monetary system.

Barring any dramatic technological breakthroughs on the food front, the widening of the gap between population growth and food production of the last several years will continue. In all too many countries, the opportunity to slow population growth with the time bought on the food front by the Green Revolution has been wasted. To be sure, there will be some further gains in output from high-yielding crop varieties, but they are not likely to match the impressive jumps registered from the mid-sixties to the mid-eighties.

The world needs to continue to strengthen agriculture in every way possible. A massive international effort is needed to protect soil, conserve water, and restore the productivity of degraded land. But the Japanese experience suggests that even doing everything feasible on the food side of the food/population equation is not enough.

A massive international effort is needed to protect soil, conserve water, and restore the productivity of degraded land.

Feeding people adequately in the nineties will depend on quickly slowing world population growth to bring it in line with the likely increase in food output. The only reasonable goal will be to try to cut the growth rate in half, essentially doing what Japan did in the fifties and what China did in the seventies. Reaching that goal depends on raising public understanding of the relationship between family size today and the quality of human existence tomorrow. Unless these goals are given top priority in national capitals, the effort will fail.[71]

Nor is a family planning effort of the magnitude needed likely to succeed if the international community does not effectively address the issue of Third World debt. The economic and social progress that normally leads to smaller families is now missing in many debt-ridden countries. Unless debt can be reduced to the point where economic progress resumes, the needed decline in fertility may not materialize.

Launching an adequate international population campaign might best be done through a U.N. conference for the world's political leaders. Such a gathering would permit an examination of the loss of momentum on the food front and a review of the shifting balance between food supplies and population. It could also lead to an examination of national and global projections so that people everywhere can better understand the consequences for the hungry—and for the world—of continuing on the current demographic path.

5

Holding Back the Sea

Jodi L. Jacobson

Quick study of a world map illustrates an obvious but rarely considered fact: much of human society is defined by the planet's oceans. And the boundary between land and water determines a great deal that is often taken for granted, including the amount of land available for human settlement and agriculture, the economic and ecological productivity of deltas and estuaries, the shape of bays and harbors used for commerce, and the abundance or scarcity of fresh water in coastal communities.

The intensified settlement of coastal areas over the past century implies tacit expectation of a status quo between sea and shore that, according to most scientific models, is about to change. On a geological time scale, the point where sea meets land is far from static. Oscillations of cooling and warming that span 100,000 years, accompanied by glaciation and melting, keep the level of the oceans in constant flux. Still, for most of recorded history, sea level has changed slowly enough to allow the development of a social order based on its relative constancy.

Global warming will radically alter this. Increasing concentrations of greenhouse gases in the atmosphere are ex-

pected to raise the earth's average temperature between 2.5 and 5.5 degrees Celsius over the next 100 years. (See Chapter 2.) In response, the rate of change in sea level is likely to accelerate from thermal expansion of the earth's surface waters and from a rapid melting of alpine and polar glaciers and of ice caps. Although the issue of how quickly oceans will rise is still a matter of debate, the economic and environmental losses of coastal nations under most scenarios are enormous. One thing is clear: no coastal nation, whether rich or poor, will be totally immune.[1]

Accelerated sea level rise, like global warming, represents an environmental threat of unprecedented proportion. Yet most discussions of the impending increase in global rates obscure a critical issue—in some regions of the world, the local sea level is already rising quickly. Egypt, Thailand, and the United States are just a few of the countries where extensive coastal land degradation, combined with even the recent small incremental changes in global sea level, is contributing to large-scale land loss. These trends will be exacerbated in a greenhouse world.

A preliminary assessment of the likely

effects of global and relative sea level rise done by the United Nations Environment Programme (UNEP) in 1989 identified the eight regions and 27 countries at greatest risk. While pointing out that potential losses from rising seas are far greater in some areas than others, the report warned that a large majority of nations will be affected to some degree by higher global rates since only 30 countries in the world are completely landlocked.[2]

As sea level rises, coastal communities face two choices: retreat from the shore or fend off the sea.

Low to middle-range estimates by the U.S. Environmental Protection Agency (EPA) indicate a warming-induced rise by 2100 of anywhere from a half-meter to just over two. A one-meter rise by 2075, well within the projections, could result in widespread economic, environmental, and social disruption. G.P. Hekstra of the Ministry of Housing, Physical Planning, and Environment in the Netherlands asserts that such a rise could affect all land up to five meters elevation. Taking into account the effects of storm surges and saltwater intrusion into rivers, he estimates that 5 million square kilometers are at risk. Although only a small percentage of the world total—about 3 percent—this area encompasses one third of global cropland and is home to a billion people.[3]

As sea level rises, coastal communities face two fundamental choices: retreat from the shore or fend off the sea. Decisions about which strategy to adopt must be made relatively soon because of the long lead time involved in building dikes and other structures and because of the continuing development of coasts. Yet allocating scarce resources on the basis of unknown future conditions—how fast the sea will rise and by what date—entails a fair amount of risk.

Questions arise about how far nations should go in safeguarding and insuring investments already made in coastal areas. Protecting beaches, homes, and resorts can cost a country with a long coastline billions of dollars, money that is only well spent if current assumptions about future sea level are borne out. Assessing the real environmental costs is difficult because traditional economic models do not reflect the fact that structural barriers built to hold back the sea often hasten the decline of ecosystems important to fish and birds. Moreover, protecting private property on one part of the coast often contributes to higher rates of erosion elsewhere, making one person's seawall another's woe.

International equity is another important issue. Low-lying developing countries stand to lose the most land from accelerated sea level rise yet can least afford to build levees and dikes on a grand scale. These regions face consequences grossly disproportionate to their relatively small contribution to the greenhouse effect. At the same time, however, development projects now in progress are putting enormous pressure on regional ecosystems while aggravating the current and likely consequences of sea level rise.

GLOBAL CHANGES, LOCAL OUTCOMES

Worldwide mean sea level depends primarily on two variables. One, the shape and size of ocean basins, involves geological changes over many millions of years. The other, the amount of water in the oceans, is influenced by climate, which can have a more rapid impact.[4]

Ocean basins change their shape and size in a process not dissimilar to the buildup of land recorded in stratified rock. The sea floor builds out from ocean ridges via the accumulation of lava, which forms multiple layers. The weight of new layers causes the earth's crust to settle and subside. If subsidence occurs more rapidly than new volcanic rock is formed, the basin deepens and the water level falls (assuming a constant volume of water). If the production of new rock exceeds subsidence, on the other hand, the basin's volume decreases and the water level rises.[5]

Sea water volume may change much more quickly than basin size and shape. A higher global average temperature can alter sea level in four ways: The density can decrease through the warming and subsequent expansion of sea water, which increases volume. And the volume can be raised by the melting of alpine glaciers, by a net increase in water as the fringes of polar glaciers melt, or by more ice being discharged from ice caps into the oceans.

Glaciers and ice shelves, such as those in Antarctica and Greenland, freeze or melt in a cycle on the order of every 100,000 years. In the last interglacial period, average temperatures were 1 degree Celsius warmer and sea level was two to six meters (6.5–20 feet) higher. At the end of the Wisconsin glaciation 15,-000–20,000 years ago, enough ocean water was collected in glaciers to drop the sea off the northeastern U.S. coast 100 meters below its levels today.[6]

Globally and locally, sea level also fluctuates day to day and year to year as a result of short-term meteorological and physical variables that may also be affected by global warming. Tidal flows, barometric pressure, the actions of wind and waves, storm patterns, and even the earth's rotational alignment all influence sea level.[7]

The slight variations in global climate of the last 5,000 years are responsible for correspondingly small fluctuations in sea level, on the order of 1–10 centimeters every century. Over the past 100 years, however, global sea level rose 10–15 centimeters (4–6 inches), a somewhat faster pace. Scientists continue to debate the cause of this rise, many arguing that there is no evidence that it is due to human-induced warming, while others are not so sure.[8]

Faster global mean sea level rise is not the only threat to coastal areas. Nor are changes in the earth's atmosphere the only consequences of human activity likely to accelerate this trend. Discussions that focus only on the global mean mask important differences in relative, or local, sea level. Although the two are fundamentally different, global sea level rise can be compounded by local fluctuations in land elevation and geological processes such as tectonic uplift or subsidence in coastal areas. Local rates of sea level rise in turn depend in large part on the sum of the global pattern and local subsidence.

Land subsidence is a key issue in the case of river deltas, such as the Nile and Bengal, where human activities are interfering with the normal geophysical processes that could balance out the effects of rising water levels. These low-lying regions, important from both ecological and social standpoints, will be among the first lost to inundation under global warming.

Under natural conditions, deltas are in dynamic equilibrium, forming and breaking down in a continuous pattern of accretion and subsidence. Subsidence in deltas occurs naturally on a local and regional scale through the compaction of recently deposited river-borne sediments. As long as enough sediment reaches a delta to offset subsidence, the area either grows or maintains its size. The Mississippi River Delta, for example, was built up over time by sediments deposited during floods and laid down by the river along its natural course to the sea. If sediments are stopped along the way, continuing compaction and erosion cause loss of land relative to the sea, even if the absolute level of the sea remains unchanged.

Low-lying regions, important from both ecological and social standpoints, will be among the first lost to inundation under global warming.

Large-scale human interference in natural processes has had dramatic effects on both relative rates of sea level rise and on coastal ecosystems in several major deltas. Channeling, diverting, or damming of rivers can greatly reduce the amount of sediment that reaches a delta, as has happened in the Indus, the Mississippi, and many other major river systems, resulting in heavier shoreline erosion and an increase in local relative water levels. Furthermore, the mining of groundwater and of oil and gas deposits can raise subsidence rates. In Bangkok, local subsidence has reached 13 centimeters per year as the water table has dropped because of excessive withdrawals of groundwater over the past three decades. (See Chapter 3.)[9]

These factors can dramatically affect the local outcome of global changes. Subsidence can result in a local sea level rise in some delta regions that is up to five times more than a global mean increase. Under a 20-centimeter worldwide mean increase, for example, local sea level rise may range from 33 centimeters along the Atlantic and Gulf coasts of the United States to one meter in rapidly subsiding areas of Louisiana

and in parts of California and Texas.[10]

Uncertainties abound on the pace of all the possible changes expected from global warming. The most immediate effect will probably be an increase in volume through thermal expansion. The rate of thermal expansion depends on how quickly ocean volume responds to rising atmospheric temperatures, how fast surface layers warm, and how rapidly the warming reaches deeper water masses. The pace of glacial melt and the exact responses of large masses such as the Antarctic shelf are equally unclear. (See Table 5–1.) Over the long term, however, glaciers and ice caps will make the largest contribution to increased volume if a full-scale global warming occurs. (Melting of the Arctic Ocean ice pack would have no effect on sea level since the ice is floating, displacing an amount of water roughly equal to that in the submerged ice.)[11]

Over the past five years, a number of scientists have estimated the possible range of greenhouse-induced sea level rise by 2100. Gordon de Q. Robin projects an increase of anywhere from 20 to 165 centimeters. Computations by other scientists yield projections as high as two to four meters over the next 110 years. Widely cited EPA estimates of global mean sea level rise by 2100 range from 50 to 200 centimeters (1.6 to 6.5 feet), depending on various assumptions about the rate of climate change. Most models do agree that initial rates of increase will be small relative to the much more rapid acceleration expected from 2050 on. (See Table 5–2.) After 2100, the rate is anybody's guess. In any case, even the low range of estimates portends a marked increase over the current global pace.[12]

If global warming runs its course unabated, resulting in average temperatures toward the higher end of the predicted range, the earth may eventually be awash in seawater. In theory, the world's total remaining ice cover con-

Table 5-1. Varying Estimates of Sea Level Rise by 2100, by Source

| | | Melting | | | |
Study	Thermal Expansion	Alpine Glaciers	Greenland	Antarctica	Total
	(centimeters)				
National Academy of Sciences (1983)	30	12	12		70
Environmental Protection Agency (1983)	28–115				56–345
National Academy of Sciences (1985)	10–30	10–30	−10 to +100	50–200	
Thomas (1985)				0–200	
Hoffman et al. (1980)	28–83	12–27	6–27	12–220	57–368

SOURCE: J.S. Hoffman et al., "Future Global Warming and Sea Level Rise," in Per Brun, ed., *Iceland Symposium '85* (Reykjavik: National Energy Authority, 1986).

tains enough water to raise sea level over 70 meters. Some early reports, taking this fact to its extreme, predicted changes of similar magnitude within a brief period of time. But such an increase is more science fiction than fact since complete melting of all ice packs would take several thousand years.[13]

The most likely candidate for melting by the early twenty-second century is the West Antarctic ice sheet. Unlike the Greenland and East Antarctic ones, which are based on land, the West Antarctic sheet is grounded in the ocean (on top of submarine mountains and ridges), making it relatively unstable. A significant increase in sea level, as from thermal expansion, could lift it off its base and break it free. Summertime temperatures in this region average about minus 5 degrees Celsius; a 5-degree increase could start this massive sheet on its way to destruction. A greenhouse-induced meltdown of the whole ice sheet would take from 200 to 500 years, raising sea level an additional six meters.[14]

Biochemist Eric Golanty claims that the resulting meltwater alone would lead to inundation of 1.5–2.1 percent of the total U.S. land area. Heavily populated areas—including one fourth to one third of Florida and Louisiana; a quarter of Delaware; the Texas cities of Galveston, Beaumont, Port Arthur, and Corpus Christi; Savannah in Georgia and Charleston in South Carolina; and low-lying regions of Atlantic City, New York City, Boston, and Washington, D.C.—would be inundated.[15]

What is important about the sea level rise expected from global warming is the pace of change. The rate expected at the global level in the foreseeable future—one meter by 2075 is certainly plausible—is unprecedented on a human time scale. Unfortunately, with today's level of population and investment in coastal areas, the world has much more to lose from sea level rise than ever before.

LANDS AND PEOPLES AT RISK

From the atmosphere to the ocean, humans are proving themselves forceful—if unintentional—agents of change. By

State of the World 1990

Table 5-2. Estimates of Sea Level Rise, 2000–2100

Range	2000	2025	2050	2075	2100
			(centimeters)		
Low	4.8	13	23	38	56
Low to mid	8.8	26	53	91	144
Mid to high	13.2	39	79	137	217
High	17.1	55	117	212	345

SOURCE: John S. Hoffman et al., *Projecting Future Sea Level Rise* (Washington, D.C.: U.S. Environmental Protection Agency, 1983).

and large, the costs of higher seas tomorrow will be determined by patterns of development prevalent in river systems and coastal areas today.

Intense population pressures and economic demands are already taking their toll on deltas, shores, and barrier islands. Rapid rates of subsidence and coastal erosion ensure that many areas of the world will experience a one-meter increase in sea level well before a global change of the same magnitude. As a result, countless billions of dollars worth of property in coastal towns, cities, and ports will be threatened and problems with natural and artificial drainage, saltwater intrusion into rivers and aquifers, and severe erosion of beaches will become even more commonplace.

The ebb and flow of higher tides will cause dramatic declines in a wide variety of coastal ecosystems. Wetlands and coastal forests, which account for most of the world's land area less than a meter above mean tide, are universally at risk. Loss of coastal wetlands in Louisiana today provides a good case study for the future.

Deterioration of the Mississippi river delta began early in the nineteenth century, shortly after levees (embankments to prevent flooding) became extensively used. Subsidence and land loss accelerated after 1940 with an increase in river diversions and the tapping of fossil fuel and groundwater deposits. Combined with sea level rise, these processes are now drowning Louisiana's coastal marshes at rates as high as 130 square kilometers per year, giving that state the dubious distinction of losing more land to the sea on an annual basis than any other region in the world.[16]

Coastal swamps and marshes are areas of prodigious biological productivity. Louisiana's marshes, for example, cover 3.2 million hectares and constitute 41 percent of all wetlands in the United States. The region supplies 25 percent of the U.S. seafood catch and supports a $500-million-a-year recreational industry devoted to fishing, hunting, and birding. The ecological benefits derived from these same wetlands have not been estimated yet. Nearly two thirds of the migratory birds using the Mississippi flyway make essential use of this ecosystem, while existing marshlands and barrier islands buffer inland areas against devastating hurricane surges. Marshes not only keep the Gulf of Mexico's salt water from intruding into local rivers and aquifers, they are a major source of fresh water for coastal communities, agriculture, and industry.[17]

What was laid down over millions of years by the slow deposit of silt washed off the land from the Rockies to the Appalachians may deteriorate in little over a century. The combination of global sea level rise and subsidence could overrun Louisiana's famous bayous and marshland by 2040, by allowing the Gulf of Mexico to surge some 53 kilometers (33 miles) inland. With the delicate coastal marsh ecology upset, fish and wildlife harvests would decline precipitously and a ripple effect would flatten the coastal economy. Communities, water supplies, and infrastructure would all be threatened. Most of these trends are already apparent in Louisiana and are becoming evident in other parts of the United States.[18]

According to EPA estimates, erosion, inundation, and saltwater intrusion could reduce the area of present-day U.S. coastal wetlands up to 80 percent if current projections of future global sea level are realized. Not only the Mississippi Delta, but the Chesapeake Bay and other vital estuarine and wetland regions would be irreparably damaged. Dredged, drained, and filled, coastal wetlands in the United States are already under siege from land and sea. Were it not for the enormous pressure that human encroachment puts on them, these swamps and marshes might have a chance to handle rising seas by reestablishing upland. But heavy development of coastal areas throughout the country means that few wetlands have leeway to "migrate."[19]

The extent of wetland loss will depend on the degree to which coastal towns and villages seek to protect shorelines under different scenarios of sea level rise. An EPA analysis showed that some 46 percent of all U.S. wetlands would be lost under a one-meter rise (from global sea level rise and local subsidence) if shorelines were allowed to retreat naturally. Building bulkheads and levees that block the path of wetland migration would entail higher losses. Fully 66 percent of U.S. wetlands would disappear if all shorelines were protected. If only currently developed mainland areas and barrier islands were safeguarded, the toll could be kept to 49 percent. Loss of up to 80 percent of the country's wetlands is envisioned under a more rapid rate of sea level rise.[20]

In any case, there will be severe reductions in food and habitat for birds and juvenile fish. No one has yet calculated the immense economic and ecological costs of such a loss for the United States, much less extrapolated them to the global level. Yet as global sea level rises, these problems will surely become more severe and widespread in ecosystems around the world.

The combination of global sea level rise and subsidence could overrun Louisiana's famous bayous and marshland by 2040.

A one-meter rise in sea level would wipe out much of England's sandy beaches, salty marshes, and mud flats, according to a 1989 study by the Natural Environment Research Council in London, for example. The most vulnerable areas lie in the eastern part of the country, including the low-lying fens and marshes of Essex and north Kent. More than half of Europe's wading birds winter in British estuaries, and they are destined to lose this vital habitat.[21]

Highly productive mangrove forests throughout the world will also be lost to the rising tide. Mangroves are the predominant type of vegetation on the deltas along the Atlantic coast of South America. On the north coast of Brazil, active shoreline retreat is less of a problem because little human settlement exists; the mangroves may be able to adapt. In the south, however, once-extensive mangroves have already been cleared or hemmed in by urban growth, especially near Rio de Janeiro. No more than 100 square kilometers of mangroves remain where thousands once stood. As sea level rises, these remaining areas will disappear too.[22]

Eric Bird of the University of Melbourne in Australia notes that mangrove-fringed coastlines have become much less extensive in Australia, Africa, and Asia in recent decades as a result of fishpond construction and land reclamation for mining, settlement, and waste disposal. Where they remain, mangroves stand on the frontlines between salt marshes and freshwater vegetation. Bird argues that submergence will kill off large areas of the seaward mangroves, especially where human developments abutting mangrove forests prevent their landward retreat. In Asia, for example, the land behind mangroves is often intensively used for fishponds or rice fields. Thus as sea level rises, it will threaten not only the mangrove species that cannot reestablish upland, but also the economic value of products derived from rice fields and brackish-water fishponds within the flood zone.[23]

In the Bight of Bangkok, the mangrove fringe has already largely been cleared and converted into fish and shrimp ponds and salt pans. Landward canals have been built to irrigate rice fields. A one-meter sea level rise would threaten to submerge all existing mangroves and an additional zone up to 300 meters landward, wiping out the fish farms. This is likely to happen on the southwestern coast of Bangladesh as well, where 6,000 square kilometers of mangroves, locally known as "sundarbans," are at risk. A maze of heavily forested waterways that is both economically and ecologically valuable, this area shields the heavily settled region behind it from the sea.[24]

Worldwide, erosion of coastlines, beaches, and barrier islands has accelerated over the past 10 years as a result of rising sea level. A survey by a commission of the International Geographical Union demonstrated that erosion had become prevalent on the world's sandy coastlines, at least 70 percent of which have retreated during the past few decades.[25]

Changes on beaches vary with the amount of sand supplied to and lost from the shore due to wave activity. The U.S. Army Corps of Engineers found that of the 134,984 kilometers of American coastline, 24 percent could be classified as "seriously" eroding. Over the past 100 years the Atlantic coastline has eroded an average of 60–90 centimeters (2–3 feet) a year; on the Gulf coast, the figure is 120–150 centimeters. Relatively few of the most intensively developed resorts along the U.S. coast have beaches wider than about 30 meters at high tide. Projections of sea level rise over the next 40–50 years suggest that most recreational beaches in developed areas could be eliminated unless preventive measures are taken.[26]

Increased erosion would decrease natural storm barriers. Coastal floods associated with storm surges surpass even earthquakes in loss of life and property damage worldwide. Apart from greater erosion of the barrier islands that safeguard mainland coasts, higher seas will increase flooding and storm damage in coastal areas because raised water levels would provide storm surges with a higher base to build upon. And the higher seas would decrease natural and artificial drainage.[27]

A one-meter sea level rise could turn a moderate storm into a catastrophic one. A storm of a severity that now occurs on average every 15 years, for example, could flood many areas that are today only affected by truly massive storms once a century. Oceanographer T.S. Murty states that as cultivation and habitation of newly formed low-lying delta land continues, "even greater storm surge disasters must be anticipated."[28]

Murty's study shows that losses are nowhere more serious than in the Bay of Bengal. About 60 percent of all deaths due to storm surges worldwide in this century have occurred in the low-lying agricultural areas of the countries bordering this bay and the adjoining Andaman Sea. Murty puts the cost of damage from storm surges in the Bay of Bengal region between 1945 and 1975 at $7 billion, but warns that this number "scarcely expresses the impact of such disasters on developing countries."[29]

Bangladesh—where storm surges now reach as far as 160 kilometers upriver—accounts for 40 percent of this toll. In 1970, this century's worst storm surge tore through the countryside, initially taking some 300,000 lives, drowning millions of livestock, and destroying most of Bangladesh's fishing fleet. The toll climbed higher in its aftermath. As the region's population mounts, so does the potential for another disaster.[30]

Studies indicate a dramatic increase in the area vulnerable to flooding in the United States as well. A one-meter rise would boost the portion of Charleston, South Carolina, now lying within the 10-year floodplain from 20 to 45 percent. A 1.5-meter rise would bring that figure to more than 60 percent, the current area of the 100-year floodplain. Effectively, once-a-century floods would then occur on the order of every 10 years. In Galveston, Texas, the 100-year floodplain would move from 58 percent of the low-lying area to 94 percent under a rise of just 88 centimeters.[31]

Sea level rise will also permanently affect freshwater supplies. Miami is a case in point. The city's first settlements were built on what little high ground could be found, but today most of greater Miami lies at or just above sea level on swampland reclaimed from the Everglades. Water for its 3 million residents is drawn from the Biscayne aquifer that flows right below the city streets. That the city exists and prospers is due to what engineers call a "hydrologic masterwork" of natural and artificial systems that hold back swamp and sea.[32]

Against a one-meter rise in ocean levels, Miami's only defense would be a costly system of seawalls and dikes. But that might not be enough to spare it from insidious assault. Fresh water floats atop salt water, so as sea levels rise the water table would be pushed nearly a meter closer to the surface. The elaborate pumping and drainage system that currently maintains the integrity of the highly porous aquifer could be overwhelmed. The higher water table would cause roads to buckle, bridge abutments to sink, and land to revert back to swamp. Miami's experience would not be unique. Large cities around the world—Bangkok, New Orleans, Taipei, and Venice, to name a few—face similar prospects.

A storm of a severity that now occurs every 15 years could flood many areas today only affected by truly massive storms once a century.

A study by the Delaware River Basin Commission indicates that a rise of 13 centimeters by the end of this decade would pull the "salt front" on that river from two to four kilometers further inland if there were a drought similar to one in the sixties that contaminated Philadelphia's water supply. A rise of 1–2.5 meters would push salt water up to 40 kilometers inland under drought conditions. The resulting contamination of fresh water would exceed New Jersey's health-based sodium standard 15 to 50 percent of the time.[33]

Countries bordering the Mediterranean are likely to suffer significant economic losses. Greece and Italy, for example, face threats to their tourism industries and to specialized agricultural industries, as well as to important harbors. A 1989 UNEP report points out that, though they constitute only 17 percent of the total land area of the Mediterranean region, the alluvial and coastal plains of most countries bordering this sea have considerable demographic and economic importance. The coast is home to 37 percent of the region's population, some 133 million people. The report cautions that, while serious environmental problems—from water pollution and salinization to shoreline erosion and loss of habitat—already exist in the region, due to agricultural and industrial practices, tourism, and urbanization, "sea level rise will affect considerably the economy and well-being of many countries especially because many low coasts will increasingly experience physical instability" resulting from subsidence and reduced sedimentation.[34]

MOST VULNERABLE, LEAST RESPONSIBLE

Social and environmental costs of sea level rise will be highest in countries where deltas are extensive, densely populated, and extremely food-productive. In these countries, most of which are in the Third World, heavy reliance on groundwater and the completed or proposed damming and diversion of large rivers—for increased hydropower and agricultural use, for flood control, and for transportation—have already begun to compound problems with sea level rise. Almost without exception, the prognosis for these vulnerable, low-lying countries in a greenhouse world is grim.

The stakes are particularly high throughout Asia, where damming and diversion of such systems as the Indus, Ganges-Brahmaputra, and Yellow rivers

has greatly decreased the amount of sediment getting to deltas. The sediments feeding Asia's many great river deltas account for at least 70 percent of the total that reaches the oceans, and they replenish agricultural land with the fertile silt responsible for a large share of food produced in those nations.[35]

As elsewhere, the deltas reliant on these sediments support sizable human and wildlife populations while creating protective barriers between inland areas and the sea. Large cities, including Bangkok, Calcutta, Dacca, Hanoi, Karachi, and Shanghai, have grown up on the low-lying river banks. These heavily populated areas are almost certain to be flooded as sea level rise accelerates.[36]

The U.N. Environment Programme's 1989 global survey represents the first attempt to analyze systematically the regions most vulnerable to sea level rise. An overall lack of data posed severe constraints on the assessments of potential impacts. In defining "vulnerability," for example, UNEP sought to evaluate population densities for the total area worldwide lying between 1.5 and 5 meters above mean sea level. At the global level and at most local sites, however, detailed topographic maps are not available for such low elevations.[37]

On a country-by-country basis, four main criteria were used to determine vulnerability. The first two—the share of total land area between zero and five meters above mean sea level and the density of coastal populations—were used to assess the likely demographic impacts. Identified as most vulnerable were areas where coastal population density exceeded 100 people per square kilometer.

Potential economic and ecological losses were gauged by the other two criteria: the extent of agricultural and of biological productivity within low-lying areas. First, UNEP isolated countries where lowland agricultural productivity grew on average more than 2 percent a year between 1980 and 1985. Second, it added the regions with the largest inventories of coastal wetlands and tidal mangrove forests.

Under these guidelines, 10 countries—Bangladesh, Egypt, The Gambia, Indonesia, the Maldives, Mozambique, Pakistan, Senegal, Surinam, and Thailand—were identified as "most vulnerable." These 10 share many characteristics, including the fact that they are, by and large, poor and populous. (See Table 5-3.) Not insignificantly, as a group they also contribute relatively little to the current buildup of greenhouse gases.

UNEP identified both primary and secondary impact areas as important in each of these countries. The primary impact area consists of the coastal region between zero and 1.5 meters elevation, which would be completely lost under a 1.5-meter rise. The secondary area (1.5–3 meters above today's mean) is vulnerable not only to a rise in seas of equivalent measure, but to the many pressures—such as an influx of environmental refugees, and increased regional demand for food, housing, and other resources—that would arise from inundation of the land closer to the sea.[38]

The sediments feeding Asia's many great deltas replenish agricultural land with the fertile silt responsible for a large share of food produced in those nations.

Detailed information on the land area, population, and economic output likely to be affected by a rise of up to three meters was unattainable for all but Bangladesh and Egypt. For data on these two countries, UNEP drew on a 1989 study by John Milliman and his colleagues at the Woods Hole Oceano-

Table 5-3. Ten Countries Most Vulnerable to Sea Level Rise

Countries	Population	Per Capita Income
	(million)	(dollars)
Bangladesh	114.7	160
Egypt	54.8	710
The Gambia	0.8	220
Indonesia	184.6	450
Maldives	0.2	300
Mozambique	15.2	150
Pakistan	110.4	350
Senegal	5.2	510
Surinam	0.4	2,360
Thailand	55.6	840

SOURCES: United Nations Environment Programme, *Criteria for Assessing Vulnerability to Sea-Level Rise: A Global Inventory to High Risk Areas* (Delft, Netherlands: Delft Hydraulics Laboratory, 1989); income and population data from Population Reference Bureau, *1989 World Population Data Sheet*, Washington, D.C., 1989.

graphic Institution in Massachusetts. Their study showed the combined effects of sea level rise and subsidence on the Bengal and Nile delta regions, where the homes and livelihoods of some 46 million people are potentially threatened.[39]

The river delta nations of the Indian subcontinent and southeast Asia depend heavily on ocean resources and coastal areas for transportation, mariculture, and habitable land. Bangladesh is no exception. The Bengal Delta, the world's largest such coastal plain, accounts for 80 percent of Bangladesh's land mass and extends some 650 kilometers from the western boundary with India to the Chittagong hill tracts. Milliman observes that because the delta is so close to the sea (most of the area is only a meter or two above that level now), an increase in sea level rise accompanied by higher rates of coastal storm erosion is likely to have a greater effect here than on any other delta in the world.[40]

Residents of one of the poorest and most densely populated nations in the world, Bangladeshis already live at the margin of survival. Most people depend heavily on the agricultural and economic output derived from land close to the sea and currently subject to annual floods both from rivers and ocean storm surges. Subsidence is already a problem in this region; the Woods Hole study indicates that as global warming sets in, relative sea level rise in the Bengal Delta may well exceed two meters by 2050. Because half the country lies at elevations below five meters, losses to accelerated sea level rise will be high.[41]

Bangladeshis depend heavily on the agricultural and economic output derived from land close to the sea.

UNEP estimates based on current population size and density show that 15 percent of the nation's land area, inhabited by 15 million people, is threatened by total inundation from a primary rise of up to 1.5 meters. Secondary increases of up to three meters would wipe out over 28,500 square kilometers (20 percent of the total land area), displacing an additional 8 million. These projections do not account for the ongoing increase in Bangladesh's population nor for continuing settlement of the delta area, and so clearly understate the potential number of environmental refugees.[42]

By the end of the next century, Bangladesh as it is known today may virtually have ceased to exist. Pressures to develop agriculture have quickened the pace of damming and channelling on the three giant rivers—the Brahmaputra, the Ganges, and the Meghna—that feed the

delta. As a result, subsidence is increasing.

This situation is being made worse by the increasing withdrawal of groundwater. Milliman of Woods Hole notes a sixfold increase in the number of wells drilled in the country between 1978 and 1985, raising subsidence to perhaps twice the natural rate. The researchers concluded that interference in the delta ecosystem today may make a far larger area and population susceptible to sea level rise, causing dislocation of more than 40 million people.[43]

Egypt—almost completely desert except for the thin ribbon of productive land along the Nile and its delta—can also ill afford the likely costs of sea level rise. The country's millions crowd on to the less than 4 percent of the land that is arable, leading to a population density in the settled area of Egypt of 1,800 people per square kilometer.[44]

In the Nile Delta, extending from just east of the port city of Alexandria to west of Port Said at the northern entrance of the Suez Canal, local sea level rise already far exceeds the global average due to high rates of subsidence. The construction of the first barrages or dams on the Nile in the 1880s cut massively the amount of sediments that nourished the delta, a situation made worse by the building of the Aswan Dam in 1902 and its enlargement in 1934. Extensive diversion of water for irrigation and land reclamation projects since then has closed down a number of the Nile's former tributaries, greatly reducing the river's outward flow.[45]

Even so, approximately 80–100 million tons of sediment were delivered annually to the Nile Delta until 1964, when the opening of the Aswan High Dam virtually eliminated the silt getting through. High rates of relative sea level rise and the accompanying acceleration in subsidence and erosion have resulted in a frightening rate of coastal retreat, reaching 200 meters annually in some places.[46]

Milliman's study suggests that local sea level rise will range from 1 to 1.5 meters by 2050, rendering up to 19 percent of Egypt's already scarce habitable land unlivable. By 2100, an expected rise of between 2.5 and 3.3 meters may inundate 26 percent of the habitable land—home to 24 percent of the population and source of an equal share of the country's economic output.[47]

To feed a population growing nearly 3 percent annually, the government has followed a strategy of land reclamation and development of lagoon fisheries on the delta banks. The principal existing natural defenses against transgression by the sea are a series of dunes and the freshwater (but increasingly brackish) lakes that fall behind them. According to James Broadus of Woods Hole, these lakes—Burullos, Idku, Manzalah, and Maryut—are the major source of the nation's approximately 100,000 tons of annual fish catch, 80 percent of which are freshwater fish. Ironically, the lakes and surrounding areas now slated for development in the regions of Port Said and Lake Maryut will likely be inundated some time in the next century.[48]

Unless steps are taken now to slow sea level rise, Egyptians can also look forward to damage to ports and harbors, increasing stress on freshwater supplies due to saline encroachment, and the loss of beaches that support tourism, such as those in Alexandria.

Extending these scenarios on Bangladesh and Egypt to the eight other most endangered nations presents a sobering picture. Despite the lack of data, preliminary findings show the situation to be equally grave. In another study, Milliman notes that when the impact of the global rise is added to that of regional subsidence and of damming and diversion, Indian Ocean deltaic areas may register a relative subsidence of at least several meters, leading to coastal regression of several tens of kilometers by the twenty-second century.[49]

At least 40 percent of Indonesia's land surface is classified as vulnerable to sea level rise. In terms of both size and diversity, the country is home to one of the world's richest and most extensive series of wetlands. Here, too, population pressures are already threatening these fragile ecosystems. Transmigration programs have resettled millions of people in the past several years from the overpopulated islands of Java and Bali to the tidal swamps of Sumatra and Kalimantan, a policy decision that may be much regretted should these lands give way to the sea. Although studies remain to be done on how many people will eventually be affected by the ocean's incursion, the numbers are certain to be high.[50]

A one- to two-meter rise in sea level could be disastrous for the Chinese economy as well. The Yangtze Delta is one of China's most heavily farmed areas. Damming and subsidence have contributed to a continuing loss of this valuable land on the order of nearly 70 square kilometers per year since 1947. A sea level rise of even one meter could sweep away large areas of the delta, causing a devastating loss in agricultural productivity for China.[51]

PAYING BY THE METER

China's 2,400-kilometer-long Great Wall is considered the largest construction project ever carried out, but it may soon be superseded in several countries by modern-day analogues: Great Seawalls. Assuming a long-run increase in rates of global sea level rise, societies will have to choose some adaptive strategies. Broadly speaking, they face two choices: fight or flight. Many governments see no alternative to building jetties, seawalls, groins, and bulkheads to hold back the sea. Yet the multibillion-dollar price tags attached to these may be higher than even some well-to-do countries can afford, especially when accounting for the long-term ecological damage such structures can cause.

Along with the intensified settlement of coastal areas worldwide over the past century has come a belief that human ingenuity could tame any natural force—an attitude that coastal geologists such as Orrin Pilkey and some of his colleagues deplore. As a result of this short-sighted approach, people have been inclined to build closer and closer to the ocean, investing billions of dollars in homes and seaside resorts and responding to danger by confrontation.[52]

Nowhere in the world is the battle against the sea more actively engaged than in the Netherlands. Hundreds of kilometers of carefully maintained dikes and natural dunes keep the part of the country that is now well below sea level—more than half the total—from being flooded. As Dutch engineers know, the ocean doesn't relinquish land easily. In early 1953, a storm surge that hit the delta region caused an unprecedented disaster. More than 160 kilometers of dikes were breached, leading to the inundation of 1,000 square kilometers of land and more than 1,800 deaths. In response, the government put together the Delta Plan, a massive public works project that took two decades and the equivalent of 6 percent of the country's gross national product each year until finally completed in 1986.[53]

The Dutch continue to spend heavily to keep their extensive system of dikes and pumps in shape, and are now protected against storm surges up to those with a probability of occurring once in 10,000 years. But the prospect of accelerated sea level rise implies that maintaining this level of safety may require additional investments of up to $10 billion by 2040.[54]

Large though these expenditures are, they are trivial compared with what the United States, with more than 30,000 kilometers (19,000 miles) of coastline would have to spend to protect Long Island, North Carolina's Outer Banks, most of Florida, the Bayous of Louisiana, the Texas coast, the San Francisco Bay area, and the Maryland, Massachusetts, and New Jersey shores.[55]

Preliminary estimates by EPA of the total bill for holding the sea back from U.S. shores—including costs to build bulkheads and levees, raise barrier islands, and pump sand, but not including the money needed for replacing or repairing infrastructure such as roads, sewers, water mains, and buried cables—range from $32 billion to $309 billion for a one-half- to two-meter rise in sea levels. (A one-meter rise would cost $73 billion to $111 billion.) Extending the projections to impoverished coastal areas of Africa, Asia, and South America underscores the futility of such an approach under a scenario of rapidly rising seas.[56]

Nevertheless, in most industrial countries at least, property owners in coastal areas have become a powerful interest group supportive of defenses that will save their land, even if only for the short term. Many countries have made vast investments reclaiming land from the sea for use by large coastal populations: witness the efforts in Singapore, Hong Kong, and Tokyo. In most cases, governments have encouraged and continued to support this constituency. Heavy investments in roads, sewers, and other public services and insurance against disasters have largely been subsidized by taxpayers, many living far from any coast.

Political pressures to maintain these lands through dikes, dams, and the like will be high. "The manner in which societies respond to the impact of rising sea levels," observes G.P. Hekstra, "will be determined by a mix of conditions [including] the vested interests that are threatened, the availability of finance . . . employment opportunities, political responsibilities and national prestige." Eric Bird argues, for example, that state capitals and other seaside towns in Australia and resorts in Africa and Asia will probably be maintained by beach nourishment programs—literally "feeding" the beach with sand transported from elsewhere—no matter the cost.[57]

Political support for subsidizing coastal areas may be undercut by competing fiscal demands over the long run. In the United States, where a burgeoning budget deficit has vastly reduced expenditures on repairs and construction of bridges and roads, for example, the Federal Highway Administration estimates that bringing the nation's highway system up to "minimum engineering standards" would cost a mind-boggling $565 billion to $655 billion over the next 20 years. Today, that agency's budget is a meager $13 billion and fiscal paralysis keeps it from growing any larger. With increasing competition for scarce tax dollars, property owners in the year 2050 may find the general public reluctant to foot the bill for seawalls.[58]

Moreover, what may seem like protection often turns out to be only a temporary palliative. While concrete structures may divert the ocean's energies from one beach, they usually displace it onto another. And by changing the dynamics of coastal currents and sediment flow, these hard structures interrupt the natural processes that allow wetlands and beaches to reestablish upland, causing them to deteriorate and in many cases disappear.[59]

Beach nourishment is a relatively benign defensive strategy that can work in some cases. And comparing the costs and benefits illustrates that it is not usually as prohibitively expensive as other approaches. Sand or beach nourish-ment, for example, can cost $620,000 per kilometer, but these costs are often justified by economic and recreational use of the areas. A recent study of Ocean City, New Jersey, found it would cost about 25¢ per visitor to rebuild beaches to cope with a 30-centimeter rise, less than 1 percent of the average cost of a trip to the beach. A more rapid rate of sea level rise would, of course, raise the ante. And the fact remains that most beach replenishment is temporary at best, indicating the need for continuous investment.[60]

Today's property owners face becoming tomorrow's proprietors of marshes and bogs.

Advocates of coastal protection strategies argue convincingly that coastal development should be limited. Most assert, for example, that future disasters may be prevented or the harm done by them lessened if highly vulnerable areas were protected. And in recent years, the tide of public opinion in several states regarding conservation of wetlands has tended to support the view that some property should be allowed to return to its natural state. Attitudes could change, of course, as today's property owners face becoming tomorrow's proprietors of marshes and bogs.

The legal definitions of private property and of who is responsible for compensation in the event of natural disasters are already coming into question. As sea level rise accelerates, pushing up the costs of adaptation, these issues will likely become part of an increasingly acrimonious debate over property rights and individual interests versus those of society at large.

Enforcement of the coastal protection law in South Carolina in the aftermath of the recent Hurricane Hugo is a good example of the types of conflicts that can arise. On September 21, 1989, Hurricane Hugo came ashore at Charleston one day after it ravaged several islands in the Caribbean. The storm, creating an ocean surge that reached 20 feet at its highest point, killed 29 people on the mainland and caused an estimated $4 billion worth of damage in the United States. It also sparked a controversy over South Carolina's new beachfront protection law. The statute completely prohibits any new seawalls and regulates commercial and residential construction in a setback area along the coast. Because the storm ate up so much of the existing beach, 159 plots of land, on which houses were destroyed, all became part of the "dead zone" where new buildings are prohibited. Even before Hugo hit, several homeowners filed suit against the state for "taking property without just compensation." The states of Maine, Maryland, North Carolina, and Texas also have enacted coastal protection laws.[61]

Site-specific studies of several towns in the United States suggest that incorporating projections of sea level rise into land-use planning can save money in the long run. Projections of costs in Charleston, South Carolina, show that a strategy that fails to anticipate and plan for the greenhouse world can be expensive. Depending on the zoning and development policies followed, including the amount of land lost and the costs of protective structures built, the costs of a one-meter sea level rise may exceed $1.9 billion by 2075—an amount equal to 26 percent of total current economic activity in this area. If land use policies and building codes are modified to anticipate rising sea levels, this figure could be reduced by more than 60 percent. Similar studies of Galveston, Texas, show that economic impacts could be lowered from

$965 million to $550 million through advanced planning.[62]

Obviously, heavily developed areas, such as the island of Manhattan, much of which is less than two meters above high tide, will not be left to be swallowed by the sea. An accounting method is needed to establish priorities and assess the costs and benefits of protection strategies versus the costs of inundation. Several analysts are attempting to develop such a model. Gary Yohe, an economist at Wesleyan University in Connecticut, is developing a method of comparing the costs of not holding back the sea with those of protecting coasts on a year-to-year basis. His economic model is a first step toward "measuring the current value of real sources of . . . wealth that might be threatened . . . if a decision to forego any protection from rising seas were made." In his preliminary analysis, using Long Beach Island, New Jersey, as a case study, Yohe focuses on estimating the market price of threatened structures, the worth of threatened property, and the social value of threatened coastline.[63]

A truly representative model should account for all the costs and benefits—economic, ecological, and social—of protection against other options. One cost not explored in Yohe's assessment is the loss in coastal ecological wealth as a side effect of protection. In keeping with the figures for the United States as a whole, for example, researchers have estimated that a 1.5-meter rise would eliminate about 80 percent of Charleston's wetlands with current barriers in place. If additional developed areas are protected by bulkheads and levees, a 90-percent loss is envisioned. The South Carolina beachfront protection law seeks to prevent this large-scale destruction, but its political viability is still in question.[64]

Protecting wetlands requires a trade-off as well. Taking shore and wetland conservation measures basically implies a willingness to relinquish to the sea some land area now in use or potentially available for social activities, such as farming and home building. A study of coastal land loss in Massachusetts by Graham Giese and David Aubrey of Woods Hole Oceanographic Institution illustrates these processes and estimates the amount of land likely to be lost in Massachusetts under three scenarios.[65]

Giese and Aubrey distinguish between upland (relatively dry terrain that is landward of wetland and not altered much by waves and tides) and wetland itself, including coastal bluffs, dunes, beaches, and marshes that are affected by these forces. Wetlands replace uplands as they migrate landward, resulting in loss of total upland area. Where wetlands are protected by law (as they are in Massachusetts) against being drained or filled, they gain at the expense of uplands, essentially protecting the ecological over the purely economic value of the land.[66]

Relative sea level in Massachusetts has been rising some three millimeters annually since 1950. Under the first scenario in Giese and Aubrey's study, which assumes a continuation of current trends from 1980 through 2025, the sea along the coast of Massachusetts would rise 36 centimeters. The state would therefore lose 23 hectares a year, or nearly 1,200 hectares over that period. The second scenario assumes a higher global rise by 2025 (EPA's low to mid-range estimate), which when combined with subsidence leads to a total land loss in Massachusetts of some 3,000 hectares by 2025. Finally, the third case, assuming a rise of 48 centimeters, costs Massachusetts nearly 4,200 hectares of commercially usable area.[67]

Whatever the strategy, industrial countries are in a far better financial position to react than are developing nations. Bangladesh, for example, cannot afford to match the Dutch kilometer for kilometer in seawalls. But its danger is no less real. Debates over land loss may be a moot point in poorer countries like Bangladesh, where evacuation and abandonment of coastal land may be the only option when submergence and erosion take their toll and when soil and water salinity increase. As millions of people displaced by rising seas move inland, competition with those already living there for scarce food, water, and land may spur regional clashes. Ongoing land tenure and equity disputes within countries will worsen. Existing international tensions, such as those between Bangladesh and its large neighbor to the west, India, are likely to heighten as the trickle of environmental refugees from the nation that is awash becomes a torrent.

PLANNING AHEAD

The threats posed by rapidly increasing sea level raise questions that governments and individuals must grapple with today. If the world moves quickly onto a sustainable path, the effects of global warming and sea level rise can be mitigated. Minimizing the impacts of climate change will require that a number of strategies be put in place right away. (See Chapter 2.) An unprecedented level of international cooperation on agricultural, energy, forestry, and land use policies will be required. More important, perhaps, is to develop a method for comparing the costs of measures to avert global warming and its consequences against the costs of adaptation. But for now, preparing to experience some degree of global and regional change in sea level is a rational response.

How can the world move away from the seemingly universal human tendency to react in the face of disaster but to ignore cumulative, long-term developments? An active public debate on coastal development policies is needed, extending from the obvious issues of the here and now—beach erosion, river damming and diversion, subsidence, wetland loss—to the uncertainties of how changes in sea level in a greenhouse world will make matters far worse. Raising public awareness on the forthcoming changes, developing assessments that account for all future and present costs, and devising sustainable strategies based on those costs are all essential.

An active public debate on coastal development policies is needed, from the obvious issues of the here and now to the uncertainties of a greenhouse world.

Taking action now to safeguard coastal areas will have immediate benefits while preventing losses from soaring higher in the event of an accelerated sea level rise. Limiting coastal development is a first step, although strategies to accomplish this will differ in every country. Governments may begin by ensuring that private property owners bear more of the costs of settling in coastal areas. A more systematic assessment is needed of the value of creating dead zones to be left in their natural state versus the economic and ecological costs that ongoing development and the subsequent need for large-scale protection will entail.

A new concept of property rights will have to be developed. Unbridled development of rivers and settlement of vulnerable coasts and low-lying deltas mean that more and more people and property will be exposed to land loss and potential disasters arising from storm surges and the like. Governments that plan over the long term to limit development of endangered coasts and deltas can save not only money, but resources as well. Wherever wetlands and beaches are not bordered by permanent structures, they will be able to migrate and reestablish further upland, allowing society to reap the intangible ecological benefits of biodiversity.

Of course, protection strategies will inevitably be carried out where the value of capital investments outweigh other considerations. But again the key is to plan ahead. As the Dutch discovered, more money can be saved over the long term if dikes and drainage systems are planned for before rather than after sea levels have risen considerably.

A cap on problematic dam-building and river diversion projects in large deltas would lessen the ongoing destruction of wetland areas and prevent further reductions in sedimentation, thereby minimizing subsidence as well. It is unlikely that the damage done by large-scale dams, like the Aswan, can be remedied. As with past projects, however, analyses of many dams now in the pipeline do not reflect the often massive present and future environmental or external costs. Better water management and increased irrigation efficiencies can both increase crop yields and save water. (See Chapter 3.) Exploring the potential gains from conservation may preclude the need for many more large-scale dams. The same can be said of curtailing development of additional dams for hydropower by encouraging energy efficiency and conservation.

Additional money is needed to do more research on sea level globally and regionally. Funds are needed to support studies of beach and wetland dynamics, to take more frequent and widespread measurements of global and regional sea level, and to design cost-effective, environmentally benign methods of coping with coastal inundation.

The majority of developing nations most vulnerable to sea level rise can do little about global warming independently. But they have a clear stake in reducing pressures on coastal areas by taking immediate actions. Among the most important of these is slowing population growth and, where necessary, changing inequitable patterns of land tenure in interior regions that promote coastal settlement of endangered areas. Furthermore, the governments of Bangladesh, China, Egypt, India, and Indonesia, to name just a few, are currently promoting river development projects that will harm delta ecosystems in the short term and hasten the date they are lost permanently to rising seas.

The issue of how to share the costs of adaptation equitably may well be among the hardest to resolve. Industrial countries are responsible for by far the largest share of the greenhouse gases emitted into the atmosphere. And no matter what strategies poorer nations adopt to deal with sea level rise, they will need financial assistance to carry them out. Problems with coastal protection, environmental refugees, changes in land and water allocation, and a host of other issues will plague poor coastal nations. The way industrial countries come to terms with their own liability in the face of accelerated sea level rise will play a significant role in the evolution of international cooperation.

6

Clearing the Air

Hilary F. French

At a time when issues like global warming and ozone depletion dominate the headlines, air pollution and acid rain can sound like yesterday's problems. Unfortunately, evidence to the contrary is widespread. In the West, efforts to combat air pollution since the first outpouring of concern in the early seventies have been only marginally successful. In Eastern Europe, the Soviet Union, and much of the developing world, air pollution is only beginning to be recognized as worthy of serious attention.

Severe health problems related to air pollution span continents and levels of development: In the United States, some 150 million people breathe air considered unhealthy by the Environmental Protection Agency (EPA). In greater Athens, the number of deaths is six times higher on heavily polluted days than on those when the air is relatively clear. In Hungary, a recent report by the National Institute of Public Health concluded that every twenty-fourth disability and every seventeenth death in Hungary is caused by air pollution. In India, breathing the air in Bombay is equivalent to smoking

An expanded version of this chapter appeared as Worldwatch Paper 94, *Air Pollution and Acid Rain: A Strategy for the Nineties.*

10 cigarettes a day. And in Mexico, the capital city is considered a hardship post for diplomats because of its unhealthy air, and some governments advise women not to plan on having children while posted there.[1]

The environmental impacts of air pollution are equally grave. Acid rain and air pollution are devastating forests, crops, lakes, and buildings over wide areas of Europe and North America. Indications are that the Third World may be next in line for widespread damage.

The one constant in the air pollution story over the years has been its growing complexity. Just as one problem seems to be coming under control, a new one arises to replace it. In the West, combating smoke and soot emissions from coal-fired factories and heating greatly improved urban air quality, at least until auto-induced pollution emerged as the primary barrier to clean air. Similarly, building tall smokestacks that spew pollutants far away from cities seemed at first an unqualified boon for air quality. But the solution to one problem was the genesis of another: it turned out that sulfur dioxide (SO_2) and nitrogen oxides (NO_x) could be transformed in the atmosphere into acid-forming particles that fall to earth far from their source in what is popularly known as acid rain.

Now industrial nations are waking up to another front in the battle for clean air—airborne toxic chemical emissions. Though the full dimensions of this problem are as yet unknown, recently released data suggest that billions of tons of harmful chemicals are commonly emitted into the air by industries. Once out the stack, these chemicals can be swept by the wind hundreds and even thousands of kilometers from their source before eventually falling to earth.[2]

Acid rain concerns transformed what had once been a national or even a local issue into an international one. Now the need to slow the buildup of heat-trapping greenhouse gases has added a supranational dimension to the air debate. Because both air pollution and global warming stem largely from common roots in energy, transportation, and industrial systems, the two problems are properly considered together when examining policy options. (See also Chapter 2.) It would be particularly foolhardy to adopt air pollution strategies that might undermine efforts to slow global warming.

On this twentieth anniversary of Earth Day, it is useful to look at what has been accomplished around the world in the realm of air pollution control—and what remains to be done. The conclusions apply not only to the West, but also to Eastern Europe and the developing world, which have the enviable opportunity to learn from the mistakes of others.

THE GLOBAL HEALTH THREAT

That air pollution could cause serious health problems first became evident during the industrial revolution, when many cities in Europe and the United States were covered with black shrouds of coal-derived smoke. On days with weather conducive to pollution, sickness and even death were omnipresent. A particularly acute episode of "black fog" in London in 1952 claimed 4,000 lives and left tens of thousands ill.[3]

In would be foolhardy to adopt air pollution strategies that might undermine efforts to slow global warming.

Incidents such as this prompted many governments to enact legislation to combat the primary pollutants of the day, sulfur dioxide (SO_2) and particulate emissions from stationary sources such as power plants, industries, and home furnaces. Both SO_2 and particulates—either alone or in combination—can raise the incidence of respiratory diseases such as coughs and colds, asthma, bronchitis, and emphysema. Particulate matter (an overall term for a complex and varying measurable mixture of pollutants in minute solid form) can carry heavy metals into the deepest, most sensitive part of the lung.[4]

With the aid of pollution control equipment and improvements in energy efficiency, many industrial countries have made major strides in reducing emissions of these harmful pollutants. The United States, for example, cut sulfur oxide emissions by 28 percent between 1970 and 1987 and that of particulates by 62 percent. (See Figure 6–1.) In Japan, sulfur dioxide emissions fell by 39 percent from 1973 to 1984. Many West European countries also reduced significantly their emissions in both categories from power plants, industries, and heating of buildings. In certain

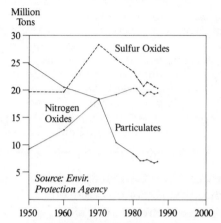

Figure 6-1. Emissions of Selected Pollutants in the United States, 1950–87

European cities, however, the widespread introduction of diesel-fueled vehicles, which emit large quantities of particulates and some SO_2, threatens to negate earlier gains.[5]

In Eastern Europe and the Soviet Union, hasty industrialization after World War II powered by the natural endowment of high-sulfur brown coal led to air quality reminiscent of London's "black fog." The lack of a market price signal prevented these countries from realizing the impressive gains in energy efficiency registered in the West after the oil shocks of the seventies, and financial constraints still make investments in pollution control a rarity.[6]

Though emissions data for developing countries are scarce, air quality in many cities is already abysmal. Rapid plans to expand energy and industrial production and the lack of adequate pollution control regulations mean that air quality could get even worse. China, for example, increased coal output more than 20 times between 1949 and 1982, and plans to continue in this direction. By the end of this decade, annual coal-burning will double if current targets are realized. In India, SO_2 emissions from coal and oil have nearly tripled since the early sixties. Growing urbanization in much of the Third World means that ever increasing numbers of people are being exposed to polluted urban air.[7]

A recent report by the United Nations Environment Programme (UNEP) and the World Health Organization (WHO) gives the best picture to date of the global spread of SO_2 and particulate pollution. (See Table 6–1.) In terms of average annual concentrations, 27 of the 54 cities with data available on sulfur dioxide for 1980–84 were in excess of or on the borderline of the WHO health standard. High on the list were Seoul, Tehran, and Shenyang, as well as Madrid, Paris, and Milan, indicating that SO_2 problems are by no means over yet for all industrial countries. Indeed, Milan topped the list of average annual concentrations, with a reading more than three times the WHO norm. Though conditions are gradually improving in most of the cities surveyed, several in the Third World reported continuing deterioration.[8]

Suspended particulate matter poses an even more pervasive threat, especially in the developing world. Fully 37 of the 41 cities monitored for this averaged either borderline or excessive levels. Annual average concentrations were as much as five times the WHO standard in Kuwait, Shenyang, Xian, New Delhi, and Beijing. The extraordinary levels noted in some Third World cities can be partially explained by natural dust; other culprits include the black, particulate-laden smoke spewed out by diesel-fueled vehicles lacking even the most rudimentary pollution controls, and motor scooters equipped with highly polluting two-stroke engines. WHO and UNEP concluded that nearly 625 million people around the world are exposed to unhealthy levels of sulfur dioxide and more than a billion—one in five people in the world—to excessive levels of suspended particulates.[9]

Table 6-1. Violations of Sulfur Dioxide and Suspended Particulate Matter Standards, Selected Cities[1]

City	Sulfur Dioxide[2]	Particulates[3]
	(number of days above standard)	
New Delhi	6	294
Xian	71	273
Beijing	68	272
Shenyang	146	219
Tehran	104	174
Bangkok	0	97
Madrid	35	60
Kuala Lampur	0	37
Zagreb	30	34
São Paulo	12	31
Paris	46	3
New York	8	0
Milan	66	n.a.
Seoul	87	n.a.

[1]Averages of readings at a variety of monitoring sites from 1980 to 1984. [2]WHO standard is a daily average of 150 micrograms per cubic meter, not to be exceeded more than seven days per year. [3]WHO standard is a daily average of 230 micrograms per cubic meter for suspended particulate matter and 150 micrograms per cubic meter for smoke, also not to be exceeded more than seven days per year. For Madrid, São Paulo, and Paris, the readings are of smoke rather than particulates.
SOURCE: United Nations Environment Programme and World Health Organization, *Assessment of Urban Air Quality* (Nairobi: Global Environment Monitoring System, 1988).

Nations that built tall stacks to improve local air quality may have simply sent their health problems elsewhere: though acid precipitation is best known as an ecological threat it is also suspected of having grave health repercussions. Sulfur dioxide can be chemically transformed into fine sulfate particles that mix with water in the air, liquefy, and become aerosols that can penetrate the deepest, most delicate tissues of the lungs, bringing toxic metals and gases along with them. Some researchers believe the mix may be responsible for as many as 50,000 deaths in the United States every year—2 percent of total annual mortality.[10]

Acid deposition threatens human health indirectly as well. It can make several dangerous metals—including aluminum, cadmium, mercury, and lead—more soluble than usual. The metals can then leach from soils and lake sediments into underground aquifers, streams, and reservoirs, potentially contaminating water supplies and edible fish. Acidic water can also dissolve toxic metals from the pipes and conduits of municipal or home water systems, poisoning drinking water. In portions of both the United States and Sweden, elevated levels of certain metals have been found in the water supplies of areas receiving acid precipitation.[11]

Pollutants that stem predominantly from cars have been fought around the world even less successfully than sulfur dioxide and particulates. Ozone—a gas formed when hydrocarbons (a product of imperfect combustion in vehicle engines and a by-product of many industrial processes) react with nitrogen oxides (produced by both cars and power plants) in the presence of sunlight—is one of the most worrying pollutants in many areas. Recent U.S. research suggests that ozone causes short-term breathing difficulty and long-term lung damage in lower concentrations than previously believed. Other pollutants of concern emitted in substantial quantity by automobiles include carbon monoxide, nitrogen dioxide by itself, lead, and toxic hydrocarbons such as benzene, toluene, xylene, and ethylene dibromide. (See Table 6–2.)[12]

Ozone has become a seemingly intractable health problem in many parts of the world. In the United States, the summer of 1988—one of the hottest and driest on record—was the worst year for ground-level ozone in over a decade. Ac-

cording to the Natural Resources Defense Council, New York City air violated the federal health standard on 34 days—two to three times a week all summer long. In Washington, the standard was topped every third day, on average, during the summer months. And in Los Angeles, ozone levels during the 1988 smog season surged above the federal standard on 172 days. All told, the American Lung Association estimates that 382 counties, home to more than half of all Americans, are currently out of compliance with the EPA ozone standard.[13]

Western Europe had its summer of 1988 in 1989, when unusually sunny conditions in some countries led to high

Table 6-2. Health Effects of Pollutants from Automobiles[1]

Pollutant	Health Effect
Carbon Monoxide	Interferes with blood's ability to absorb oxygen, which impairs perception and thinking, slows reflexes, causes drowsiness, and can cause unconsciousness and death; if inhaled by pregnant women, may threaten growth and mental development of fetus.
Lead	Affects circulatory, reproductive, nervous, and kidney systems; suspected of causing hyperactivity and lowered learning ability in children; accumulates in bone and other tissues, so hazardous even after exposure ends.
Nitrogen Oxides	Can increase susceptibility to viral infections such as influenza, irritate the lungs, and cause bronchitis and pneumonia.
Ozone	Irritates mucous membranes of respiratory system; causes coughing, choking, impaired lung function; reduces resistance to colds and pneumonia; can aggravate chronic heart disease, asthma, bronchitis, emphysema.
Toxic Emissions	A broad category including many different compounds that are suspected of causing cancer, reproductive problems, and birth defects; benzene, for one, a known carcinogen.

[1]Automobiles are a primary source, but not the only source, of these pollutants.
SOURCES: National Clean Air Coalition, *The Clean Air Act: A Briefing Book for Members of Congress* (Washington, D.C.: 1985); John D. Graham et al., "The Potential Health Benefits of Controlling Hazardous Air Pollutants," in John Blodgett, ed., *Health Benefits of Air Pollution Control: A Discussion* (Washington, D.C.: Congressional Research Service, 1989).

ozone levels. In the normally cloudy United Kingdom, for example, readings above WHO standards on several days prompted calls for a "smog alert" system. In the rest of Europe, a monitoring network run by the Organisation for Economic Co-operation and Development (OECD)—though not complete—indicates that ozone levels in excess of WHO suggested levels occur at least occasionally over much of the continent.[14]

Information is scarce on ozone conditions in Eastern Europe, the Soviet Union, and the developing world. As vehicle fleets grow, however, conditions are likely to worsen in many areas. Latin America appears to have a grave problem, due to a combination of conducive weather conditions and rapidly growing fleets of vehicles lacking pollution controls. In Mexico City, the relatively lenient government standard of a one-hour ozone peak of 0.11 parts per million, not to be exceeded more than once daily, is topped more than 300 days a year—over twice as often as Los Angeles violates its much stricter standard.[15]

Emissions of carbon monoxide and nitrogen oxides remain a problem even where catalytic converters have been introduced on cars—Australia, Austria, Canada, Japan, Norway, South Korea, Sweden, Switzerland, and the United States—not to mention those where they have not. The recent WHO/UNEP report estimated that 15–20 percent of urban residents in North America and Europe are exposed to unacceptably high levels of nitrogen dioxide and 50 percent to unhealthy carbon monoxide concentrations.[16]

By fortuitous coincidence, cars equipped with catalytic converters cannot tolerate leaded gas. Where they are mandatory, the use of lead-free gas has resulted in impressive emissions reductions, with unequivocal benefits to human health. In the United States, lead emissions fell by 96 percent between 1970 and 1987 as a result of this relationship, in conjunction with national legislation requiring steep reductions in the lead content of gas burned in old cars not equipped with converters. As a result, the average lead level in Americans' blood dropped by more than one third between 1976 and 1980.[17]

In Mexico City, 7 out of 10 newborns were found to have lead levels in their blood in excess of WHO norms.

Most of the world, however, mandates neither converters nor the use of unleaded gas. The WHO/UNEP report estimated that a third of North American and European city dwellers are exposed to either marginal or unacceptable concentrations of lead in the air. Paris topped the list of a limited sample, with an annual average concentration far in excess of the WHO guideline. In a study in Mexico City, 7 out of 10 newborns were found to have lead levels in their blood in excess of WHO norms.[18]

In the United States, long-overdue attention is finally being paid to the health threat posed by airborne toxic chemicals from industrial sources. Such chemicals—which can cause cancer and birth and genetic defects—have often fallen through the regulatory crack. The current wave of concern has been fueled by EPA's recent announcement that 2.7 billion pounds of hazardous pollutants were released into the air by industries in 1987, including 235 million pounds of known cancer-causing chemicals. Even this number is an underestimate, as it omits sources such as waste dumps, dry cleaners, and cars, and releases into soil or water that enter the air through evaporation. The true number may be in excess of 4.8 billion pounds annually, EPA admits.[19]

In a 1987 study, the agency concluded that air toxic emissions may cause 2,000 cancer deaths a year. Because the study tallied the effects of only a third of the carcinogenic chemicals, it was likely an underestimate. If all the hundreds of chemicals emitted into the air had been included and their synergistic effects considered, the number would probably be far higher. Other studies have found high cancer rates in communities near certain types of factories in West Virginia and Louisiana. For example, in West Virginia's Kanawha Valley—home to roughly 250,000 people and 13 major chemical plants—state health department records show that between 1968 and 1977 the incidence of respiratory cancer was more than 21 percent above the national average. According to EPA statistics, a lifetime of exposure to the airborne concentrations of chloroform, ethylene oxide, and butadiene in this valley could cause cancer in one resident in 1,000.[20]

Unfortunately, similar data are not available for other countries. Wherever uncontrolled polluting industries such as chemical plants, metals industries, and paper manufacturers exist, however, emissions levels are undoubtedly high. Measurements of lead and cadmium in the soil of the upper Silesian towns of Olkosz and Slawkow in Poland, for instance, are among the highest ever recorded anywhere in the world. The government is considering a ban on growing vegetables in several Silesian towns due to concentrations of lead, zinc, cadmium, and mercury that are 30–70 percent higher than World Health Organization norms.[21]

Airborne toxic chemical emissions are likely to rise rapidly in developing countries as polluting industries are built. The Third World's share of world iron and steel production rose from 3.6 per-

cent in 1955 to 17.3 percent in 1984. In India, pesticide production increased from 1,460 tons in 1960 to 40,680 in 1980—nearly thirtyfold. Production of dyes and pigments grew at a comparable pace over the same period, reaching 30,-850 tons by 1980. Most of these countries have few pollution controls and little environmental regulation.[22]

Recent evidence suggests that toxic air emissions—like acid deposition—can be carried great distances before falling to the ground. This explains in part why DDT, presumably blown in from Central America or Mexico, is being found in the Great Lakes years after the pesticide was banned in the United States and Canada. And why researchers at McGill University in Montreal have found DDT and polychlorinated biphenyls throughout the Eskimos' food chain, including in bears, fish, berries, and snow.[23]

Putting a dollar value on all these health costs of air pollution is difficult, as it involves judgments about the worth of good health and human life. The few guesses made suggest a very high price. Thomas Crocker of the University of Wyoming, for one, estimates that air pollution costs the United States as much as $40 billion annually in health care and lost productivity. A study conducted for Los Angeles officials concluded that the proposed air quality management plan for the region (described later in this chapter) would save $9.4 billion a year in health care expenses—more than three times what it is projected to cost.[24]

THE ECOLOGICAL TOLL

Though concern for human health was the motivating factor behind the world's first control laws, the last 20 years have demonstrated that air pollution poses an equally grave threat to the natural and built environments. Ecosystems are actually often damaged at lower levels of pollution than humans are, a fact not yet adequately reflected in air pollution standards. Fishless lakes and streams, dying forests, and faceless ancient sculptures all provide sad testament to the destruction that industrialization has wrought.

The first warning came in the late sixties, when scientists in Scandinavia began to suspect that SO_2 emissions from the more urban and industrialized countries of Europe, such as the United Kingdom and West Germany, might be responsible for declining fish stocks in their lakes. American scientists who soon did extensive studies of acidity at the Hubbard Brook Experimental Forest in New Hampshire found similar omi-nous indications. Extensive investigations in the seventies revealed that acid deposition was indeed acidifying water and killing fish and other aquatic organisms in geologically susceptible areas of Scandinavia and North America. (See Table 6–3.)[25]

In recent years, large areas of the world have been found to fall into this worrying category of "geologically susceptible." In the United States, much of the Great Lakes region, several southeastern states, and many of the mountainous areas of the West appear to be at risk, in addition to the Northeast, where damage has been evident for some time. Half the 700,000 lakes in the six eastern provinces of Canada are deemed extremely acid-sensitive, as are large areas in all western provinces, the Yukon, the Northwest Territories, and Labrador. In

Table 6-3. Evidence of Acidified Lakes, Selected Countries

Country	Evidence
Canada	Over 14,000 lakes strongly acidified, and 150,000 in the East (one in seven) suffering biological damage.
Finland	Survey of 1,000 lakes indicates that those with a low acid-neutralizing capacity are distributed across the country; 8 percent of these lakes have no neutralizing capacity; most strongly acidified ones are located in southern Finland.
Norway	Fish in waters covering 13,000 square kilometers eliminated, and affected in waters over a further 20,000 square kilometers.
Sweden	14,000 lakes unable to support sensitive aquatic life and 2,200 nearly lifeless.
United Kingdom	Some acidified lakes in southwestern Scotland, in western Wales, and in the Lake District.
United States	About 1,000 acidified lakes and 3,000 marginally acidic ones, per Environmental Defense Fund; 1984 EPA study found 552 strongly acidic lakes and 964 marginally acidic ones.

SOURCES: Jim Ketcham-Colwill, "Acid Rain: Science and Control Issues," Environmental and Energy Study Institute Special Report, July 12, 1989; U.N. Economic Commission for Europe, "Current Geographical Extent of Acidification in Rivers, Lakes, and Reservoirs in the ECE Region" (draft), June 15, 1988.

Europe, similarly vulnerable parts of the Netherlands, Belgium, Denmark, Switzerland, Italy, West Germany, and Ireland have been added to the better known areas of Scandinavia and the United Kingdom. Vast parts of Asia, Africa, and South America are also acid-sensitive.[26]

Canadian biologist David Schindler warns that the ecological consequences are great. Though concern initially focused on damage to fish, many other aquatic organisms can also be affected. To investigate the biological impact of this process, Schindler and his colleagues deliberately acidified two lakes in northwest Ontario. When the first lake went from a pH of 6.5 to 5.0 over eight years, approximately a third of the species studied were eliminated. Extrapolating from these findings, Schindler estimates that in the Adirondack and the Poconos-Catskills regions, more than half the sensitive species such as mollusks, leeches, and crustaceans have been wiped out.[27]

Fishless lakes and streams, dying forests, and faceless ancient sculptures all provide sad testament to the destruction industrialization has wrought.

Recent evidence suggests that the threat to coastal waters may be of similar magnitude. A 1988 study by the Environmental Defense Fund concluded that acid deposition is a major contributor to the degradation of the Chesapeake Bay. By a process known as eutrophication, nitrogen from a variety of sources "fertilizes" algae to the extent that it chokes off the oxygen supply and blocks the sunlight required by other aquatic plants and animals. The study found that airborne nitrates account for fully 25 per-

cent of the total nitrogen load currently entering the Chesapeake. If emissions continue their upward climb, this share will rise to 42 percent by the year 2030. Other research has shown that 27 percent of the Baltic Sea's nitrogen load comes from the air.[28]

In the early eighties, concern about the possible effects of acid deposition spread from lakes and streams to forests. Signs of widespread damage attributed to acid deposition first arose in West Germany. The share of forests there showing signs of damage rose from 8 percent in 1982, the first year a survey was done, to 34 percent in 1983, to 50 percent by the following year. It peaked in 1986, at 54 percent, and has since declined slightly—to 52 percent in 1988. Because dead trees are not included in the surveys, however, slightly lower percentages do not necessarily mean an improving situation.[29]

Wide public debate about the causes and consequences of forest decline followed its discovery. Though the exact mechanisms are still not precisely understood, most scientists believe a complex mixture of pollutants—including acid deposition, ozone, and heavy metals—renders forests susceptible to a range of natural stresses, such as droughts, extremes of heat and cold, and blights, that combine to cause the decline. Though initial fears of massive forest death throughout the continent have not been borne out, a high economic and ecological toll has already been paid.

Since the initial West German alarm, concern about forest damage has spread throughout the world. In Europe, several countries have initiated annual surveys. The results are now brought together in an annual assessment of European forest damage published by the U.N. Economic Commission for Europe. The 1988 report found at least preliminary signs of damage in each of the 26 areas surveyed. Twenty-two of the surveys—most of them national— found 30 percent or more of their forests to be damaged; for eight, the figure was half or more. Across the continent, nearly 50 million hectares have been identified as damaged, representing 35 percent of Europe's total forested area. (See Table 6–4.)

North America may be next in line. Though the damage documented to date has been mostly restricted to isolated, high-altitude environments, many fear it is only a matter of time before it spreads to other areas. In the United States, some declines, such as those of Jeffrey and ponderosa pines in California and of eastern white pines, have been conclusively linked to high concentrations of ground-level ozone. In others, such as those of the red spruce, Fraser fir, yellow pines, and sugar maple, researchers are less certain about the role of air pollution. Even with these species, however, high levels of recorded pollution often correspond to areas of extensive damage, suggesting a link.[30]

On the summit of Mount Mitchell in North Carolina—where essentially all the red spruce and Fraser fir trees are now dead—ozone levels more than half the time exceed those at which tree damage has been proved to occur in controlled laboratory studies. Cloud samples taken on this mountain during 1986 showed a pH varying from 5.4 (slightly acidic) to as low as 2.2 (about the same as vinegar), with a mean of 3.4. "It's plain," says plant pathologist Robert Bruck, who has been studying damage in this area for years, "that no one has proved, or ever will, that air pollution is killing the trees up here. But far more quickly than we ever expected, we've ended up with a highly correlated bunch of data—high levels of air pollution correlated to a decline we're watching in progress."[31]

The economic toll to the forestry and tourism industries is potentially enormous. Environmental scientists in Poland predict that forest loss will cost the country $1.5 billion by 1992. Economists Werner Schulz and Lutz Wicke estimate that forest damage will cost West Germany between 5.5 and 8.8 billion deutsche marks ($2.98–4.77 billion) annually over the next 70 years depending upon how strictly emissions are controlled. The losses are not only monetary: Fichtelberg Mountain along the East German and Czechoslovakian border once attracted many visitors who scaled it to marvel at the view. "Now they weep," writes Mike Leary of the *Philadelphia Inquirer*. "What a horrible tragedy," said one woman gazing from the summit at the view below: a huge expanse of dead trees and broad brown patches where forests once stood.[32]

In addition to forests, air pollution also threatens crops. Ozone is the primary concern, although SO_2, nitrogen oxides, and sulfates and nitrates are also thought to be potentially harmful. The most comprehensive studies on this have been conducted in the United States. A 1987 government report by the National Acid Precipitation Assessment Program concluded that current levels of ozone were reducing crop yields by 1 percent or less for sorghum and corn, by about 7 percent for cotton and soybean, and by more than 30 percent for alfalfa. Total crop losses were estimated to be in the range of 5–10 percent of production. According to one estimate, this represents an economic loss of some $5.4 billion.[33]

Reports of similar damage in the Third World are starting to be heard. Damage in China's southwest forests is being increasingly linked by scientists to air pollution and acid rain caused by a heavy reliance on high-sulfur coal. In Sichuan's Maocaoba pine forest, more than 90 percent of the trees have died. On Nanshan hill in Chongqing, the biggest city in southwest China, an 1,800-hectare forest of dense masson pine has

State of the World 1990

Table 6-4. Forest Damage in Europe, 1988

Country or Area	Total Forest Area[1]	Estimated Area Damaged	Share of Total
	(thousand hectares)		(percent)
Czechoslovakia	4,578	3,250	71
Greece	2,034	1,302	64
United Kingdom	2,200	1,408	64
Estonia, Soviet Union	1,795	933	52
West Germany	7,360	3,827	52
Tuscany, Italy	150	77	51
Liechtenstein	8	4	50
Norway[2]	5,925	2,963	50
Denmark	466	228	49
Poland	8,654	4,240	49
Netherlands	311	149	48
Flanders, Belgium	115	53	46
East Germany	2,955	1,300	44
Bulgaria	3,627	1,560	43
Switzerland	1,186	510	43
Luxembourg	88	37	42
Finland	20,059	7,823	39
Sweden	23,700	9,243	39
Wallonia, Belgium[2]	248	87	35
Yugoslavia	4,889	1,564	32
Spain	11,792	3,656	31
Ireland[2]	334	100	30
Austria	3,754	1,089	29
France	14,440	3,321	23
Hungary	1,637	360	22
Lithuania, Soviet Union	1,810	380	21
Bolzano, Italy	307	61	20
Portugal	3,060	122	4
Other[3]	13,474	n.a.	n.a.
Total[4]	140,956	49,647	35

[1]For areas where only conifers surveyed, "total forest area" means total forested area of conifers. For Yugoslavia, which conducted only a regional survey, "total forest area" means total area surveyed. [2]Conifers only. In Ireland, only trees less than 60 years old were assessed. [3]Includes unsurveyed portions of countries that have done regional and conifer-only surveys. [4]Does not include Turkey or any of the Soviet Union except for Estonia and Lithuania.
SOURCE: Worldwatch Institute, based on United Nations Environment Programme and United Nations Economic Commission for Europe, "Forest Damage and Air Pollution: Report of the 1988 Forest Damage Survey in Europe," Global Environment Monitoring System, 1989.

been reduced by almost half. Both these regions have highly acidic rain and elevated levels of sulfur dioxide. China's *Science and Technology Daily* reported in May 1989 that acid rain is causing serious damage in Hunan Province, including crop losses worth about 1 billion yuan ($260 million at official exchange rates) annually. Localized damage to trees and crops from air pollution, and possibly from acid deposition, has also been reported in Chile, Brazil, and Mexico.[34]

Elevated levels of acidity and ozone have been found in tropical rain forests. In Latin America, the pollution is attributed to the enormous fires that rage as cattle ranchers and settlers clear land. In Africa, it stems from fires that burn for months across thousands of kilometers of African savannas. They are set by farmers and herders to clear shrubs and permit the growth of crops and grass. Tropical areas are thought to be especially vulnerable to acidification, because their soils are naturally low in buffering agents.[35]

In contrast to damage to lakes and forests, that to the built environment is most frequently a local problem. Sulfur dioxide and its acidic chemical derivatives are believed to be the chief culprits, though nitrogen oxides, ozone, and other pollutants also contribute.[36]

Corrosion of historical monuments is particularly evident in Europe—from the Acropolis, to the Royal Palace in Amsterdam, to the medieval buildings and monuments of Krakow, Poland. Although some decay is to be expected in structures dating to antiquity, pollution is greatly speeding the process. T.N. Skoulikidis, a Greek specialist on acid corrosion, has estimated that Athenian monuments have deteriorated more in the past 20–25 years from pollution than in the previous 2,400. Damage to historical artifacts and edifices is evident throughout Italy. "Classic marble busts," says *New York Times* correspondent Paul Hofmann, are being "transmogrified into noseless and earless plaster grotesques." In the Katowice region of southern Poland, trains must slow down in certain places because the railway lines have corroded, apparently from acid pollution.[37]

In the United States, air pollution may prevent the nation's historical monuments from ever reaching the ripe old age of Europe's. Already, Independence Hall in Philadelphia, where the Declaration of Independence was signed, is experiencing damage. At the Gettysburg Civil War battlefield, every statue or tablet made of bronze, limestone, or sandstone is being slowly but inexorably eaten away. Both the Statue of Liberty and the Washington Monument are also reportedly threatened.[38]

Again, the Third World is following the example of the First. The Taj Mahal appears to be endangered by emissions from an upwind oil refinery that are eroding its marble and sandstone surfaces. Recent research has found that acid rain falling on the Yucatan Peninsula and much of southern Mexico is destroying the temples, murals, and megaliths of the Mayans. The primary source of the damaging emissions is believed to be uncapped Mexican oil wells and oilfield smokestacks near Coatzacoalcos and Ciudad del Carmen on the Gulf of Mexico. Exhaust from tour buses is also thought to contribute to the decay.[39]

The costs of materials corrosion to national economies are monumental. Though quantifying the exact toll is difficult, given scientific uncertainties about the precise mechanisms of decay and difficulties in distinguishing between natural and acid-induced erosion, some rough attempts have been made. Researchers at the Swedish Corrosion Institute estimated in 1984 that corrosion to materials of all kind cost Sweden $2.5 billion per year. A 1980 study for the Dutch Ministry of Health and Environmental Protection estimated damage to monuments, libraries, and archives in the Netherlands at $10–15 million annually. Studies conducted in the United States have suggested a multibillion-dollar price tag. When these numbers are added to the other costs exacted by air pollution and acid rain, reducing emissions begins to look cheap by comparison.[40]

REDUCTION STRATEGIES

The tremendous damage to health and environment inflicted by air pollution and acid rain has not been lost on the public or on policymakers. In the western, industrial world, the last 20 years has been a period of intensive political and scientific activity aimed at clearing the air. The approaches to date, however, have tended to be technological Band-Aids rather than efforts to address the roots of the problem: inappropriate energy, transportation, and industrial systems. Though they have achieved some success, a more comprehensive strategy will be needed to confront the air quality challenges of the nineties.

The most widespread technological intervention has been the introduction of electrostatic precipitators and baghouse filters for the control of particulate emissions from power plants. Their use is now required in virtually all OECD countries (Western Europe, the United States, Canada, and Japan). Though such technologies can reduce particulate emissions directly from the smokestack by as much as 99.5 percent, they do nothing to prevent acidic particles such as sulfates and nitrates from forming outside the smokestack from gaseous emissions. Such particles can wreak particularly great harm on both human health and the environment.[41]

The predominant technique used to reduce sulfur dioxide has been to put flue-gas desulfurization (FGD) technology, popularly known as scrubbers, on coal-burning power plants. Scrubbers can remove as much as 95 percent of a given plant's SO_2 emissions. Among members of OECD, nearly 140,000 megawatts of power plant capacity were either equipped with FGD or had it under construction at the beginning of 1988. The United States led in total "scrubbed" capacity, with 62,000 megawatts. But only 20 percent of the total U.S. coal-fired capacity was equipped with scrubbers as of January 1987, compared with roughly 40 percent in West Germany, 50 percent in Sweden, 60 percent in Austria, and 85 percent in Japan. By the end of the decade, the figures will be 70 percent in Italy, 85 percent in West Germany, 100 percent in the Netherlands, but still only 30 percent in the United States.[42]

Though scrubbers have been rare elsewhere, this is slowly beginning to change. Czechoslovakia, for example, announced a major new environmental protection initiative in early 1989, including a projected $1.3-billion investment by the mid-nineties in desulfurization equipment at nine of the country's most heavily polluting power plants. China indicated in December 1987 that it would, for the first time, equip a planned coal-fired plant with FGD.[43]

For control of nitrogen oxide emissions from power plants, countries have pursued a variety of approaches with mixed results. The most simple have been various forms of combustion modifications, which yield reductions of some 30–50 percent. To date, the United States has invested heavily in this approach, and the United Kingdom and Portugal somewhat. More expensive, but also more effective, is a process known as selective catalytic reduction

(SCR). This can reduce emissions by 80–90 percent. Japan pioneered the technology in the seventies, primarily as an antismog measure. At the end of 1986, it had 91 units in operation, which was about 90 percent of total OECD-member installations and 54 percent of the country's fossil-fuel-fired generating capacity. West Germany and Austria have more recently embarked on ambitious SCR construction programs.[44]

Also under intensive investigation are so-called clean coal technologies that lower emissions of both SO_2 and NO_x during combustion. Most prominent among them are various fluidized bed combustion technologies and a process known as integrated gasification-combined cycle, whereby coal is transformed into a gas, which is then used to run a turbine. Excess heat is tapped to produce steam, which powers a second turbine. These technologies have the advantages of offering deep cuts in SO_2 and NO_x while simultaneously burning more efficiently. Some of them are beginning to be commercialized, though others are still at the demonstration phase.[45]

In general, new power plants are being fairly tightly controlled in most western countries. The greater problem comes in retrofitting existing facilities. This is important, particularly as construction of new power plants has slowed considerably in the industrial world. To date, Denmark, the Netherlands, Sweden, the United Kingdom, and West Germany are the only OECD members to have undertaken significant retrofitting.[46]

Though the technologies just outlined provide necessary immediate reductions, they are not the ultimate solution. For one, they can create environmental problems of their own, such as the need to dispose of scrubber ash, a hazardous waste. Second, they do little if anything to reduce carbon dioxide emissions, so

make no significant contribution to solving global warming. For these reasons, they are best viewed as a bridge to the day when energy-efficient societies are the norm and when renewable sources such as solar, wind, and water power provide the bulk of the world's energy. (See also Chapters 2 and 10.)

Approaches to date have tended to be technological Band-Aids rather than efforts to address the roots of the problem.

Energy efficiency is an essential strategy for reducing emissions from power plants, though it is rarely promoted as such. The American Council for an Energy-Efficient Economy (ACEEE) has identified cost-effective, widely available conservation measures that could bring electricity demand 15 percent below utility forecasts in a region of the U.S. Midwest that is responsible for 33 percent of the nation's utility-generated sulfur dioxide emissions. Implementing this amount of conservation would cut SO_2 emissions by 7–11 percent between 1992 and 2002 and NO_x emissions by an undetermined amount.[47]

Equally important, the savings resulting from avoided power plant construction can more than offset the cost of emissions controls at existing plants. Indeed, ACEEE found that consumers in the Midwest could realize a net savings of $4–8 billion if emissions control and accelerated conservation were both pursued as part of a national effort to halve SO_2 emissions, as compared with a scenario in which neither occur. Emissions control in the absence of efficiency investments would run in the billions of dollars.[48]

Around the world, it is essential that governments put economic incentives for energy reform into place as part of their air quality strategies. One way of doing that would be to allow utilities that reduce customers' bills through efficiency investments to earn a slightly higher rate of return than otherwise. Another is by market-based power planning systems that incorporate environmental costs, thereby encouraging both efficiency and renewable energy. The State of New York is already pioneering this approach: fossil-fuel-based generating sources in two utility regions in the state must add nearly a penny per kilowatt-hour to their costs to account for air pollution and an additional half a cent per kilowatt-hour to account for other environmental externalities.[49]

Reducing urban air pollution may require a major shift away from automobiles as the cornerstone of transportation systems.

Though materials recycling has caught the public's attention as a way to save on scarce landfill space, its pollution prevention potential is equally important. Each ton of newsprint made from waste rather than wood lowers energy use by one fourth to three fifths and air pollutants by some 75 percent. Aluminum produced from recycled cans rather than virgin ore cuts emissions of nitrogen oxides by 95 percent and of sulfur dioxide by 99 percent.[50]

When it comes to reducing automotive pollution, a similar preventive approach is needed. To date, engine modifications and catalytic converters have been the principal strategies used to lower harmful emissions. However, even in countries that have mandated converters—which reduce a car's hydrocarbon emissions by an average of 87 percent, carbon monoxide emissions by

an average of 85 percent, and nitrogen oxides by 62 percent over the life of a vehicle—rising vehicle fleets are overwhelming their efficacy. In addition, converters actually slightly increase carbon dioxide emissions. Again, more fundamental additional measures are required.[51]

Most basically, reducing urban air pollution may require a major shift away from automobiles as the cornerstone of transportation systems. As congestion increases exponentially in most major cities, driving to work is becoming increasingly unattractive anyway. Convenient public transportation, increased car pooling, and facilitated bicycle commuting are for many cities the cheapest, most effective, most sensible way to proceed. (See Chapter 7.)

Indeed, in many cities around the world driving restrictions already exist. In Florence, Italy, the heart of the city has been turned into a pedestrian mall during daylight hours. Central Rome is off-limits to normal car traffic for seven hours a day, during the morning and evening rush hours. Budapest bans motor traffic from all but two streets in the downtown area. Mexican President Carlos Salinas de Gortari has asked for prohibitions on driving in the city on particularly bad days. In Santiago, one fifth of all vehicles are kept off the streets each weekday based on their license-plate number. Lagos has had a similar system in place since 1976, with odd and even plate numbers taking turns. Their experience, however, shows that such plans must be implemented with care—wealthy residents began to buy two cars, one with an odd-numbered license and the other, an even.[52]

Short of an all-out ban on private cars in downtown areas, city governments can encourage commuters to leave their cars at home by making public transportation inexpensive and convenient. Eradicating or taxing free parking benefits and collecting hefty tolls for road use are other means toward this end.

One option under consideration around the world is the widespread introduction of vehicles powered by alternative fuels, such as methanol, ethanol, natural gas, hydrogen, and electricity. Although these alternatives would reduce emissions, in some cases substantially, their widespread use runs the risk of falling into the classic anti-pollution pitfall: solving one problem by substituting another.

For example, though most analysts agree that methanol would yield some reductions in ozone formation, vehicles run on this fuel produce two to five times as much formaldehyde, a potentially carcinogenic compound, as those run on gasoline do. And if methanol is made from coal, as it might well eventually be if it were relied on in a substantial way, carbon dioxide emissions would likely be 20–160 percent higher than in cars using gasoline—an untenable idea in a greenhouse world. In addition, methanol is a highly poisonous, colorless, odorless substance that could contaminate groundwater in the event of leaking storage tanks, and that could seriously harm someone if splashed on them during refueling or even kill them if accidentally ingested.[53]

The problems posed by alternative fuels are not necessarily insurmountable. If such programs are to render real benefits, however, they must be implemented with great care as one component of a broader effort.

Less glamorous but perhaps more practical would be measures such as reducing the volatility of gasoline and introducing widespread inspection and maintenance programs to ensure that emissions control systems are functioning properly. Indeed, a recent U.S. Office of Technology Assessment (OTA) report concluded that reducing gasoline volatility would cost only $120–750 per ton of hydrocarbons reduced, and implementing inspection and maintenance programs $2,100–5,800 per ton of reductions, whereas replacing gasoline with methanol would cost $8,700–51,000 per ton reduced.[54]

A clear priority as society struggles with both air pollution and global warming is to encourage the manufacture and purchase of automobiles that are both low in emissions and high in fuel economy. There is a direct relationship between fuel economy improvements and carbon dioxide emissions reductions, but a far more ambiguous one between fuel economy and "conventional" pollutants. In the latter case, the relationship is complicated by variables such as combustion temperature and amount and type of catalysts used.

Though the automobile industry has claimed over the years that the goals of reducing emissions and improving fuel economy are in direct conflict, both EPA and OTA maintain that this is not the case. Some measures chosen to improve emissions may entail a fuel economy penalty, but this can be avoided. Indeed, an analysis by Chris Calwell of the Natural Resources Defense Council of 781 car models shows that the 50 most efficient emitted one third less hydrocarbons than the average car and roughly half as much as the 50 least efficient ones.[55]

An incentive system could help push consumers and carmakers in the right direction. Analysts at the Lawrence Berkeley Laboratories have suggested that a revolving fund could give purchasers of low-emission, fuel-efficient vehicles a rebate financed by a tax on new high-emission, gas-guzzling cars. Consumer demand would then spur the industry to produce such autos. An intermediary step would be to at least provide consumers with information as to how polluting a car is. Just as a given model's fuel economy is common knowledge, so would be its emissions rating.[56]

As with power plant and auto emissions, efforts to control air toxics will be most successful if they focus on waste minimization rather than simply on control. Such a strategy also helps prevent waste from just being shifted from one form to another. Many control technologies, such as scrubbers and filters, produce hazardous solid wastes that must then be disposed of on land. OTA has concluded it is technically and economically feasible for U.S. industries to lower production of wastes and pollutants by up to 50 percent within the next few years. Similar reduction possibilities exist in most other countries.[57]

The most effective incentive for waste reduction is strict regulation of disposal into the land, air, and water.

Probably the most effective incentive for waste reduction is strict regulation of disposal into the land, air, and water. This will force up the price of disposal, making it cost-effective for industries to reduce waste generation. Certain economic incentives could also help. For example, a "deposit-refund" system might be put in place whereby industries would be taxed on the purchase of hazardous inputs, but refunded for wastes produced from them that are recovered or recycled.[58]

Experience in the United States has shown that public access to information about what chemicals a plant is emitting can be instrumental in spurring a response: within weeks of its release of the information on emissions mandated under right-to-know legislation, Monsanto announced its intention to cut back its toxic air emissions by 90 percent by 1992.[59]

Most European countries have yet to provide for the release of information about emissions from industrial plants, although the European Economic Community (EEC) is reported to be considering a draft directive on freedom of information on environmental matters that would improve this somewhat. Glasnost is gradually improving the environmental data flow in some of Eastern Europe and in the Soviet Union, though much progress in this area remains to be made. Grassroots groups in some developing countries are also beginning to break down the secrecy barriers.[60]

MOBILIZING FOR CLEANER AIR

Though the time seems ripe for reform of the current approach to air pollution control, few policymakers are considering implementation of the comprehensive strategies that are necessary. Recent developments at the national and international levels—though steps forward—remain inadequate to the task at hand.

In the United States, Congress is currently debating major amendments to the 1970 Clean Air Act. The long-stalled process was given a boost by the Bush administration, which submitted a plan of its own for Congress to consider. This active presidential leadership, combined with strong public support for strengthened environmental regulations, has raised hopes that legislation will soon be in place that will halve emissions that cause acid rain, tighten emissions standards for automobiles significantly, and require much stricter control of toxic air pollutants.[61]

Though any legislation is an improvement over the status quo, the administration's proposal misses the opportunity to address the problem at a

fundamental level through energy efficiency, transportation reform, and waste
reduction. From this point of view, the
acid rain provisions are probably the
most promising, as they would give the
utilities their choice of control technique, including energy efficiency. Posi-
tive incentives, however, could have
helped make this the clear first choice for
utilities rather than simply one of many
options.

Because air pollution crosses European borders with impunity (see Table
6–5), international cooperation on reducing emissions is essential. For example, 96 percent of the sulfur deposited in
Norway originates in other countries.
The Norwegians can thus do little to
save their lakes without the help of troublesome neighbors.

Though the past few years have seen
important incremental advances in international fora, there remains much room
for improvement. Under the auspices of
the U.N. Economic Commission for
Europe, agreements to reduce emissions
of sulfur dioxide and nitrogen oxides
have been reached. The SO_2 protocol
calls for a 30-percent reduction in emissions or their transboundary flows from
1980 levels by 1993. The nitrogen oxide
protocol calls for a freeze on emissions
in 1994 at 1987 levels, as well as further
discussions beginning in 1996 aimed at
actual reductions.[62]

Most but not all major players have
signed on to these protocols, and some
have made commitments to go beyond
them. At least nine countries have now
pledged to bring sulfur dioxide levels

Table 6-5. Sulfur Pollution in Selected European Countries, 1988

Country	Total Emissions[1]	Total Deposition	Share of Emissions Exported[2]	Share of Deposition Imported
	(thousand tons)		(percent)	
Norway	37	210	76	96
Austria	62	181	74	91
Sweden	110	302	69	89
Switzerland	37	65	81	89
Netherlands	145	104	80	72
France	760	622	67	59
West Germany	750	628	63	56
Czechoslovakia	1,400	659	75	47
Poland	2,090	1,248	68	46
Soviet Union[3]	5,150	3,201	61	38
Italy	1,185	510	72	36
East Germany	2,425	787	75	22
Spain	1,625	590	72	22
United Kingdom	1,890	636	71	15

[1]Unless otherwise noted, emissions figures are preliminary 1988 data. Austria, Sweden, and France are
1987 data; Italy, 1986; and Spain, 1983.   [2]May be deposited either in another country or over a body of
water.   [3]Only the European part of the Soviet Union. Thus, export figure includes exports to the Asian
part of the Soviet Union.
SOURCE: Worldwatch Institute, based upon emissions data in Economic Commission for Europe, "Annual
Review of Strategies and Policies for Air Pollution Abatement" (draft), September 26, 1989, and on data
on transboundary flows supplied by the European Monitoring and Evaluation Programme.

down to less than half of 1980 levels by 1995; Austria, Liechtenstein, Sweden, and West Germany have committed to reducing them by two thirds. On nitrogen oxides, 12 West European nations have agreed to go beyond the freeze and reduce emissions by 30 percent by 1998. Work is now under way on a protocol aimed at reducing hydrocarbon emissions, an ingredient in the formation of ground-level ozone.[63]

A November 1988 EEC directive represents a binding commitment by the members to reduce acid-rain-causing emissions significantly. The directive will lower Community-wide emissions of SO_2 from existing power plants by a total of 57 percent from 1980 levels by 2003, and of nitrogen oxides by 30 percent by 1998. The amount each country will be required to cut emissions is based on each member-state's contribution to long-range transboundary pollution, level of industrial development, dependence on indigenous high-sulfur energy resources, and pollution control efforts before 1980. Belgium and West Germany, for example, would have to reduce SO_2 emissions by 70 percent by 2003, while Ireland could have 25 percent higher output and Portugal, 79 percent higher.[64]

These commitments are far better than nothing. But they are not enough. For one, it is unclear whether all signatories to the SO_2 and NO_x protocols will meet their targets. Second, the reductions envisioned are too little and too late to protect the environment adequately. Ecologists suggest that cuts on the order of 90 percent in sulfur and nitrogen oxides and 75 percent in ground-level ozone are what is really needed. How individual countries will choose to meet their given targets remains an open question. Unfortunately, energy efficiency rarely appears at the top of the lists of options.[65]

Though Europe has been quicker than the United States in addressing acid rain, it has for the most part been much slower to tackle urban air quality problems. Non-EEC countries such as Austria, Norway, Sweden, and Switzerland have had strong auto emissions control legislation in place for several years, but the Community has been unable to agree on stringent standards.

This finally changed in June 1989, when the EEC Council of Environmental Ministers ended a nearly four-year debate and approved new standards for small cars that will be as tough as those now in effect in the United States. To meet them, small cars will have to be equipped with catalytic converters. Though standards for large cars were already on the books, the EEC Environment Commissioner is considering stricter ones for medium and large cars as well. A Community-wide speed limit is also being looked at as an air pollution control measure. Although clearly a step forward, it is somewhat ironic that Europe sees its adoption of U.S. standards as a major victory at the same time that the United States is realizing these regulations do not go far enough.[66]

Los Angeles—with the worst air quality in the United States—is one of the first regions in the world to really understand that lasting change will not come by tinkering with the status quo. Under a bold new air quality plan, automobile use will be discouraged, public transportation boosted, and household and industrial activities that produce emissions tightly controlled. For example, paints and solvents will have to be reformulated, gasoline-powered lawn mowers and barbecues and fuels that require lighter fluid will be banned, and the number of cars per family limited.[67]

In Eastern Europe and the Soviet Union, air pollution has only recently emerged as a potent political issue. Fyodor Morgun, then head of the Soviet

Union's State Committee on the Protection of Nature, reported to the Nineteenth Communist Party conference in July 1988 that air pollution levels in 102 Soviet cities, affecting more than 50 million people, were often 10 times higher than the Soviet standard. This official recognition of the problem was an important step forward. Similarly, when Solidarity sat down with the Polish government in March 1989 for the Round Table discussions, environmental issues in general and air pollution in particular figured prominently on the agenda: a fact of enormous significance. More than rhetoric, however, will clearly be necessary to solve the problems.[68]

To make a dent in their pollution, Eastern Europe and the Soviet Union will need western technologies as well as domestic reform. Given current economic straits in these countries, however, "environmental aid" from the West will be necessary to help with the purchase of pollution control, energy efficiency, renewable energy, and waste reduction technologies. Such aid can be in the best interests of the purveyor, as stemming pollution elsewhere before it crosses borders is often far cheaper than taking further incremental steps at home. Concern about global warming provides added impetus for western nations to provide assistance with energy efficiency and renewable energy.

Some promising initiatives are already under way. The Natural Resources Defense Council is cooperating with the Soviet Union on energy efficiency, helping them set up the regulatory structures that should encourage it. The Rocky Mountain Institute has also been advising the Soviet government on efficiency. A Swedish-Soviet committee of representatives from both countries' paper and pulp industries has recently been formed to cooperate on cleaning up the Baltic Sea. Under the committee's plan, Swedish technology will be exchanged for Soviet natural gas and pulp.[69]

Several countries are giving environmental aid for Eastern Europe. To cite just a few of the many recent initiatives, the U.S. Congress passed legislation that would allocate $40 million in environmental assistance for Poland and Hungary, a token sum given the magnitude of the problems, but symbolically important nonetheless. The Swedish government recently announced that it plans to provide Poland with $45 million in aid over the next three years, mostly for environmental protection. East and West Germany recently initiated a three-year program to cooperate on pollution control. East Germany will contribute 120 million deutsche marks ($65 million) and West Germany, 300 million ($163 millon) that will go toward the purchase of advanced coal-burning technology for power plants and other pollution control measures. The two countries will also standardize air pollution data in order to facilitate the forecasting of photochemical smogs that waft across the border.[70]

To make a dent in pollution, Eastern Europe and the Soviet Union will need western technologies as well as domestic reform.

Air pollution is just beginning to emerge on the political agenda in Third World cities as well. Though the challenge is enormous, at least the issue is starting to be addressed. In Cubatão, Brazil, a notoriously polluted industrial city known as "the Valley of Death," a five-year-old government clean-up campaign is starting to make a dent in the problem. Total emissions of particulates were cut from 236,600 kilograms a day in 1984 to 70,782 in 1989. Mexico City, too, is embarking on an ambitious cleanup. With the support of the World Bank and the German, Japanese, and United States governments, the municipal government hopes to introduce a package of measures to cut automotive pollution dramatically over the next two to three years.[71]

Industrial countries are involved in a variety of efforts to assist developing countries with air pollution problems. The International Environmental Bureau in Switzerland and World Environment Center in New York help facilitate transfer of pollution-control information and technology to the Third World. The World Bank is currently actively exploring ways to step up their air pollution activities. One proposed World Bank-U.N. Development Programme project would help urban Asian governments confront environmental problems, including air pollution. Legislation currently before the U.S. Congress would, in the interests of slowing global warming, require the U.S. Agency for In-

ternational Development to encourage energy efficiency and renewable energy through their programs.[72]

In attempting to help Eastern Europe, the Soviet Union, and the Third World with air pollution problems, western industrial countries should not simply transfer wholesale the pollution control strategies that have not been entirely successful at home. An ill-conceived approach such as this could do more harm than good.

The experience of the past several decades may not indicate as much, but air pollution is an eminently solvable problem. Tinkering with the present system, however, will simply not be adequate. A comprehensive approach will be necessary that focuses on pollution prevention rather than pollution control. Though such a strategy will require some investment, the payback will be great. Faced with ever mounting costs to human health and the environment, the question is not how can society afford to control air pollution. How can we afford not to?

7

Cycling Into the Future

Marcia D. Lowe

In a world so transformed by the automobile that whole landscapes and lifestyles bear its imprint, a significant fact goes unnoticed. Although societies the world over define transportation in terms of engine power, the greatest share of personal transport needs is met by human power.

From the ten-speeds of Boston to the black roadsters of Beijing, the world's 800 million bicycles outnumber cars by two to one—and each year three times as many bikes as cars are produced. Bicycles in Asia alone transport more people than all the world's autos do.[1]

In developing countries—where urban workers cycle to their jobs and rural dwellers pedal two- and three-wheelers piled high with loads of goods—pedal power is an important part of national economies and the only alternative to walking that many people can afford. In industrial countries, bicycles are a practical supplement to motorized transportation.

Meanwhile, a steady rise in the amount of driving in the world's major urban areas is pushing congestion to intolerable levels. Traffic jams and the byproducts of gasoline combustion—air pollution in cities, acid rain, and global climate change—point to the need for an alternative to the automobile.

A few countries do encourage bicycling in their transportation strategies. Public policy can facilitate cycling in a number of ways. By making ordinary streets safe for bikes and providing a network of cycleways—separate cycle paths, bike lanes on streets, and wide shoulders to accommodate cyclists on major roads—physical improvements can give cyclists and drivers equal access to as many destinations as possible.

A pro-bicycle stance need not be anti-automobile. The Netherlands, the most bicycle-friendly of all industrial countries, has the western world's highest densities of both cycleways and cars, plus a high level of public transit service. Indeed, maximizing the use of bicycling and mass transit could curb the steep growth in driving before urban air becomes unbreathable and rush-hour traffic grinds to a standstill.[2]

An expanded version of this chapter appeared as Worldwatch Paper 90, *The Bicycle: Vehicle for a Small Planet*.

THE SILENT MAJORITY

Most of the world's bicycles are in Asia. China alone has roughly 300 million. And since only 1 person in 74,000 owns an automobile, Chinese commuters have little choice but to make the most of their bikes. They are also popular because, like cars in industrial countries, they offer the luxury of individual mobility and door-to-door travel without detours or extra stops for other passengers. During the last 10 years, rising incomes in China have been matched by rising bicycle purchases, and the country's fleet has nearly tripled. Domestic bike sales in 1987 reached 35 million—meaning that more people bought bikes in China than purchased cars worldwide.[3]

Elsewhere in Asia, bicycles often make up two thirds of the vehicles on city streets during rush hour. Many Asian urban transit systems are enhanced by pedal-powered "paratransit," in which three-wheeled vehicles for hire—variously called rickshaws, trishaws, pedicabs, and becaks—transport one or more passengers. These resourceful adaptations of the bicycle do much the same work cars do elsewhere. Cycle rickshaws are the taxis of Asia; heavy-duty tricycles, hauling up to half-ton loads, are its light trucks. In Bangladesh, trishaws alone transport more tonnage than all motor vehicles combined.[4]

In 1987, more people bought bikes in China than purchased cars worldwide.

The rest of the developing world lags far behind in bicycle transportation. In much of Africa and even more widely in Latin America, the prestige and power of auto ownership has made governments ignore pedal power and has led people to scorn the bicycle as a vehicle for the poor. Some colonial administrations in Africa built urban bicycle lanes or paths, but most of these have deteriorated or been abandoned in the past three decades. And many African women do not ride bicycles for reasons of propriety or religion or because of encumbering clothing.[5]

Africa and Latin America lack Asia's widespread domestic bicycle industries. The few bikes available are often of poor quality, spare parts are scarce, and maintenance skills are inadequate. Despite these constraints, the bicycle's utility and the lack of other options have led to more intensive cycling in some African settings—even if it means splicing together worn-out brake cables or filling a punctured inner tube with sand. In parts of Burkina Faso, Ghana, Zimbabwe, and a few other countries, bicycles are widely used.[6]

Latin America also has its cycling pockets, where the vehicle's practicality transcends its low social status. In Bogotá, the largest bakery traded in most of its trucks for 900 space-saving delivery trikes that now bring goods to over 60,000 local shops. Nicaragua became perhaps the first Latin American government to promote cycling actively, pledging in 1987 to supplant its war-ravaged transport sector with 50,000 bicycles, some purchased by the government and others donated from outside.[7]

Several heavily polluted areas in Eastern Europe and the Soviet Union have encouraged a revival of bicycling as part of their environmental strategies. The Lithuanian city of Siauliai launched the Soviets' first comprehensive cycling program in 1979; a new bike path system and extensive parking facilities have helped raise bicycle use there. In small Hungarian cities, roughly half the journeys to work are by bike.[8]

Industrial countries have a surprising number of bicycles per person, considering how few are typically found on most city streets. Many industrial nations have two or three times as many bikes per person as their Asian counterparts. But where people have access to automobiles, ownership does not necessarily translate into use. (See Table 7–1.) One in four Britons has a bicycle, yet only 1 transport trip out of 33 is made by bike. In the United States, just 1 in every 40 bikes is used for commuting; most of the rest are ridden for fitness, sport, or play—or are collecting dust in basements.[9]

A comparison of bicycle and car ownership illustrates each country's reliance

on bikes for actual transport. (See Table 7–2.) The United States has more than seven times as many bicycles per person as India, but because one in every two Americans owns an automobile—compared with 1 out of 500 Indians—bicycles play a much more modest role in the U.S. transportation system.

West Europeans are the industrial world's heaviest bicycle users. In several countries—among them Denmark, the Netherlands, and West Germany—bike owners outnumber nonowners. Pro-bicycle planning in the last 20 years (rather than conducive climate or flat terrain, as is often assumed) distinguishes Europe's truly "bicycle-friendly" countries. The Netherlands and Denmark lead this group, with bicycle travel making up

Table 7-1. Cycling as Share of Daily Passenger Trips, Selected Cities

City	Share of Daily Trips
	(percent)
Tianjin, China	77[1]
Shenyang, China	65
Groningen, Netherlands	50
Beijing, China	48
Delft, Netherlands	43
Dacca, Bangladesh	40[2]
Erlangen, West Germany	26
Odense, Denmark	25
Tokyo, Japan	25[3]
Moscow, Soviet Union	24[3]
Delhi, India	22
Copenhagen, Denmark	20
Basel, Switzerland	20
Hannover, West Germany	14
Manhattan, United States	8[4]
Perth, Australia	6
Toronto, Canada	3[4]
London, England	2
Sydney, Australia	1

[1]Share of nonwalking trips. [2]Trips by cycle rickshaw only. [3]Share cycling or walking to work. [4]Vehicle trips (versus passenger trips). SOURCE: Worldwatch Institute, based on various sources.

Table 7-2. Bicycles and Automobiles in Selected Countries, Circa 1985

Country	Bicycles	Autos	Cycle/ Auto Ratio
	(million)		
China[1]	300.0	1.2	250.0
India	45.0	1.5	30.0
South Korea	6.0	0.3	20.0
Egypt	1.5	0.5	3.0
Mexico	12.0	4.8	2.5
Netherlands	11.0	4.9	2.2
Japan[1]	60.0	30.7	2.0
West Germany	45.0	26.0	1.7
Argentina	4.5	3.4	1.3
Tanzania	0.5	0.5	1.0
Australia[1]	6.8	7.1	1.0
United States[1]	103.0	139.0	0.7

[1]1988. SOURCE: Worldwatch Institute, based on Motor Vehicle Manufacturers Association (MVMA), *Facts and Figures* (Detroit, Mich.: various editions); MVMA, various private communications; International Trade Centre, *Bicycles and Components: A Pilot Survey of Opportunities for Trade Among Developing Countries* (Geneva: UNCTAD/GATT, 1985); *Japan Cycle Press International*, various editions; and other sources.

20–30 percent of all urban trips—up to half in some towns.[10]

In several European countries, from 10 to 55 percent of railway patrons in suburbs and smaller towns arrive at the station by bicycle. Traffic jams and air pollution in the past decade have spurred authorities in Austria, Switzerland, and West Germany to encourage more bicycle use. Riders in Belgium, France, the United Kingdom, and southern European countries, by contrast, still get little support.[11]

Many auto-saturated cities of North America and Australia have all but abandoned the bicycle. The growth of suburbs has sent jobs, homes, and services sprawling over long distances that inhibit both cycling and mass transit. Many major cities are largely bicycle-proof, their roadways and parking facilities designed with only motor vehicles in mind. Although several cities explicitly emphasize safety for cycle commuters—including Seattle, Calgary, Melbourne, and several university towns—these are the exceptions.[12]

The world is sufficiently equipped, measured by the number of vehicles, to let bicycles take on a larger share of the transportation burden. Nearly 100 million bicycles are manufactured each year. Backed by strong domestic demand and growing export markets, major producers are sure to keep expanding their capacity, especially in Asia.[13]

VEHICLE FOR A SMALL PLANET

The automobile—which has brought industrial society a degree of individual mobility and convenience not known before—has long been considered the vehicle of the future. But countries that have become dependent on the car are paying a terrible price: each year brings a heavier toll from road accidents, air pollution, urban congestion, and oil bills. Today people who choose to drive rather than walk or cycle a short distance do so not merely for convenience, but also to insulate themselves from the harshness of a street ruled by the motor vehicle. The broadening of transport options beyond those requiring an engine can help restore the environment and human health—indeed, the very quality of urban life.[14]

Nearly everyone who lives in a large city is exposed to one hazard of the automobile era—air pollution. Cars and other motor vehicles create more air pollution than any other human activity does. Gasoline and diesel engines emit almost half the carbon monoxide, hydrocarbons, and nitrogen oxides that result from all fossil fuel combustion worldwide. (See Chapter 6.) Airborne lead, sulfur dioxide, particulates, and many other harmful if not toxic or carcinogenic pollutants also spew out of the tailpipe. On particularly hazy days in the world's most smog-filled cities—Athens, Budapest, Mexico City, and many others—officials must call temporary driving bans to avert public health crises.[15]

Car-induced air pollution does damage far beyond city limits. Vehicles are a major source—indeed, in the United States, the largest source—of the nitrogen oxides and organic compounds that are precursors to ozone. Ground-level ozone, believed to reduce soybean, cotton, and other crop yields by 5–10 percent, takes an estimated annual toll of $5.4 billion on crops in the United States alone. Nitrogen oxides can be chemically transformed in the atmosphere into acid deposition, known to destroy aquatic life in lakes and streams, and suspected of damaging forests throughout Europe and North America.[16]

As serious as these losses are, they pale in comparison to the disruptions that global warming could inflict on the biosphere. (See Chapter 2.) Motor vehicles are thought to be responsible for 17 percent of the worldwide output of carbon dioxide (CO_2)—the greenhouse gas that accounts for roughly half the warming effect. And carbon monoxide contributes indirectly to the warming by slowing the removal of methane and ozone, two other greenhouse gases, from the lower atmosphere.[17]

Excessive motorization also deepens the oil dependence that is draining national economies. Many nations' transport sectors account for more than half their petroleum consumption. Even with the recent fall in crude prices, fuel bills are particularly harsh for indebted developing nations that spend large portions of their foreign-exchange earnings on imported oil. In 1985, low-income developing countries (excluding China) spent on average 33 percent of the money they earned through merchandise exports on energy imports; many spent more than half.[18]

The oil shocks of the seventies brought home the precarious nature of petroleum as an energy source, exposing the vulnerability of car-dependent countries. Today's stable prices and supply have lulled importers into complacency. But as oil demand continues to rise, another oil crisis looms in the nineties.[19]

The world's largest gridlocked cities may run out of momentum before they run out of oil. Road traffic in many cases moves more slowly than bicycles, with average car speeds during peak times sometimes down to eight kilometers per hour. Traffic density in Taiwan's capital is 10 times that of congestion-plagued Los Angeles, and work trips take Mexico City commuters up to four hours each day. Some police units in European and North American cities—including London, Los Angeles, and Victoria—use bicycles rather than squad cars for patrolling congested urban centers. The economic and social costs of congestion, such as employee time lost through tardiness and inflated goods prices resulting from higher distribution costs, are bound to multiply if car commuting trends persist.[20]

Decision makers typically seek only technological solutions to auto-induced problems, an approach that, without alternatives to driving, is inadequate. Although exhaust controls such as the catalytic converter have dramatically reduced pollution from U.S. passenger cars since the early sixties, for example, rapid growth in the vehicle fleet and in kilometers traveled have partially offset this progress. Moreover, the catalytic converter has contributed to impressive reductions in hydrocarbon and carbon monoxide emissions in the United States, but it actually slightly increases the CO_2 buildup, which contributes to climate change.[21]

For every person who makes a trip by bicycle instead of by car there is less pollution, less fuel used, and less space taken on the road.

Similarly, the quest for petroleum alternatives has focused largely on "clean" fuels such as methanol, made from coal or natural gas, and alcohol substitutes distilled from corn and other crops. But methanol contributes to ozone formation and, if derived from coal, to climate change by emitting twice as much carbon dioxide per unit of energy as gasoline does. Using crop-based feedstocks for alternative fuels carries its own environmental side-effects, as well as posing a potential conflict with food production.[22]

Nor is building more roads the answer to congestion. Transport planners are finding that constructing new freeways just attracts more cars, as some public transit riders switch to driving and as new developments spring up along the new roads. Former Federal Highway Administration (FHWA) head Robert Farris concedes: "We can no longer completely build our way out of the congestion crisis by laying more concrete and asphalt. Time is too short, money is too scarce, and land is often not available."[23]

For every person who makes a trip by bicycle instead of by car there is less pollution, less fuel used, and less space taken on the road. Bicycle transportation, rather than replacing all motorized trips, would mainly supplant short automobile journeys, precisely those that create the most pollution (because a cold engine does not fire efficiently). A shift to bicycles for such journeys would therefore yield a disproportionately large benefit. Bicycle commuting holds great potential, since more than half of all commutes in the United States and nearly three fourths of those in the United Kingdom are eight kilometers or less.[24]

Unfortunately, dangerously high pollution levels are even more of a health threat to people who are exercising outdoors, including cyclists. Air pollution during such periods becomes, ironically, a deterrent to bicycling. Until air quality improves, cyclists need a workable alternative—ideally, mass transit—on particularly hazy days. Eventually, when more drivers have switched to bicycling, the air is likely to be safer for everyone.

In the rush to run engines on gasoline substitutes such as corn-based ethanol, decision makers have overlooked a technology that converts food directly into fuel. A cyclist can ride 5.6 kilometers on the calories found in an ear of corn—and nothing has to be distilled or refined.

Bicycles use less energy per passenger kilometer than any other form of transport, including walking. (See Table 7–3.) A 16-kilometer commute by bicycle uses 350 calories of energy, the amount in one bowl of rice. The same trip in the average American car uses 18,600 calories, or more than half a gallon of gasoline.[25]

Bicycle transportation also uses space more efficiently than automobile transport. In fact, any other transport mode, especially mass transit, can move more people per hour in a lane of a given size than cars can even at highway speed. Taking into account both space occupied and speed of travel, automobile traffic accommodates roughly 750 persons per meter-width of lane per hour, compared with 1,500 persons traveling by bicycle, 5,200 persons by bus in a separate bus lane, and 9,000 persons by surface rapid rail. Replacing short car trips with bicycling and longer ones with mass transit—especially with cycling as the way to get to the station—would save considerable space on roads.[26]

Cycling is thus an ever more attractive alternative to the daily grind of traffic congestion. Except when there is no al-

Table 7-3. United States: Energy Intensity of Selected Transport Modes, 1984

Mode	Energy Intensity
	(calories per passenger kilometer)
Automobile, one occupant	1,153
Transit Bus	570
Transit Rail	549
Walking	62
Bicycling	22

SOURCES: Mary C. Holcomb et al., *Transportation Energy Data Book: Edition 9* (Oak Ridge, Tenn.: Oak Ridge National Laboratory, 1987); President's Council on Physical Fitness and Sports, private communication, June 23, 1988.

ternative but to ride in the same traffic stream, cycling commuters benefit both themselves and their employers by being less vulnerable than their colleagues to hypertension, heart attacks, and coronary disease, and by arriving at work more alert. The proof that people enjoy cycling to keep fit is in the popularity of stationary exercise bikes; the irony is that so many people drive to a health club to ride them.[27]

Measured by its benefits both to society and the individual, the bicycle is truly a vehicle for a small planet. "Bicycling is human-scale," writes New York cycling activist Charles Komanoff. "Bicycling remains one of New York City's few robust ecological expressions . . . a living, breathing alternative to the city's domination by motor vehicles. There is magic in blending with traffic, feeling the wind in one's face, the sheer fact of traversing the city under one's own power."[28]

PEDAL-POWERED DEVELOPMENT

Most people in the Third World will never sit inside—let alone own—an automobile. Private car travel is the privilege of only a tiny elite, and thinly stretched government budgets cannot provide adequate public transit services for burgeoning populations. Bicycles and their three- and four-wheeled cousins can enhance mobility at little cost, improve access to vital services, and create a wide range of employment opportunities. Yet deeply impoverished countries pour precious export earnings into motorized transport, ignoring or even subverting human-powered options that people and governments alike could better afford.

In developing countries, an automobile can cost as much as 30 times the annual per capita income—or some 18 to 125 times as much as a basic one-speed bicycle. It is no wonder there are so few cars per person in the developing world. Motorcycle ownership is growing in many countries—usually much faster than car ownership—but is still beyond the means of the majority. Mopeds are increasingly popular in Asian cities and have promise as an intermediate vehicle. Both these motorized two-wheelers, however, pose their own pollution problems.[29]

Deeply impoverished countries pour precious export earnings into motorized transport, ignoring human-powered options that people and governments could better afford.

For many people in the developing world, the only affordable alternative to walking is a public bus, and yet even in urban areas buses can be scarce. Many Asian, African, and Latin American cities—where most people depend on buses—have less than half the coverage (in buses per capita) of U.S. and European cities, where riders usually have the option of driving anyway. The expansion of Third World transit systems can easily overwhelm government budgets burdened by other crucial demands such as housing, water supply, and sanitation. Privately run bus companies, which often help ease a government's transit burden, also fall short for lack of equipment and fuel, and are impeded by the high incidence of unpaved or narrow streets. In much of the Third World, fully meeting transport needs through mass transit is a worthy, but probably unattainable goal.[30]

Particularly in rural areas, therefore, people have little choice but to walk.

Rural transport needs in developing countries consist mainly of moving farm goods and supplies and gathering household necessities such as water and fuel. Families spend an inordinate amount of time on the latter each day, with women carrying the brunt of this hardship quite literally on their heads. A study in Kenya found that women do 89 percent of all water and firewood gathering for the family. Women and children may spend three to six hours a day fetching water for the household.[31]

Across the developing world, farmers typically need to haul loads weighing up to 150 kilograms (330 pounds) over moderately long distances. These tasks do not call for a truck or automobile, but they do require some intermediate kind of vehicle. Throughout tropical Africa and parts of Asia, even a "small" load of 30 kilograms becomes a crushing burden when carried on the head, back, or shoulders. "Headloading" and other forms of human porterage can severely injure the spine, joints, muscles, and internal organs. According to a report in Bangladesh, half the broken necks that occur are sustained during falls while the victims are carrying loads on their heads.[32]

Field studies by Intermediate Technology Transport, Ltd., a London-based agency that gives technical assistance to local development projects in the Third World, have shown that a bicycle can provide a travel capacity (a combination of speed and payload) that is at least five times that possible through walking. And attaching a trailer to a bicycle allows the rider to carry comfortably up to 200 kilograms—several times the maximum headload.[33]

Several international organizations are working with local authorities in developing countries to design sturdy load-carrying bicycles and trailers, and to improve vehicle designs for the disabled. Some have adapted bicycles to

other purposes—powering a paddy thresher, a peanut sheller, or a water pump, for example—in a way that they can be easily adjusted back again for transportation. Inter Pares, a Canadian agency, is redesigning rickshaws originally developed for six-foot Europeans into lighter, cheaper, more maneuverable vehicles for the typical five-foot Bangladeshi.[34]

Bicycles provide a livelihood for many in the Third World's informal sector—the mass of self-employed entrepreneurs who, operating with little or no capital, constitute from 20 to 70 percent of the labor force in urban areas. Mobile vendors cycle with bundles of newspapers through Tanzanian towns, bring hot lunches to factory workers in Sri Lanka, and sell fresh bread from their bicycles in Iran. Some 5,000 *tricicleros* in Santo Domingo, capital of the Dominican Republic, use heavy three-wheelers to circulate much of the city's supply of fresh food, coal, scrap metal, and various materials for recycling.[35]

Professor V. Setty Pendakur of the University of British Columbia, an authority on urban transport in South Asia, believes that the social benefits of pedal-powered transit—employment, access, and affordability—far outweigh the disadvantages, mainly the lack of efficiencies of scale possible with motorized public transit.[36]

But bias against human-powered vehicles is severe in Third World cities, and even worse in the countryside. For decades, governments have devoted rural funds to building motorable roads, neglecting the fact that few people there have cars or trucks, and that motorized public transit rarely reaches them. Transportation expert Wilfred Owen has noted that "many miles of roads in poor countries prove more useful for drying beans and peppers than for moving traffic."[37]

Although policymakers could improve

transport at little cost by supporting bi-cycles and other human-powered vehi-cles, they seldom do. City governments are often hostile to rickshaws and similar services, either discouraging them with regulations, fines, and taxes, or wiping them out entirely through bans and con-fiscations. In Jakarta, the city has confis-cated some 100,000 cycle rickshaws over the past five years and dumped them into the sea—to "reduce traffic conges-tion." Dacca, Bangladesh, recently threatened to phase out rickshaws—de-spite the fact they account for more than half the passenger trips in the city and employ an estimated 140,000 people. Rickshaws throughout South Asia, and even in some African countries, where they are less common, are also in jeop-ardy.[38]

Governments typically defend their assaults on rickshaws by declaring them unsafe or inhumane for the drivers. The more likely motive, however, is to clear the streets of vehicles they think make the city look poor and backward. Ac-cording to Pendakur: "The cycle rick-shaw provides the highest employment for a given investment and serves the poor efficiently and flexibly, reaching areas not reached by conventional buses. While there is a need to modern-ize and make the life of the rickshaw puller tolerable, the answer may not lie in costly motorization but in better de-sign of the cycle rickshaw and the streets."[39]

Governments are not the only ones reinforcing discrimination against human-powered vehicles. Michael Re-plogle, president of the U.S.-based Insti-tute for Transportation and Develop-ment Policy (ITDP), notes that in 1985 the World Bank—a principal source of urban transport funds in the developing world—published a 400-page study of China's transport sector that failed to even mention the word bicycle.[40]

ITDP, a nongovernmental organiza-tion, sends donated bicycles and parts where most needed to improve people's access to services and free the hard-working poor for productive activities. Its bikes have multiplied the effective-ness of literacy teachers in Haiti, well drillers in Bolivia, and health workers in Nicaragua and Mozambique. The Insti-tute is breaking further ground by pro-moting all-terrain bicycles—also called mountain bikes—that are more appro-priate for rugged conditions than the half-century-old British roadster model usually found in the Third World.[41]

Developments in recent years suggest that human-powered vehicles may yet penetrate official circles. A handful of national governments, lending banks, and other international agencies that have steadfastly pursued motorization policies are finding their one-track ap-proach unsustainable.

One example in progress is an inter-mediate transport project in Ghana. With advice from Intermediate Technol-ogy Transport and assistance from the World Bank, the government is now pro-moting the production and use of bicycle trailers and handcarts. The Bank is also proposing to help Ghana build low-cost rural roads meant chiefly for nonmotor-ized vehicles. Sufficient to serve an occa-sional truck or car without having to meet stricter construction standards, the proposed roads would cost about $2,400 per kilometer, or roughly 8 percent as much as a conventional rural road.[42]

That there are only slightly more de-veloping countries making bicycles than producing automobiles—even though domestic auto manufacturing demands much more capital and technical and managerial resources—epitomizes the Third World's missed opportunities with human-powered vehicles. Experi-ence in Asia has shown that starting a bicycle industry is a relatively low-risk venture for developing countries that have little industrial base. A small assembly plant and repair shop can run on about $200 worth of tools, and 100 bicycles can be manufactured for the energy and materials it takes to build a medium-sized car.[43]

Starting a bicycle industry is a relatively low-risk venture for developing countries that have little industrial base.

India has demonstrated how a nearly self-sufficient bicycle industry can be created by first assembling bicycles with imported parts, then producing frames in local workshops, and gradually establishing small factories to produce parts domestically. From a modest beginning five decades ago, India has become a major world producer. It directs more than 90 percent of its bicycle exports to other developing countries, and through joint-venture and license agreements is sharing its small-scale, labor-intensive techniques with countries throughout Asia, Africa, and the Caribbean.[44]

Engineer and urban planner Ricardo Navarro, who is nurturing a string of small workshops in El Salvador that he hopes will become a vigorous domestic bicycle industry, has succinctly described the potential economic role of pedal power: the bicycle is "a *mecanismo indispensable* for development."[45]

CYCLING SOCIETIES

What sets apart the handful of countries that, in a world seduced by automobiles, have chosen to embrace the bicycle? China, Japan, the Netherlands, Denmark: these few cycling societies are not notably different, in terms of living standards, geography, or climate, from their noncycling neighbors. A study of transport systems in 12 North American and West European countries by John Pucher confirms that wide variations in people's transport decisions are not chiefly influenced by levels of income, technology, or urbanization. The difference lies in enlightened public policy and strong government support.[46]

Chinese authorities recognized decades ago that human-powered transport could move more people more cheaply than any other option. They began investing in low-cost, mass production of bicycles when most people were still too poor to own one, and they directed infrastructure improvements to nonmotorized travel. When commercial access to villages was opened in the early sixties, it was largely over rural tracks built for people moving on foot, animal carts, or bicycles.[47]

Special bicycle avenues with five to six lanes each are common in Chinese cities. Motorized traffic is often separated from pedestrians and cyclists on three-track roads, and some cities set apart space for load-carrying bicycles. Convenient, guarded bicycle parking is plentiful, as are services for maintenance and repair. Throughout China, city governments have long used bicycles to relieve pressure on overcrowded buses by paying commuters a monthly allowance for cycling to work. China has thus provided high-quality transport to large numbers of people while postponing the need for heavy public transit investments.[48]

Japan is another cycling society in which bicycle trips for practical use far outnumber those for leisure or sport. Census figures for 1980 showed that 7.2 million commuters—approximately 15 percent of the total—rode bikes to work or to commuter rail stations. Bicycle ownership has climbed from an average of 1 per household in 1970 to 1.5 in 1988. Though recreational cycling is quickly gaining enthusiasts, most bicyclists ride for everyday shopping and commuting to work or school.[49]

Many of Japan's rail passengers pedal daily to train and subway stations. Since the sixties, Japanese transport has been dominated by commuter railroads that link urban centers to the rapidly growing suburbs. As development fanned out farther from the rail lines, many commuters switched from slow feeder buses to bikes for their daily journeys to suburban stations. By the mid-seventies, "bicycle pollution"—a phrase coined for the hundreds or even thousands of bicycles crammed in front of some railway stations—spurred the government to promote bike parking as a way out of the chaos.[50]

National legislation passed in 1980 empowers local governments in Japan to require that railways and private businesses build ample bicycle parking, an approach known internationally as "bike-and-ride." Today more than 8,600 of these official and private parking sites exist, with total capacity for 2.4 million bikes. Limited land space in urban Japan—where downtown real estate can cost over $75,000 per square meter—has inspired the construction of bicycle parking towers. Dozens of transit stations have multistory structures in which automated cranes park thousands of bikes.[51]

Much of Europe is moving toward pro-bicycle planning. Many towns in Sweden, Switzerland, and West Germany have steadily increasing shares of cyclists in their traffic, and authorities in recent years have stepped up their commitment to cycle planning. West Germany's years of *Verkehrsberuhigung* (traffic calming) have helped encourage a cycling environment by confining motorized traffic with physical barriers and reduced speed limits. Swedish towns have experimented with restraining motor vehicles since the seventies by using "traffic cells" that divide a city into zones to reroute traffic from denser areas onto main roads, making smaller streets safer for cycling and walking.[52]

The bicycle's top two European champions, the Netherlands and Denmark, owe their current lead in cycling policy to having had a head start. Bicycle tourism is a major industry, with thousands of people taking extended trips along country roads to enjoy diverse landscapes and visit picturesque towns. As elsewhere on the continent, pro-bicycle planning for utilitarian purposes in these countries is part of a recent search for alternatives to automobile dependence.[53]

The Netherlands has a long tradition of cycling both for recreation and everyday transport, though the bicycle's role declined in the fifties and sixties as more people bought automobiles. But cycling made a comeback when the repercussions of growing car use were dramatized by energy shocks and ecological crises in the seventies. Between 1975 and 1985 the government spent some $230 million to build cycleways and parking, and to increase transit access through bicycle facilities at rail stations. Highway construction expenditures, by contrast, began to decline; by the early eighties, funding for bicycle projects exceeded 10 percent of capital spending on roadways.[54]

In 1986 the Netherlands' cycle paths covered 13,500 kilometers. Perhaps even more significant than this achievement, however, are Dutch efforts to create direct, uninterrupted cycling routes—thus making riding practical, rather than simply getting cyclists out of the way of other traffic. As a result, the share of trips made by bicycle in Dutch towns and cities is typically between 20 and 50 percent.[55]

In Groningen, the largest city in northern Netherlands, cycling accounts for half of all urban trips. The city of

Delft's cycling provisions include underpasses and bridges across dangerous intersections, and innovations in traffic management. Some traffic-activated signals, for example, normally sensitive only to motor vehicles, detect bicycles as they arrive at intersections. Pavement lines give cyclists space to stop ahead of other traffic and then move ahead first. Cyclists are also permitted to ride against the traffic on some one-way streets.[56]

Rather than replace the automobile, the Netherlands has sought to balance auto transport with bicycling, public transit, and walking as a national policy goal. Pedestrian-only streets and reduced speed limits are common in Dutch towns, and car parking is prohibited on many city streets. Residential streets are often transformed into *woonerven*, or "living yards," a traffic-calming concept introduced in the seventies. In a *woonerf* the road becomes a paved courtyard. All means of transport are allowed, but bicyclists and pedestrians have priority and cars enter only as "guests."[57]

A cycling society can emerge out of one already hooked on the automobile.

A current car boom, however, is challenging Dutch officials to put their energies into simply maintaining the present popularity of cycling. In 1988, Dutch Transport Minister Nellie Smit-Kroes announced a bold plan in which taxes will increase the costs of buying and driving a car by about half, while public transit will receive an extra $5.7 billion per year. A new electronic system will log the number of kilometers each car travels, and "excessive drivers" will then pay additional taxes.[58]

The Netherlands' closest peer in cycling is Denmark, where as Copenhagen's Mayor of Traffic notes, poets have celebrated the bicycle and sculptors have made statues of cyclists. Danish authorities have encouraged cycling in much the same way Dutch officials have. The share of bicycles in Denmark's traffic counts, though slightly lower than that in the Netherlands, is still impressive: 20 percent of personal trips in Danish cities and towns are made by bicycle. Danish rail stations typically accommodate several hundred bikes at a time.[59]

North America's closest approach to a cycling society is the bicycle-friendly university town. Two such communities in northern California, Palo Alto and Davis, vie for the title of U.S. biking capital. Of the two, Davis has the higher cycling rate—25 percent of total trips in the community of 44,000 are made by bike—and cycle trailers filled with groceries or children are not an unusual sight. The town has some 48 kilometers of bicycle lanes for 161 kilometers of streets, and roughly 32 kilometers of separate cycle paths.[60]

Palo Alto has spent, since 1980, roughly $1 million—mostly from state grants—on bicycle lockers and racks, bike bridges, and lighted cycle paths. All road patching in town must adhere to high smoothness standards. The centerpiece of Palo Alto's 65-kilometer system of bikeways is its bicycle boulevard, a 3.2-kilometer stretch in the middle of town where bikes are the only through-traffic allowed. A 1983 zoning ordinance requires new buildings beyond a certain size to provide secure bicycle parking and showers for employees, and several large employers in the area add their own incentives. Amenities at Xerox, for example—which include a towel service in the shower room—help explain why 20 percent of the company's local employees cycle to work, one of the highest bicycle commuter rates nationwide.[61]

These examples give ample proof that a cycling society can emerge out of one already hooked on the automobile. Palo Alto Council member Ellen Fletcher, who cycles to city meetings and has become known nationally for her bicycle advocacy, knows how to fulfill the bicycle's potential. "All you have to do is make it easier to ride a bike than drive a car," she says. "People will take it from there."[62]

GETTING THERE FROM HERE

In a 1988 *Newsweek* interview on the U.S. gridlock predicament, former FHWA traffic systems chief Lyle Saxton remarked: "We have built our society around the automobile and we have to deal with it." But as the world's cycling societies demonstrate, neither governments nor citizens need accept the status quo. With public policy support and private initiatives on many levels—from international institutions to the individual commuter—bicycle transportation can become an everyday alternative.[63]

National governments can make automobile owners pay more of the hidden expenses of driving. The costs of these items—including road building and maintenance, police and fire services, accidents and health care—are borne by all taxpayers, drivers and nondrivers alike. In the United States, the hidden price of car driving may total as much as $300 billion a year.[64]

One way to counteract this enormous subsidy is for governments to tax auto ownership and use, and to invest the revenues in mass transit and in cycling and pedestrian facilities. In Pucher's recent study of 12 countries in Western Europe and North America, it is clear that when drivers are made to pay the costs of automobile travel through taxa-tion of ownership and use, total mileage driven tends to decline.[65]

Pucher found that Denmark and the Netherlands—the two with some of the highest taxes—also have among the lowest figures for per capita kilometers of automobile travel. The United States comes last in taxation of both auto sales and gasoline, and is second only to Sweden—the country with the next-to-lowest gas tax—in kilometers of auto travel per person. Moreover, while the United States puts these tax monies into highway expenditures, European countries put most of the proceeds into general revenues.[66]

Taxes and user fees can be extended to other amenities such as space in parking lots and roads. In the United States, 75 percent of all commuters have free parking provided by employers, a tax-free fringe benefit for employees and a tax-deductible expense for businesses that provide it. In Japan and much of Europe, in contrast, public policy makes parking both more expensive and less available. Singapore charges private cars carrying fewer than four passengers "congestion fees" for entering the downtown area during rush hours. Since 1975 the scheme has raised Singapore's average downtown traffic speeds by 20 percent, reduced traffic accidents by 25 percent, and trimmed fuel consumption by an estimated 30 percent.[67]

As the world's cities rapidly grow larger, the expansion of mass transit systems will become increasingly important. Third World governments straining to extend public transit service to outlying settlements can maximize access by fostering bike-and-ride. In the industrial world, bike-and-ride is an alternative to increased car commuting both between city centers and suburbs and from one suburb to another. Transit authorities have typically tried to attract passengers by building automobile parking lots, which require at least 20 times as much space as bicycle parking. For a tiny fraction of that cost, secure bicycle facilities can increase convenient access to transit stations by a radius of at least one to three kilometers, enlarging the total area served roughly ninefold.[68]

Leaders in the developing world can make national transportation policies more effective by considering the mobility and access needs of their impoverished majorities, providing credit where necessary to help people purchase bicycles and other low-cost vehicles. Particularly in countries that discriminate against rickshaw drivers and others, a first step is to correct bias against human power and focus on improving rickshaws, not banning them. Governments may need to balance motorized transit investments with less costly bicycle subsidies and cycling infrastructure.

Governments can also foster domestic bicycle industries, while avoiding heavy import taxes and other policies that in the past have overprotected them, leading to poor-quality products and, eventually, to factory closings. It is important for authorities to encourage the flow of technologies from elsewhere in the developing world, and to support research in adapting low-cost vehicles to local conditions.[69]

Multilateral banks and other development organizations can help spur these changes. The World Bank, for example, now emphasizes bicycles and low-cost roads in several projects that are part of the Sub-Saharan Africa Transport Program. This joint undertaking with the United Nations Development Programme, the United Nations Economic Commission for Africa, and several bilateral donors began in 1987 to reevaluate the ineffective, motor-biased transport policies of the region. The Bank's series of nonmotorized transport projects seeks to address the problems of rural access, focusing on trips that take place off the road network and on the transport needs of women.[70]

Local governments in developing and industrial countries alike can promote cycling more effectively by creating an official bicycle advisory council that reports to the mayor, city council, or equivalent authority—following the example set by Toronto and Washington, D.C., among others. This advisory body can help decision makers ensure that all transport improvements consider the needs of cyclists. Building codes and ordinances can specify that new developments beyond a given size must include bicycle parking and showers for commuters, and that new or rebuilt roads and bridges include safe bicycle access. A specific portion of downtown parking space could be devoted to bicycles and a percentage of all transport spending allocated to cycling facilities.[71]

Official support for cycleways in much of the United States has been undermined by a deep divide among bike advocates. On one side are those who oppose any special cycleways or routes that are separate from motor vehicle traffic, believing them unsafe and discriminatory. Other cyclists advocate building special bikeways where practical. The controversy stems from the all-too-common situation in which a separate facility for bicycles is inferior to that for motor traffic—particularly if use of a poorly designed or deteriorating cycle path is made mandatory.

Experience has shown, however, that if separate paths are of high quality, cyclists will readily use them whether mandatory or not. No single type of cycleway is appropriate for every place or for every rider. Differences in moving speeds, volume of traffic, and skill levels all should be taken into account when deciding on a bicycle lane, bike path, or route markings on regular roads.

Still, the debate among cycling advocates has led many local authorities to dismiss requests for new bikeways on the grounds that cyclists themselves do not want them. Meanwhile, millions of Americans annually use 4,350 kilometers of abandoned rail corridors that have been converted into cycling and hiking trails largely through the efforts of the Rails-To-Trails Conservancy, a private organization.[72]

Traffic management is also important in balancing the needs of all types of vehicles, particularly where limited space makes physical separation impossible. For decades the Netherlands has led the way in calming traffic with *woonerven*, and West Germany is trying to create the same effect with *Verkehrsberuhigung* and reduced speed limits. With such steps, urban design and traffic restraint can help different types of transport coexist safely.

Despite progress in recent years with better traffic management and road designs, cyclists are still more vulnerable than drivers. Safety education for all road users and law enforcement for offenders are crucial—for riders and drivers alike. In Manhattan, a decline in bicycle-pedestrian accidents each year since 1985 is credited in part to increased police vigilance in issuing summonses to cyclists, and in part to the emphasis by cycling organizations on mutual responsibility and cooperation.[73]

The long-term interests of cities will be best served if metropolitan areas of 50,000 people or more adopt regional planning and create growth plans that limit urban sprawl. Innovative land use plans cluster homes, shopping, and workplaces, and they base layouts as much as possible on cycling and walking distances, with an emphasis on public transit to complete the transport balance.[74]

Government action—from national taxation policies to local provision of bicycle racks—is not likely to happen with-

out a concerted effort from cycling advocacy groups and individuals acting to counter strong automobile lobbies. Local bike clubs with a handful of members and international groups with tens of thousands may be the most powerful forces for this broadening of transport alternatives. When Mayor Edward Koch announced in 1987 that three main avenues in New York City would be closed to bicycles on weekdays, cycling advocacy groups demonstrated against the ban, took legal action to reverse it on a technical flaw, and finally persuaded the mayor not to try banning bicycles again.[75]

Millions of Americans annually use 4,350 kilometers of abandoned rail corridors that have been converted into cycling and hiking trails.

In June 1989, the League of American Wheelmen, a national cycling organization founded in 1880, held the first National Congress of Bicyclists since cycling's Golden Age in the 1890s. The congress established several goals, from urging local officials to enforce traffic laws that affect cyclists to persuading the U.S. Department of Transportation to include cycling in national transportation policy.[76]

Perhaps the greatest potential for change lies with the individual cyclist. Pressing employers and local authorities to provide cycling facilities—and simply using bikes whenever possible—can have enormous impact. Some cyclists can make all the difference in simply leading by example: Argentine President Carlos Menem recently urged citizens to ease the shock of soaring gasoline prices by riding bicycles, and is a cyclist himself.[77]

Each time a driver makes a trip by bicycle instead of by car not only the cyclist but society as a whole reaps the benefits. One great irony of the twentieth century is that around the globe vast amounts of such priceless things as land, petroleum, and clean air have been relinquished for motorization—and yet most people in the world will never own an automobile.

As author and cyclist James McGurn writes, "The bicycle is the vehicle of a new mentality. It quietly challenges a system of values which condones dependency, wastage, inequality of mobility, and daily carnage . . . .There is every reason why cycling should be helped to enjoy another Golden Age."[78]

8

Ending Poverty

Alan B. Durning

"Until the lions have their historians," declares an African proverb, "tales of hunting will always glorify the hunter." Likewise, the historians of the world's fortunate class—those billion-odd people who inhabit industrial lands—have already labeled the twentieth century a time of economic miracles.[1]

The history of the wealthy is impressive. Since 1900, the value of goods and services produced each year worldwide has grown twentyfold, the products of industry fiftyfold, and the average distance travelled by the well-to-do perhaps a thousandfold. As the century enters its final decade, commoners of the world's affluent nations live like the royalty of yesteryear, and elites literally live like gods.[2]

Yet the poor have a different tale to tell. The disparities in living standards that separate them from the rich verge on the grotesque. In 1989, the world had 157 billionaires, perhaps 2 million millionaires, and 100 million homeless. Americans spend $5 billion each year on special diets to lower their calorie consumption, while 400 million people

around the world are so undernourished their bodies and minds are deteriorating. As water from a single spring in France is bottled and shipped to the prosperous around the globe, nearly 2 billion people drink and bathe in water contaminated with deadly parasites and pathogens.[3]

The histories of rich and poor diverged particularly sharply in the eighties. For industrial nations, the decade was a time of resurgence and recovery after the economic turmoil of the seventies. For the poor, particularly in Africa and Latin America, the eighties were an unmitigated disaster, a time of meager diets and rising death rates.

Destitution in the modern world is perpetuated by mutually reinforcing factors at the local, national, and international levels that form a global poverty trap. Poverty's profile, furthermore, has become increasingly environmental. The poor not only suffer disproportionately from environmental damage caused by those better off, they have become a major cause of ecological decline themselves as they have been pushed onto marginal land by population growth and inequitable development patterns. Economic deprivation

An expanded version of this chapter appeared as Worldwatch Paper 92, *Poverty and the Environment: Reversing the Downward Spiral.*

and environmental degradation reinforce one another to form a maelstrom—a downward spiral that threatens to pull in ever more.

For the poor, the eighties were an unmitigated disaster, a time of meager diets and rising death rates.

Still, great hope for poverty alleviation exists around the world: poor people have formed hundreds of thousands of grassroots organizations to help themselves gain what official development programs fail to provide. These organizations have found scores of innovative ways to break out of the poverty trap, and to reverse the downward ecological spiral. The challenge ahead is to turn these hopeful beginnings into a full-fledged mobilization to end poverty.

AND THE POOR GET POORER

Poverty is far more than an economic condition. Though traditionally measured in terms of income, poverty's horror extends into all aspects of a person's life: susceptibility to disease, limited access to most types of services and information, lack of control over resources, subordination to higher social and economic classes, and utter insecurity in the face of changing circumstances. Flowing from these physical dimensions is poverty's psychological toll—the erosion of human dignity and self-respect.

Unfortunately, even the most basic poverty indicator—income—is little monitored. It is possible to know precisely how much money is in circulation in Haiti, how much steel is produced in Malaysia, and how many automobiles

there are in the Congo. But the numbers of people living in misery is largely a matter of conjecture.

Economists define a poverty line by calculating the income in cash or kind a family requires to meet its basic needs either solely for food, or for food, clothing, and shelter. Ideally, governments carry out household surveys or censuses to determine what share of their people have incomes below that threshold. Unfortunately, the procedure is marred by the variety of poverty lines employed and the lack of true house-to-house monitoring. In this chapter, absolute poverty signifies the lack of sufficient income in cash or kind to meet the most basic biological needs for food, clothing, and shelter. The absolute poverty income threshold varies widely between $50 and $500 per year, depending on such things as prices, access to subsistence resources, and availability of public services.

In the early eighties, World Bank and U.N. Food and Agriculture Organization (FAO) estimates of the numbers of people living in absolute poverty ranged between 700 million and 1 billion. Most indicators suggest that in 1989 poverty increased dramatically in sub-Saharan Africa and Latin America, as well as in parts of Asia, swamping reductions in China and India. Unfortunately, absent direct monitoring, tracking poverty's course requires inference from trends in average incomes, wages, prices, unemployment rates, income distribution patterns, and health indicators.[4]

Since 1950, average income per person worldwide has doubled to $3,300 in real terms, but the fruits of global economic growth have almost all gone to the fortunate. Grouping the world's nations into four classes based on their per capita income in 1985 brings the disparities into sharp focus. (See Figure 8–1.) Wealthy nations almost tripled their per capita income since mid-century, but

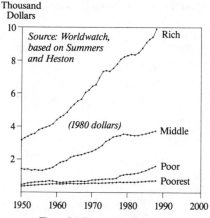

Thousand
Dollars

Source: Worldwatch,
based on Summers
and Heston

Rich

(1980 dollars)

Middle

Poor
Poorest

1950   1960   1970   1980   1990   2000

**Figure 8-1. Income Per Person,
Four Economic Classes of Countries, 1950–88**

that figure in the poorest countries has remained effectively level. (Throughout this chapter, income per person or average income is defined as annual per capita gross domestic product in 1980 U.S. dollars and—unlike commonly used World Bank figures—adjusted to reflect widely divergent prices for basic goods in different countries.)[5]

Wide as the gap between rich and poor appears when measured in average income, the real situation is worse. Averages disguise the gross disparities in income distribution that characterize the majority of countries: 60–70 percent of the people in most countries earn less than their nation's average income. Almost nowhere do the poorest fifth of households collect even 10 percent of national income, while the richest fifth commonly receive half. (See Table 8–1.) Different income distributions mean the poor may be fairly well off even where average income is low, or may suffer greatly where average income is high. In 1985, for example, Egypt's per capita income was about half of Peru's, but because Egypt is more equitable, poor Egyptians earned one third more than poor Peruvians.[6]

The world as a whole is probably less

equitable than any nation. The fifth of humanity living in the richest countries have average incomes 15 times higher than the fifth living in the poorest. Were sufficient data available to group the world's people by their true incomes rather than national average incomes, however, the richest fifth might be found to earn 30 or 40 times what the poorest do.

For the poor of Africa, Latin America, and parts of Asia, the eighties were a time of cruel reversals, a period when the global economy seemed to conspire against them. On top of the runaway population growth and accelerating environmental decline that were already dragging down living standards across the Third World, prices for poor nations' exports plummeted, and international debt siphoned a growing share of their income into the hands of foreign financiers.

Since 1950, the gap between rich and poor nations has grown mostly because the rich got richer. But since 1980, in many developing countries the poor have been getting poorer too. Forty-three developing nations probably finished the decade poorer, in per capita terms, than they started it. The 14 most devastated—including Zambia, Bolivia, and Nigeria—have seen per capita income plunge as dramatically since their troubles began as the United States did during the Great Depression. Indeed, the term developing nation has become a cruel parody: many countries are not so much developing as they are disintegrating.[7]

The human impacts of the decade's economic crisis in Africa, Latin America, and parts of Asia have been ruinous. Malnutrition is documented to be on the rise in Burma, Burundi, The Gambia, Guinea-Bissau, Jamaica, Niger, Nigeria, Paraguay, the Philippines, Nicaragua, El Salvador, and Peru, and is undoubtedly increasing elsewhere as well, particularly

Table 8-1. Approximate Income Distribution of Most Populous Nations and of World, Most Recent Available Year[1]

Country	Year	Households' Share of National Income		Equity Ratios[2]
		Poorest 20 percent	Richest 20 percent	
		(percent)		
China[3]	1984	12	31	3
Soviet Union	1980	9	36	4
Japan	1979	9	37	4
West Germany	1978	8	40	5
United Kingdom	1982	7	40	6
Bangladesh	1982	7	45	7
Italy	1977	6	44	7
Indonesia	1976	7	49	8
Egypt	1974	6	48	8
Pakistan	1985	—	47	—
India	1976	5	50	10
Philippines	1985	5	53	10
Thailand	1976	5	57	11
United States	1986	4	46	12
France	1975	4	50	13
Turkey	1973	4	57	16
Mexico	1977	3	54	18
Brazil	1982	2	64	28
World[4]	1985	4	58	15

[1]Data for Vietnam, Nigeria, and Iran are unavailable. [2]Equity ratios (share of richest to share of poorest) were calculated based on unrounded income share figures and may therefore appear inconsistent with income shares listed. [3]Cities only. [4]Estimated by grouping nations into five classes based on per capita income.

sources: World Bank, *World Development Report 1988* (New York: Oxford University Press, 1988); World Resources Institute, *World Resources 1988–89* (New York: Basic Books, 1988); Agha M. Ghouse, "Urban Poverty and Unemployment in Pakistan," *Pakistan & Gulf Economist,* December 31, 1988; Joel Bergsman, *Income Distribution and Poverty in Mexico,* Staff Working Paper 395 (Washington, D.C.: World Bank, 1980); Robert D. Hamrin, "Sorry Americans—You're Still Not 'Better Off'," *Challenge,* September/October 1988; Soviet Union estimated from William U. Chandler, *Changing Role of the Market in National Economies,* Worldwatch Paper 72 (Washington, D.C.: Worldwatch Institute, September 1986).

among societies' least fortunate members. The World Bank reports that from 1979 to 1983 life expectancy fell in nine African countries, and that more than 100 million Africans lack sufficient food to sustain themselves.[8]

Poverty's most savage toll is measured in the lives of children. In Zambia, twice as many children died from malnutrition in 1984 as in 1980. The infant mortality rate in Brazil rose in 1983 and 1984 for the first time in decades, and rose most steeply in the country's poorest regions. Similar trends are afoot in much of the

Third World, leading UNICEF, the United Nations Children's Fund, to conclude in its 1989 annual report that "at least half a million young children have died in the last twelve months as a result of the slowing down or the reversal of progress in the developing world."[9]

Given the scarcity of consistent data that directly measure poverty, the global poverty rate cannot be determined precisely. Nonetheless, some sense can be gathered of the rate in each developing country by examining a wide variety of social and economic indicators, including the few recently performed surveys of household income, from dozens of published and unpublished sources and from the internal documents of multilateral institutions. Although the margins of error are wide for each country, totaling these individual assessments on a regional basis balances out inaccuracies, providing the best sketch of the extent of absolute poverty possible with current data. (See Table 8-2.)[10]

Rising poverty rates in Africa, Latin America, and parts of Asia appear to have swamped poverty reductions in India and China, with the result that in 1989 approximately 1.2 billion people

Table 8-2. People Estimated to be Living in Absolute Poverty, 1989

Region	Number of People[1]	Share of Total Population
	(million)	(percent)
Asia	675	25
Sub-Saharan Africa	325	62
Latin America	150	35
North Africa and Middle East	75	28
Total	1,225	23

[1]Estimates are best thought of as midpoints of ranges that extend 10 percent above and 10 percent below listed figures.
SOURCE: Worldwatch Institute.

lived in absolute poverty. In part because of population growth, that total is much larger than ever before. Perhaps more significant is what this means overall. World Bank figures suggest that the global poverty rate stood at 22.3 percent in 1980, after declining steadily but gradually since mid-century. Our own estimate of 1.2 billion people translates to a rate of 23.4 percent. During the eighties, in other words, the global poverty rate not only stopped falling, it rose—despite substantial reductions in the number of impoverished in the two most populous countries on earth.[11]

The poor are overwhelming illiterate, and therefore lack access to information and ideas that could help them escape poverty.

These 1.2 billion people are not a homogeneous group, but a few generalizations help answer the question, Who are the poor? Despite rapid urbanization and growing urban poverty in much of the world, four fifths of those in absolute poverty still live in rural areas. Only in Latin America do a large share of the poor—nearly half—live in cities.[12]

The poor are overwhelmingly illiterate, and therefore lack access to information and ideas that could help them escape poverty. They are often distinct in race, tribe, or religion from dominant wealthy groups. They are slightly more likely to be female than male, particularly in urban areas, leading some analysts to speak of a global "feminization of poverty."[13]

Women's burdens multiply endlessly. They are paid less than men but they work more. They are less well educated, but bear greater responsibility for the health of children. They are expected to give birth to, raise, and feed numerous—

preferably male—offspring and consequently grow weak and ailing as their bodies are exhausted by the cycle of repeated pregnancy and childbirth. They are commonly abused and beaten at home, but have few legal rights and fewer property rights. For poor women, as one Brazilian woman says, "the only holiday . . . is when you are asleep."[14]

Numerically, the group most plagued by poverty is children. As income declines, family size increases. Consequently, perhaps two thirds of the world's absolute poor are under the age of 15, and the prospects for these young people are even worse than for their parents. Wracked by disease, lacking sufficient nourishment and clean water, one third of these youngsters die before their fifth birthday. Many of those who survive are physically stunted and mentally impaired as a result of chronic hunger during the critical age of six months to two years.[15]

The work of the poor is concentrated at the margins of the global economy. Most are landless agricultural laborers, sharecroppers, marginal farmers, or, if they live in cities, unskilled laborers in the underground economy. What work they can find is usually piecemeal and unstable, not to mention backbreaking and tedious, yet they work hard day in and day out.[16]

The utter deprivation of absolute poverty is rare in affluent lands, yet it is worth noting that the less severe form of poverty found there was also on the rise during the eighties, particularly in the United States, the United Kingdom, and Eastern Europe. One fifth of Soviet citizens reportedly lived below the official poverty line of 75 rubles a month ($117, at mid-October 1989 official rate).[17]

In the United States, the eighties found more people living below the poverty line—just over $12,000 per year for a family of four in 1988—than at any time since the War on Poverty was initiated in the mid-sixties. In 1979, the equity of income distribution began to deteriorate rapidly; by 1986, disparities in earnings were the worst on record. Two years later, some 32 million Americans—and one fifth of American children—lived below the official poverty threshold.[18]

Thus with widening income disparities around the globe, it will not be long before there is a Third World within the First World, and a First within the Third. The inevitable question is, Why does poverty continue to spread in an age incomparably more prosperous than any in history?

THE GLOBAL POVERTY TRAP

Uruguayan historian Eduardo Galeano once wrote, "The division of labor among nations is that some specialize in winning and others in losing." The global division of labor has relegated the poor to the role of perpetual loser. They are caught by forces at local, national, and global levels that combine to form a three-tiered trap.[19]

At the local level, elements of the poverty trap include skewed patterns of access to land and other assets, physical weakness and heightened susceptibility to disease, population growth, and powerlessness against corrupt institutions. These are reinforced at the national level by innumerable policies—from tax laws to the structure of development investment—that neglect or discriminate against the poor. And at the global level, the poor are held down by the devastating combination of oppressive debt burdens, falling export prices, and rising capital flight. All these factors and forces, like diabolical counterparts to Adam Smith's invisible hand, strengthen each other as they interlock.

The first of the poverty trap's four local parts is the lack of productive assets: the poor are poor not only because they earn so little but because they own so little too. In developing societies, where three of four people earn their living in agriculture, the most crucial asset is land. Yet ownership of farmland is concentrated in the hands of a fortunate few. Latin America has the worst record. The skewed landownership there is a legacy of colonial times, when Spanish and Portuguese rulers established vast plantations; 1 percent of landlords commonly own more than 40 percent of the arable land. Crowded Asian nations are somewhat better, while in Africa land is less scarce and collective tribal landholding arrangements moderate inequality.[20]

Combined with the momentous force of population growth, maldistribution of land pushes ever more of the poor into the vulnerable position of being farmers without land. As millions of poor families divide already small farms between their children, plots become too small to provide subsistence. The mechanization of agriculture in some regions has displaced millions more, as commercial operators expel sharecroppers, squatters, and smallholders.

The typical poor person is no longer a subsistence farmer but a dispossessed laborer; indeed, the number of landless rural households in the world has never been higher. Estimated in 1981 at 167 million households—938 million people—the landless and near-landless are expected to increase to nearly 220 million households by the end of this decade. They account for a large share of rural families in many developing countries, ranging as high as 92 percent in the Dominican Republic. (See Table 8–3.) Reflecting patterns of land concentration, landlessness is most pronounced in Latin America, somewhat less prevalent in Asia, and still new to Africa, though even there it is rising in more fertile regions.[21]

Because of the dispossessed state of most of the poor, their day-to-day calculus of survival now has three variables: the availability of work, the wages offered, and the price of food. In each case, the prospects are not good. Wages and prices went against the poor during much of the eighties. Labor force growth alone means that nearly 40 million people enter the job market each year. Already, unemployment and underemployment, though hidden, are rampant in both urban and rural areas of the Third World.[22]

A debilitating but frequently overlooked dimension of the poor's lack of assets is their extreme vulnerability to unforeseen expenses and emergencies, such as natural disasters, crop failures, illnesses, and the costs of legal battles. To meet these needs a poor family must often borrow at usurious rates from moneylenders, piling debt on deprivation, or must sell or mortgage whatever they have, frequently at distress sale prices. Vulnerability leads to poverty ratchets, events that suddenly make poor people poorer, diminishing at a stroke both their assets and their income.[23]

The second part of the poverty trap at the local level is physical weakness and illness. Lacking consistent nourishment, clean water, basic medical care, and sufficient housing space to avoid rapid spread of infection, the poor are chronically weakened by disease. Physical weakness can combine with low earnings to form a vicious circle: for lack of food, the poor have no energy to work; for lack of work, they have no money to buy food.[24]

Population pressures are the third component of the local trap. Rapid growth in numbers drives wages down to the survival level as the poor compete with each other for scarce work. It

**Table 8-3. Landless and Near-Landless Agricultural Households,
Selected Countries and Africa, Mid-Seventies**

Country/Region	Near-Landless	Landless	Combined
	(percent of rural households)		
Dominican Republic[1]	48	44	92
Guatemala	47	38	85
Ecuador	52	23	75
Peru	46	29	75
Brazil	10	60	70
Philippines	34	35	69
Colombia	24	42	66
El Salvador[1]	—	65	65
Honduras	46	18	64
Bangladesh[1]	33	29	62
Costa Rica	11	44	55
India	13	40	53
Mexico	33	18	51
Malaysia	35	12	47
Africa	30	10	40

[1]Data from late seventies to mid-eighties.
SOURCES: William C. Thiesenhusen, ed., *Searching for Agrarian Reform in Latin America* (Boston: Unwin Hyman, 1989); Milton J. Esman, *Landlessness and Near-Landlessness in Developing Countries* (Ithaca, N.Y.: Center for International Studies, Cornell University, 1978); Joel Bergsman, *Income Distribution and Poverty in Mexico,* Staff Working Paper 395 (Washington, D.C.: World Bank, 1980); M. Riad El Ghonemy, ed., *Development Strategies for the Rural Poor,* Economic and Social Development Paper 44 (Rome: Food and Agriculture Organization, 1984).

stretches investment resources thin and raises the number of children for whom each worker must provide. It overtaxes natural resources, diminishing their productivity. Yet poor couples have large families because they know that some of their children are likely to die, and because they lack access to family planning. Having many children is part of a strategy for economic security: when times are bad for some, they may be better for others. This "strength in numbers" strategy lowers the chances of pulling the whole family out of poverty, but also reduces the risk of falling deeper—into starvation.[25]

Local poverty also stems from powerlessness. Unable to read, the poor are sometimes misled or intimidated into signing away their rights to land or accepting debt repayment terms that verge on extortion. Local officials employ well-intended laws and regulations to hassle the less fortunate to the point of paralysis. Legal systems are often a dead end for the poor, riddled as they commonly are with procedural delays and corruption. In places, the exploitation of poor people does not even wear the trappings of legality. The active human rights networks of Brazil and India both report killings and brutality against the rural poor almost daily. When the wealthy go to court against the illiterate poor there is little competition. When they go to war, there is none.[26]

Inequalities in Third World villages are replicated and reinforced at the national level, where tax laws, budgeting, and voluminous regulations everywhere favor the few over the many. National investments have long been tilted sharply toward the fortunate inhabitants of cities and away from the rural poor. In the worst cases, pharaonic extravagance fuels the diversion of resources. President Houphouët-Boigny of Côte d'Ivoire recently completed construction of a $200-million air-conditioned cathedral that dwarfs even St. Peter's in Rome. Trimmed with nine acres of French stained glass and Italian marble, the church sits in an impoverished country where only one tenth of the population is Catholic.[27]

Côte d'Ivoire's new cathedral is an apt symbol for normal practices of development investment in much of the Third World. In dozens of nations, governments have suppressed prices paid to farmers, siphoning millions of dollars into grandiose industrialization projects in the capital. The resulting high-technology facilities, inappropriate to local circumstances of scarce capital and abundant unskilled labor, now stand as monuments to futility and inequity: they either stand idle or, when funded, continuously redistribute wealth from poor farmers and laborers to a few hundred salaried workers and managers. Either way, governments infatuated with construction of industrial temples have doomed the hinterlands to worsening poverty.

Government budgets are inequitably distributed even within the sectors ostensibly aimed at helping the poor—education and health. In sub-Saharan Africa, fewer than half the school-age children are enrolled, while large shares of education budgets provide highly subsidized university education, disproportionately to children of the urban wealthy. What African governments spend putting one student through four years of college could pay for six years of primary education for a class of 35 poor, rural children.[28]

Likewise, subsidized urban hospitals take the lion's share of public health budgets, but help those who least need assistance. Community-based health workers and rural clinics are consistently underfunded. Some 70–85 percent of public health spending in developing countries overall goes into expensive curative care while community health and preventive care combined receive only 15–30 percent. The World Health Organization believes that increasing annual investment in preventive care by only 75¢ per person in the Third World would save 5 million lives each year.[29]

The pivotal question is which way resources are moving across the North-South boundary.

At the international level, debt, trade protectionism in industrial countries, capital flight, and plummeting world prices for the products developing nations export add further layers to the poverty trap. The poor world's debt to the rich is so colossal that discussions of the issue sound surreal. In 1989, the Third World owed $1.2 trillion—nearly half its collective gross national product—to industrial-country banks and governments. If people in Zambia were to give every penny they earned to their nation's foreign financiers beginning January 1, 1990, they would not have paid their debt until May of 1993.[30]

Staggering as the absolute quantities of debt are, however, the pivotal question is which way resources are moving across the North-South boundary. Until 1984, industrial countries gave more to developing countries in loans each year than they took back in interest and principal payments. At that point, the flow become a backward torrent; by 1988, the poor were paying the rich $50 billion a year. The massive diversion of resources to the North has taken a toll not only on the people of developing lands, but on the land itself. Forests have been recklessly logged, mineral deposits carelessly mined, and fisheries overexploited, all to pay foreign creditors.[31]

Adding insult to injury, as debt piled on top of developing economies, export earnings collapsed from under them. Most developing countries remain heavily dependent on exports of such raw materials as copper, iron ore, and timber, and of such cash crops as sugar, cotton, and coffee. Yet between 1980 and 1987, the prices of 33 such commodities fell by 40 percent on average, catching the Third World between the blades of rising debt and falling earnings. (See Figure 8–2.)[32]

Rising trade barriers in rich countries are one cause of declining export prices for poor lands. The European Economic Community, for example, levies a tariff four times as high against cloth imported from poor, heavily indebted nations as from rich ones. All told, World Bank figures suggest that each year industrial-country trade barriers cost developing countries $50–100 billion in lost sales and depressed prices.[33]

Beyond the disastrous trends in bank loans and international trade, economic turmoil has spooked First World investors and incited national elites to bundle their fortunes off to foreign banks. The dimensions of this exodus of private money are gargantuan: capital flight from Latin America amounted to about $250 billion by late 1987. Venezuelans' overseas holdings are valued at $58 billion; the nation's external debt is $37 billion.[34]

As countries foundered in the inhospitable world economy of the eighties, the World Bank and International Monetary Fund promoted controversial economic reforms called structural adjustment. Yet no impoverished nation can reform its economy sufficiently in the short term to compensate for massive debt burdens (aggravated by high real interest rates in New York) and falling prices for its goods (exacerbated by trade policies in the European Economic Community).

The economy that most needs adjustment is the global one. Until that takes place, structural adjustment will continue to be little better than, in the words of UNICEF, "a rearranging of the furniture inside the debtors' prison."[35]

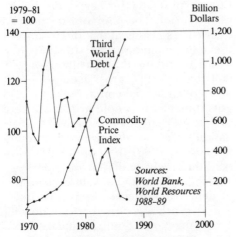

Figure 8-2. Commodity Price Indexes and
Total Third World Debt, 1970–87

POVERTY AND THE ENVIRONMENT

Most of the world's looming environmental threats, from groundwater contamination to climate change, are byproducts of affluence. But poverty drives ecological deterioration when desperate people overexploit their resource base, sacrificing the future to salvage the present. The cruel logic of short-term needs forces landless families to raze plots in the rain forest, plow steep slopes, and shorten fallow periods. Ecological decline, in turn, perpetuates poverty, as degraded ecosystems offer diminishing yields to their poor inhabitants. A self-feeding downward spiral of economic deprivation and ecological degradation takes hold.

In the rural Third World, human reliance on ecosystems goes unmediated by the long chains of commerce, industry, and civil infrastructure that shape life in rich countries. For the have-nots, food comes from the soil, water from the stream, fuel from the woods, fodder from the pasture, and fruit from the trees around the hut. Poor people know well that to endanger any of these things is to imperil themselves and the lives of their children. Farmers with secure rights to a piece of land tend to care for it meticulously, taking a long-term view and forgoing current benefits for dependable future gains.

The central pole around which the downward spiral turns is the lack of resources—the first element of the local poverty trap. Poor but secure smallholders rarely overburden their land; dispossessed and insecure rural households often have no choice but to do so. Access to a resource without control over it can be equally harmful. Nothing incites people to deplete forests, soils, or water supplies faster than fear they will soon lose access to them. Neither hired workers, nor hired managers, nor tenant farmers care for land as well as owners do. In Thailand's forests, squatters given long-term rights to use their plot care for the land better than squatters with no legal standing, but not as well as those who own their plots outright.[36]

As the global poverty trap tightens and the world's poor become increasingly insecure and dispossessed, the conditions for ecological degradation spread to more of the earth's fragile lands. Two of the most common sequences by which poor people fall, or are pushed, into the downward spiral are illustrated by the situations in Nepal and Costa Rica.

Nothing incites people to deplete forests, soils, or water supplies faster than fear they will soon lose access to them.

Nepal exemplifies the way sheer growth of human numbers fuels the spiral, as human practices at a given level of technology exceed a local environment's carrying capacity. As population swells, peasants in highland valleys are forced to expand their plots onto steep forested hillsides, extending women's journeys to gather fuel and fodder. Over the past decade, during which forests have receded to half their original extent, women's daily journey has increased by more than an hour, their workday in the fields has shortened, family incomes have fallen, and diets have deteriorated.[37]

Shubh Kumar and David Hotchkiss of the International Food Policy Research Institute (IFPRI) in Washington report not only that food consumption in the region has fallen by 100 calories per person on average, but that in village after village childhood malnutrition rates and deforestation rates are closely coupled. In the hills of Nepal, in other words, the health of a village's children can be read in the retreating tree line on surrounding slopes.[38]

Where population density is less acute than in Nepal, skewed landownership patterns have the same effect—large holdings are underattended, small holdings overburdened, and millions left landless. In Costa Rica, a country often praised for its enlightened environmental policies, the poor have been driven into this desperate state largely by a cattle boom that has swept the nation since the early sixties. Costa Rica was once almost completely cloaked in tropical forest, holding within its small confines perhaps 5 percent of all plant, animal, and insect species on earth. By 1983, after two decades of explosive growth in the cattle industry, pastures covered roughly half the nation's arable land, only 17 percent of the original forest remained, and soil erosion was rampant.[39]

Aside from its ecological toll, the cattle boom has been an economic debacle. It was driven not so much by market forces as by heavy flows of international capital to, and special treatment for, the 2,000 politically powerful ranching families who control the industry. The Costa Rican poor, for their part, have only gotten hunger out of the deal. Smallholders have been pushed from their land to make way for four-legged competitors. Rural employment opportunities dried up because ranching is the least labor-intensive agricultural activity in the country. The rising tide of landlessness has spilled over into the expanding cities, onto the fragile slopes, and into the forests, where families left with little choice accelerate the treadmill of deforestation.[40]

The poverty trap patterns seen in Nepal and Costa Rica reappear on every continent, and the net effect is universal: the poor are increasingly concentrated in fragile regions where land is least productive and tenure least secure—arid and semiarid lands, mountain slopes, tropical forests, and sprawling shanty-towns around overcrowded cities. The geographic concentration of poverty in inhospitable lands is driven partially as people move in, but more so by the heightened population growth rates that poverty itself brings. Inevitably, the poor

are pulled into the downward spiral in growing numbers. Jeffrey Leonard of the World Wildlife Fund–Conservation Foundation in Washington has attempted to roughly quantify this trend using data from the World Bank, the United Nations, and IFPRI.[41]

Leonard's estimates focus on the poorest fifth of people in developing countries, a subset including about 780 million of the absolute poor. He calculates that 370 million of these—47 percent of the total—live on marginal or fragile lands, where at today's population densities the downward spiral is almost inevitable. If this ratio holds true not just for the poorest fifth of people in the Third World, but for the entire class of people in absolute poverty, then 580 million people are now trapped in the downward spiral. Many others suffer from environmental degradation, but the impact on their living standards is more limited so far.[42]

The same forces that have pushed 580 million people into the downward spiral—lack of secure control over resources, population growth, inequity, and misguided policies—are largely responsible for the deterioration of marginal, nonagricultural land. Throughout history, poor people have drawn on a resource base far larger than their own land. They gather fruits, nuts, fish, game, roots, leaves, fibers, reeds, dung, and wood from so-called common property resources—hillside forests and lowland swamps; rivers, riverbanks, and floodplains; lakes and lakeshores; open rangeland and dense thickets. These areas are a critical safety net for the poor, providing a fallback means of survival. In dryland regions of India, the landless gather one fifth of their annual income, along with numerous nonmarketed goods, by harvesting the natural products of common areas.[43]

Over generations, most agricultural and fishing villages and pastoral bands evolved intricate codes of conduct to avoid depleting the commons upon which they depended. Yet in vast areas of the Third World, common lands—and the management systems that poor people used on them—have been overrun: privatized by the fortunate, overburdened by the unfortunate, or appropriated by the state.

In a study of Indian dryland commons, N.S. Jodha of the International Crops Research Institute for the Semi-Arid Tropics in Patancheru, India, found that over the past three decades more than half the common property distributed through the government agrarian reform program has ended up in the hands of those who already own significant quantities of land. And in Botswana, communal pasture in the savannah has been fenced by large ranchers, with support from the World Bank. Many so-called land reforms in Latin America have simply exported the rural underclass to a fragile rain forest frontier, where new settlers lay waste to ecosystems previously harvested for generations by tribal people. Almost without exception, the net effect of state land policies has been to curtail drastically common property resources open to the poor without expanding their private property resources commensurately.[44]

The deterioration of common areas is sometimes most directly caused by the breakdown of traditional management regimes that takes place as governments nationalize the resource base or intervene obtrusively in its use. Since the beginning of this century, newly independent nations freeing themselves from colonial bonds have followed their colonizers' path, codifying the earlier policy that common resources are the domain of the sovereign. Contradictory legal rights and overlapping jurisdictions, often combined with weak or corrupt ministries and courts, have undercut the authority and legitimacy of traditional management regimes—and thereby poor people's control over their resource base—without providing a viable substitute. Predictably, these failed government policies have left the rich richer, the poor poorer, and the environment denuded.

In the northern Sahel region of Mali, for example, villages traditionally enforced local restrictions on tree cutting for sheep and goat fodder by confiscating and slaughtering the best breeding male of the offender's flock. This punishment, swift and harsh, kept forests intact over centuries. When government foresters began to implement the previously declared nationalization of woodlands in the late sixties, however, villagers who enforced the traditional rules became liable for breaking the law. Forests have since been severely depleted.[45]

Even where poverty is not a cause of environmental decline, it is a ticket to suffer environmental abuses caused by others. The poor of the mammoth cities of the Third World face all the environmental perils of underdevelopment along with those of overdevelopment. Their cookstoves fill their hovels with old-fashioned smoke, multiplying the danger of respiratory diseases such as tuberculosis that run rampant through crowded slums. In the streets, they breathe air polluted in concentrations that rival the worst recorded in industrial lands. (See Chapter 6.) Lacking adequate sewerage and water supplies, they drink and bathe in water contaminated with both human and chemical wastes.[46]

The shantytowns of the poor are found in areas eschewed by the better off: in floodplains, on perilous slopes, around—sometimes in—garbage dumps containing unknown quantities of toxic materials, and near hazardous industrial zones. The fact that Bhopal's victims were overwhelmingly poor was no coincidence; industrial accidents worldwide take the lives of those who cannot afford to live far from the belching stacks.[47]

The disproportionate exposure of the poor to pollution and hazardous materials has gone almost unmeasured in the Third World, but some idea of its magnitude can be gathered by examining the situation in the United States. Neighborhood-by-neighborhood comparisons of income level, race, and toxic waste site location reveal a disturbing but not so surprising pattern. The poorer the neighborhood, and the darker the skin of its residents, the more likely it is to be near a toxic waste dump. Three fourths of hazardous waste landfills in the American Southeast are in low-income, black neighborhoods, and more than half of all black and Hispanic Americans live in communities with at least one toxic waste site. As in the United States, so in the Third World: the rich get richer, and the poor get poisoned.[48]

In the United States, the poorer the neighborhood, and the darker the skin of its residents, the more likely it is to be near a toxic waste dump.

The environmental victimization of the poor extends beyond local hazards. The potential poverty impacts of global climate change are incalculable. Low-lying countries such as Bangladesh, Egypt, Indonesia, and Thailand could lose large areas of cropland as seas rise and storms become more severe in the next 30–40 years. (See Chapter 5.) Many of the world's most productive agricultural areas will suffer higher heat, reduced rainfall, or both, increasing the frequency of massive crop failures. (See Chapter 4.) Storms, floods, droughts,

heat waves, famines, and social upheaval—all likely to be commonplace in a greenhouse world—are precisely the events, the poverty ratchets, that tend to make poor people poorer.

In addition, climate change could divert resources away from the poor on a scale unimaginable today. The debt burden of the eighties—on which UNICEF blames the death of a half-million children per year—could be dwarfed by the climate burden of the twenty-first century. Health, education, and antipoverty budgets would probably be slashed as governments poured hundreds of billions of dollars into sea walls, new irrigation systems, flood prevention, and countless other projects. Throughout economic history, hard times have almost always widened the gap between rich and poor. There is no reason to believe uncontrolled climate change will be a story with a different ending.[49]

With perhaps 580 million poor people trapped in the crushing downward spiral of economic and environmental deterioration, the future of poverty alleviation—and of environmental protection—will contain unprecedented challenges. Hundreds of millions already suffer from deforestation, desertification, fuelwood scarcity, erosion on mountain slopes, and urban air pollution. If, along with uncontrolled climate change, these other trends continue along their current trajectories, the poverty rate will almost certainly skyrocket, perhaps doubling worldwide by the second half of the next century.

In other words, following a business-as-usual course into the future could doom half of humanity to absolute poverty by sometime between 2050 and 2075. Whatever the numbers, if destructive environmental trends are not controlled, the downward spiral will accelerate out of control, pulling in many millions more.

REVERSING THE DOWNWARD SPIRAL

National development, while benefiting the middle and upper classes handsomely, has either helped the poor little or actually hurt them. Yet the means and methods are at hand to alleviate the worst aspects of absolute poverty. Around the world, public and private initiatives have created hundreds of innovative ways to pry open—if only slightly—the jaws of the poverty trap. Turning this potential into reality, however, will require national governments and international agencies to redefine development goals. It will also require decisive international action.

True development puts first those that society puts last. In the words of Mahatma Gandhi: "Whenever you are in doubt . . . apply the following test. Recall the face of the poorest and the weakest man whom you may have seen, and ask yourself if the step you contemplate is going to be of any use to him. Will he gain anything by it? Will it restore him to a control over his own life and destiny?"[50]

For development to help the poor, it must put them first not only as intended beneficiaries, but as active participants, advisors, and leaders. True development does not simply provide for the needy, it enables them to provide for themselves. This implies a great deal of humility on the part of outsiders. As nutrition analyst Paulus Santosa put it, "It would be very hard to find professional nutrition workers in Indonesia today who can raise a family of five with U.S. $0.50 per day and stay healthy." On poverty, the only true experts are the poor.[51]

The groundwork for true development is expanding impressively in many regions as poor people organize themselves to fight poverty and environmental decline. Indeed, the accelerating proliferation of self-help groups is the most heartening trend on the poverty front. Grassroots action has grown steadily, if unevenly, since mid-century, as an expanding latticework of religious and nonreligious independent development groups have organized among the poor. Self-help organizations spread explosively in the eighties, as economic and environmental conditions deteriorated. Grassroots environmental and anti-poverty groups probably number in the hundreds of thousands, and their collective membership in the hundreds of millions. Spotlighting a few successful efforts helps reveal the opportunities for dramatic progress when the poor are enabled to overcome old barriers.[52]

Information is one of the most important assets, and innovative educational programs have found ways to bring it to the poor. In Bolivia's rugged Yungas region, hundreds of impoverished peasants take high-school-level courses in market towns when they come to sell their produce. The curriculum, designed by a dedicated independent group called CETHA to be relevant to local conditions, also offers intensive, week-long vocational courses during the agricultural slack season. Most important, the effort has managed to enroll nearly as many women as men. Studies on every continent show that as female literacy rates rise so do income levels, nutrition levels, and child survival rates; at the same time, population growth slows, as women gain the self-confidence to assert control over their bodies.[53]

A new generation of loan programs has found a direct and highly effective way to increase poor people's access to productive assets. The world's premier loans-for-the-poor program was initiated in 1976 by Bangladeshi Mohammad

Yunus. He began distributing tiny loans, around $65 apiece, to landless villagers so that they could buy basic assets, such as rice hullers, cows, and raw materials. The project has spread like wildfire ever since. In 1983, Yunus established the Grameen (Village) Bank, which by early 1989 had a half-million borrowers, most of them women. Repayment rates are higher than for any commercial lending program, and Grameen's clients have augmented their incomes dramatically. One study found borrowers earning 43 percent more than their counterparts in nearby unaffected villages. Income gains were largest, furthermore, for the poorest borrowers.[54]

For development to help the poor, it must put them first, as active participants, advisors, and leaders.

Trees are another asset that can liberate the poor. In the Indian state of West Bengal, a government program has enabled villagers to turn a halfhearted land reform into a route out of the downward spiral. In the Midnapore District, with a population of 7 million, an agrarian reform in the late seventies assigned small plots in degraded commons to landless families. These infertile allotments remained vacant until 1981 when the newly founded Group Farm Forestry Program began distributing free tree seedlings. Many villagers enrolled and covered their land with the hardy trees. Villagers in the Nagina area, mostly lower caste and tribal people, used the bulk of their earnings to purchase small irrigated fields in the fertile valleys from absentee landlords. Distribution of degraded commonland, followed by tree cultivation, thus became self-help land reform.[55]

Returning control over local resources to villagers can have dramatic effects. In the devastated Aravali hills of the west Indian state of Rajasthan lies the community of Seed, whose tribal inhabitants have registered their village under the little-used state Gramdan Act. The neglected act, inspired by Gandhi, transfers all common lands in the village to the local council. The results of empowering the community are plainly visible: the village, write Indian environmental analysts Anil Agarwal and Sunita Narain, "is lush green and full of grass, like an oasis in the denuded Aravalis." Even during the severe drought of 1987, when cattle died by the thousands in neighboring areas, Seed's villagers filled 80 bullock carts with grass.[56]

Poor people's physical weakness from malnutrition and illness can be alleviated through low-cost, participatory provision of clean drinking water and basic health care. In 1981, Tanzanian health authorities began meeting with dozens of communities to discuss the best strategy for an all-out assault on child mortality. Kicked off in one region with UNICEF assistance in 1983, the program is a model of social mobilization from the international to the village level, involving government, the media, and freshly organized village health committees. At regularly scheduled Village Health Days, parents and newly trained health workers weigh and vaccinate children, treat common ailments, and discuss health and nutrition. By 1988, the incidence of severe malnutrition had fallen by nearly two thirds, and the child death rate by a third.[57]

Powerlessness is perhaps the most tenacious local aspect of the poverty trap. Land reforms, welfare systems, employment programs, minimum wages, basic infrastructure investments, and health and nutrition assistance are commonly delayed, distorted—sometimes plundered—by landowners, intermediaries, and petty officials. A decade ago, when the Bangladesh Rural Advancement Committee (BRAC), an independent development organization, undertook an emergency famine aid effort in a border region, it quickly discovered government relief should have been more than sufficient. But local landlords and moneylenders had formed a "complex net of co-operative connections linking them into a seemingly irresistible network of corruption." Interviewing landless people in 10 villages, BRAC took records of land grabs, unfair loans, embezzlements, and bribes. As the pervasiveness of the scam was revealed, the poor organized themselves to reclaim property wrongfully taken and to get their due from government offices.[58]

The inspiring antipoverty efforts described above give a sense of the force unleashed when development starts with the poorest. In the final analysis, however, ending poverty requires more than a piecemeal approach; it requires a government that explicitly directs policy toward aiding the least fortunate. In countries of all ideological hues, poverty has retreated dramatically only when ambitious programs have been in place to create the foundations of equitable development.

The state of Kerala, India, provides a model of incremental but comprehensive poverty alleviation under democratic rule. This southern Indian state, with a population as large as Canada's in an area the size of Switzerland, has been putting the poor first for decades. Peasants and laborers there are exceptionally well organized and vocal, thanks to a history of social activism among the educated classes, and have been able to elect governments committed to helping them.[59]

Kerala's villages have access to health, education, and transportation services at a level unknown anywhere else in India: a recent survey of basic services in a half-million villages ranked Kerala first in 15 of 20 categories. Immunization campaigns cover most of the population. Schools and clinics are spread throughout the state, along with "fair price shops" that sell basic goods to the poor at low cost. Safe drinking water supplies and family planning services are available to a large and growing share of the population.[60]

A comprehensive land reform program begun in 1969 gave 1.5 million tenants and laborers rights to the land they tilled or to their homes and gardens. Despite per capita income less than two thirds the all-India average, Kerala's adult literacy rate is almost twice the national mark, its people typically live 11 years longer, its birth rate is one third lower, its infant death rate is two thirds lower, and inequalities between sexes—and castes—are less pronounced than in any other state.[61]

The ingredients of Kerala's success are a common thread running through effective efforts to dismantle the poverty trap worldwide: Literacy, especially for women, gives access to information and leads to higher incomes, improved health, and smaller families. Secure land rights allow the poor to increase income and economic security. Local control over common resources helps arrest the ecological downward spiral. Credit lets the poor get productive assets such as livestock and tools. Clean drinking water and primary health care reduce the incidence of debilitating diseases that kill children (inducing parents to have larger families) and sap adults' strength. Family planning gives poor women means to effectively control their fertility, space births at a healthy interval, and improve their own health. Grassroots organizations enable the poor to direct the development process and to accelerate progress by working together.

Implementing poverty alleviation programs locally and nationally, however, will come to little without complementary reforms at the international level, where the debt burden remains undiminished and the threat of climate change hangs like Damocles' sword. If World Bank and International Monetary Fund structural adjustment packages incorporated some of the antipoverty priorities just outlined they would often help both the unfortunate and the economy overall, by enabling poor people to generate more income. Alternately, structural adjustment might simply focus on removing economic distortions and inefficiencies that benefit primarily the rich. Because adjustment plans are often quickly drawn, they tend to remove obvious subsidies to the poor, such as those on food, without touching invisible subsidies to the rich, such as protection of industrial sectors.[62]

Reversing the downward spiral means negotiating deep cuts in the debt burden in exchange for policy reforms to help the poor out of the cycle of environmental impoverishment. Moves to cancel some African nations' debts to western governments are a good start, as is U.S. Treasury Secretary Brady's call for reducing Latin nations' debt to private banks. But more comprehensive action will be required. Since early in the debt crisis, banks have formed an almost united front, but debtor nations have never joined together to coordinate their bargaining positions. Reaching a final solution may require roundtable international negotiations between debtors and creditors, an international debt reduction agreement, and an appropriate impartial mechanism.[63]

Beyond debt relief, unlocking the global poverty trap will require lowering industrial-country trade restrictions against imports from poor nations, helping developing nations diversify away from dependence on volatile commodity markets, and working toward international cooperation to track and tax capital flight.

Finally, poverty alleviation must look to the threat of global climate change, which if allowed to run its course could overwhelm all other progress against destitution. Over the long run, fossil fuel use in industrial countries may be the most important determinant of the global poverty rate. Every year that passes without an international accord to stabilize and then reduce greenhouse gas emissions effectively dooms millions more to live as paupers. At the same time, insofar as deforestation contributes to global warming, and poverty to deforestation, poverty alleviation has a place in any comprehensive climate protection plan.

No one can question that there is something morally bankrupt about a world where, amid extravagant abundance, one quarter of the human family is doomed to an unremitting battle for survival. But poverty is more than a moral issue. Failure to launch an all-out assault on poverty will not only stain the history of our age, it will guarantee the destruction of much of our shared biosphere. For although environmental damage penalizes the poor more consistently and severely than it does the wealthy, the circle eventually becomes complete.

When the poor destroy ecosystems in desperation, they are not the only ones who suffer. Sheets of rain washing off denuded watersheds flood exclusive neighborhoods as surely as slums. Potentially valuable medicines lost with the extinction of rain forest species are as unavailable to the rich in their private hospitals as they are to the poor in their rural clinics. And the carbon dioxide released as landless migrants burn plots in the Amazon or the Congo warms the globe as surely as do the fumes from automobiles and smokestacks in Los Angeles or Milan. The fate of the fortunate is immutably bonded to the fate of the dispossessed through the land, water, and air: in an ecologically endangered world, poverty is a luxury we can no longer afford.

9

Converting to a Peaceful Economy

Michael Renner

Since biblical times, people have been admonished to beat swords into plowshares. Never has such advice been more appropriate. The relentless pursuit of military might has brought humanity to the brink of annihilation. Perhaps worse, the priorities of the global arms race have also kept us from meeting pressing social and environmental needs.

Crushing poverty, rampant disease, and massive illiteracy characterize the lives of hundreds of millions in developing countries. And in many industrial nations, a growing underclass has emerged; some inner-city areas resemble war zones, not places for people to live. Nations devoting a large portion of their wealth to the military have done so at the expense of economic vitality. All of humanity—rich or poor, militarily strong or weak—confronts the specter of unprecedented environmental devastation. Military spending may not cause all these problems. But the devotion of large-scale resources to building military power has stopped us from addressing them adequately.[1]

The world now has a remarkable opportunity to redirect society's priorities. Following a decade that started with a rash of new wars and confrontation, diplomacy and overtures toward disarmament are renewing hopes for a less violent world. The cold war that has gripped us for almost half a century is waning. The world witnessed a dramatic sign of this when the Berlin Wall crumbled. Several hot spots in the Third World—the Persian Gulf, Nicaragua, Namibia, Angola, and Western Sahara—appear to be cooling off.

Major arms control and disarmament treaties are in the offing that, if concluded, would release a great deal of resources. The proposed Strategic Arms Reductions Talks (START) agreement would cut U.S. and Soviet strategic nuclear weapons by roughly one third. Fueled by pervasive health and environmental problems associated with the superpowers' nuclear weapons industries, a negotiated ban on the production of all nuclear weapons materials has been proposed. In Europe, strong momentum can be found for denuclearization and a sharp reduction of conventional forces. And U.S. and Soviet negotiators are edging toward a treaty mandating the destruction of chemical weapon stockpiles, though the Bush administration insists on continued production.

These developments are accompanied by strong budgetary pressures on virtually all governments to rein in military spending. The U.S. military budget has begun to level off in real terms, with further reductions expected, and the Soviets have announced significant cutbacks. Arms export sales have declined—from a peak of about $57 billion in 1984 to $47 billion in 1987—due largely to economic crises in many developing nations. The global arms budget, an estimated $1 trillion per year, may indeed have peaked.[2]

Clearly, the world's armies and military industries will still remain large, even with these recent developments. Wars still rage in many areas. And a number of developing countries are actually expanding their capacity to manufacture arms. Nevertheless, there is a growing sense that the world is at a historic juncture, the beginning of a new era in international relations in which reliance on force is increasingly counterproductive.

In an age of weapons of mass destruction and delivery systems of global reach, arms no longer provide the security we expect; more important, the pursuit of military prowess is undermining our economic health. Yet few governments, communities, or military contractors have the expertise and the institutions to reverse the armament process without social and economic upheaval. Accomplishing this difficult task will require a careful program of economic conversion—releasing resources now tied up in the military sector and planning for their use in health and educa-

tion programs, environmental protection, and other areas of need.

INITIAL OBSTACLES

Converting a great deal of a society's productive wealth to civilian use is no doubt an ambitious undertaking, but there should be no insurmountable problems. The major barriers are not technical but political, ranging from the power and agendas of vested interests to the widespread misconception that military spending makes good economic sense.

Military contractors have little incentive to move out of defense work: They enjoy low-risk operations, generous cost-plus contracts, and large profits. Conversion would mean a loss of power and privilege. Conversion advocates Lloyd Dumas and Suzanne Gordon observe that it "is seen as a fundamental challenge to management's important prerogatives—the ability to shut down operations whenever and wherever it desires and to produce whatever it wants, whenever, wherever, and however it wants." Indeed, according to Jonathan Feldman of the National Commission for Economic Conversion and Disarmament, a private group in Washington, D.C., "in the past, top military managers have resisted planning for conversion to the point that they have sacrificed facilities to permanent closure rather than propose their reuse for civilian purposes through conversion."[3]

Without a convincing economic alternative to defense jobs, military industry employees see disarmament as a threat to their livelihoods. Most engineers are reluctant to transfer to civilian jobs, which tend to be less well paid and involve less exotic tasks. This distinction is not so pronounced for the average defense production worker. But as long as military orders are received and employment appears secure, most workers are apparently willing to stick with arms production.[4]

The myth of military-led prosperity has obscured the fact that civilian spending creates significantly more jobs.

National defense establishments are predictably hostile to the prospect of genuine conversion planning because it would jeopardize their command over a sizable portion of society's resources and, thus, a good measure of political power. The politics of the military pork barrel not only provides the Defense Department with a lever to press for continued high military spending, it is a serious impediment to a conversion program: every military-dependent community represents a formidable constituency for maintaining the status quo.

In the United States, for example, the Pentagon has quite consciously attempted to spread military work across as many congressional districts as possible, to hedge against the possibility that Representatives and Senators will vote against controversial weapons projects. Even where their political philosophy inclines them otherwise, members of Congress are likely to vote for military programs vital to livelihoods in their districts.[5]

The myth of military-led prosperity—a conviction rooted in the U.S. experience of the forties, when war spending pulled the economy out of the Depression—remains a major obstacle to conversion. In truth, the military pork barrel turns out to be empty for most communities. A recent study by Employment Research Associates in Lansing,

Michigan, found that 321 out of the 435 U.S. congressional districts in effect pay a "net Pentagon tax": They pay more in defense-obligated federal taxes than is returned to them in military salaries and contract money. The few winners are mostly concentrated along coastal and southern states, in what has been dubbed the "gun belt."[6]

Experience in the United States and elsewhere demonstrates that military-related employment rises and falls with the government procurement cycle and the vagaries of international relations. (See Figure 9–1.) During the Carter-Reagan buildup, jobs in the military industry doubled—to almost 3.4 million—between 1976 and 1987, a postwar level surpassed only during the conflict in Korea. Strong budgetary pressures, however, have already caused a decline from 1987's peak and portend a severe retrenchment in the years ahead.[7]

West Germany's military industry employment has followed a similar roller-coaster even though, unlike the United States, the country was not involved in any wars. The number of jobs rose rapidly to peak at almost 350,000 in 1961, but fell to 200,000 by 1973. Since then, employment has increased to roughly

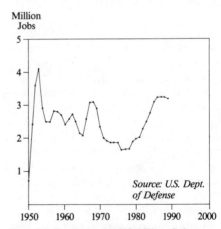

Figure 9-1. Employment in U.S. Military Industry, 1950–89

280,000, but it is likely to decline again as the procurement cycle comes to an end.[8]

Apart from the ups and downs of procurement cycles, arms production is steadily becoming more capital-intensive, reducing the number of jobs generated per military dollar. In the United Kingdom, for example, military industrial employment declined by 25 percent (from roughly 1 million) between 1963 and 1978. Even though the early years under Prime Minister Thatcher brought a major rearmament program, jobs created by military outlays fell by a further 10,000 between 1978 and 1986.[9]

Continued belief in the myth of military-led prosperity has obscured the fact that civilian spending creates significantly more jobs. In the United States, for example, spending $1 billion (in 1981 dollars) on guided missile production creates about 9,000 jobs; when producing military aircraft, it creates 14,000 jobs. But spending the same amount on local transit would yield 21,500 jobs; on educational services, 63,000 jobs; and on air, water, and solid waste pollution control, 16,500 jobs. According to Employment Research Associates, a $40-billion conversion program could bring a net gain of more than 650,000 jobs.[10]

These findings are corroborated by studies in other countries. For example, 1 billion deutsche marks ($570 million at 1988 exchange rates) saved in the West German military budget and spent on alternative programs would create at least 800 more jobs, and possibly up to 6,500 more, than would be lost in the military sector. In India, $13,500 is needed to generate one job in an ordnance factory, compared with $3,800 in industry generally, and a mere $80–90 in road construction or agriculture.[11]

The lost economic opportunities that military spending entails are echoed in investment trends, productivity, and inflation. Undoubtedly, society would benefit if the resources now absorbed by defense were used in the civilian realm. Yet because the military sector in many countries is of such enormous size, without proper preparation the rechanneling could lead to economic disruption and social dislocation. With a conversion strategy in place, however, reversing the arms buildup becomes an economic opportunity—not a penalty.[12]

WHAT IS CONVERSION?

Conversion goes beyond a mere reshuffling of people and money. It involves a political and institutional transformation. Although the military sector could take over some civilian responsibilities, such an approach falls short of the degree of public accountability and democratization of decision making that conversion promises. Moreover, since the war-making capacity of the defense departments would remain relatively unconstrained, a shift away from military priorities could easily be reversed.

"Diversification" is often offered as an escape from dependence on military contracts and an alternative to conversion. A defense contractor may lessen reliance on military orders simply by acquiring other companies, or a defense-dependent community may seek greater economic balance by attracting new firms to the area. Yet such strategies are not concerned with finding alternative uses for the work force and resources previously tied up in military production. In fact, they often go hand in hand with plant closures and massive layoffs. Diversification, as primarily a financial strategy, may be good for a company's balance sheets, but its value for society is questionable.[13]

The most immediate task in any conversion undertaking is to identify the re-

sources involved in the military sector—
essentially to make an inventory of existing skills and equipment at all military-
related facilities, from research labs to
arms production sites to military bases—
and to assess the likely impact of any
disarmament measures on the work
force and the local community.

Enough information is publicly available to sketch a rough global picture of
these resources. Worldwide, the military
sector absorbs between one quarter and
one third of all R&D expenditures, capital investment, and scientists and engineers employed. Although this industry
is highly capital-intensive, in some countries it nevertheless accounts for a significant share of the industrial work
force.[14] (See Table 9–1.)

National averages, however, understate the dependence of certain regions
and communities on defense spending.
The military industry is not only geo-
graphically concentrated, it is also heavily oriented toward a handful of industrial sectors and disproportionately
reliant on engineers, scientists, and
managers. Thus what seems negligible
on the aggregated national level may
pose grave adjustment problems during
disarmament. In part because of a strong
desire for secrecy, most governments
provide too little information—that is,
not detailed enough in geographical,
sectoral, and occupational terms—for a
firm assessment of the specific problems
and promises of conversion.

Although better data are needed at the
local level, we can assess roughly how
many people and what resources global
disarmament would make available for
alternative civilian purposes. Four possi-

Table 9-1. Share of Industrial Work Force Employed in Military-Serving Industry, Selected Countries, Mid-Eighties

Country	Employment in Military Industry	Share of Industrial Employment[1]	Country	Employment in Military Industry	Share of Industrial Employment[1]
	(thousand)	(percent)		(thousand)	(percent)
Israel	90	22.6	Norway	15	2.7
Malaysia	3	18.0	Singapore	11	2.7
United States	3,350	11.1	Italy	160	2.3
China	5,000	10.0	Turkey	40	2.3
Soviet Union	4,800	9.7	Sweden	28	2.2
United Kingdom	700	9.0	Spain	66	2.0
France	435	6.3	Austria	16	1.4
Argentina	60	5.0	Netherlands	18	1.3
Egypt	100	4.1	Pakistan	40	0.8
India	280	3.0	Brazil	75	0.7
West Germany	290	2.8	South Korea	30	0.7

[1]Employment in military industry measured against total employment in mining, manufacturing, gas, electricity, water, and construction industries.
SOURCES: Total industrial employment from *Statistical Yearbook 1985/86* (New York: United Nations, 1988); military industry employment from Worldwatch Institute, based on Peter Wilke and Herbert Wulf, "Man-power Conversion in Defense-Related Industry," Disarmament and Employment Program, Working Paper No. 4, International Labour Organisation, Geneva, 1986, on Michael Brzoska and Thomas Ohlson, "Trade in Major Conventional Weapons: The Changing Pattern," *Bulletin of Peace Proposals*, Vol. 17, No. 3–4, 1986, and on other sources.

ble scenarios of future worldwide military spending illustrate the effect on military-related employment over 25 years. (See Figure 9–2.) Based on an estimated total of 45 million military employees (29 million in the armed forces and 16 million in arms-producing industries), the scenarios assume that spending cuts translate proportionately into job losses.[15]

The first two underscore that, even short of disarmament, employment in the military industry is in jeopardy at a time of tight budgets. The first scenario assumes that global military spending keeps pace with inflation, but that productivity gains mean 3 percent fewer workers are needed every year. The second assumes that spending will not compensate for annual inflation of 5 percent. The last two scenarios assume spending cuts on top of inflation: under one, phased-in cuts increase from 2 to 10 percent; under the other, with immediate spending cuts of 10 percent, the fall in employment would be rapid. Over 25 years, between 9 million and 44 million jobs would be lost.

These scenarios, of course, only give a rough indication of the number of jobs that would have to be replaced with civil-ian employment. The actual savings and employment effects depend strongly on the specifics of future disarmament agreements (such as the extent to which treaties constrain future arms production) and the way governments tailor their armed forces to comply with any treaties.

A few estimates have been made of the savings possible each year from future arms agreements. U.S. savings from a START treaty, for example, could be as high as $14 billion during the decade ahead if currently planned modernization programs are shelved. (They could be as low as $7 billion if these programs proceed.) Reduced warhead production and operations costs could bring in an additional $3–5 billion. The superpowers' conventional forces in Europe will probably be trimmed during the nineties: a 50-percent cut could translate into U.S. savings of $15 billion and Soviet savings on the order of $30 billion. Removing all U.S. ground and air forces from Europe would save an estimated $34 billion. A naval agreement reducing forces across the board by 20 percent could yield U.S. savings of $15 billion and Soviet savings of perhaps $10 billion. Unfortunately, little parallel work has been done yet on the possible impact of arms treaties on jobs.[16]

Once an inventory of skills and equipment and an assessment of the economic impact of arms treaties has been undertaken, the conversion process proper can begin. Alternative products need to be selected and developed, engineering studies on producing them with existing capabilities prepared, machinery and production layouts refashioned where needed, and local and regional marketing studies done to see which products are in sufficient demand.

These tasks are best tackled in a decentralized manner—by those most familiar with local opportunities and obstacles. Alternative use committees,

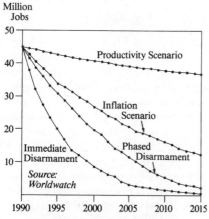

Figure 9-2. Job Losses Under Four Disarmament Scenarios, 1990–2015

composed of representatives from management and workers of a military firm, would be entrusted with preparing a conversion "blueprint," a plan to be implemented once disarmament begins and military contracts are terminated. Because it usually takes about two years to consider properly the alternative employment of people, equipment, and buildings, advance planning is crucial to the success of any such undertaking.[17]

The deeply ingrained work habits of military-industry employees pose a special challenge. Production and clerical workers can be relatively easily transferred to civilian employment, but retraining is crucial for scientists, engineers, and managers. Successful conversion requires a reorientation in management practices, which now are geared to the demands of a single customer and which regard cost considerations as secondary. Military industry engineers tend to be overspecialized, accustomed to choosing unusually costly materials and to product specifications and performance requirements of a complexity unnecessary and unacceptable in the civilian sphere.[18]

The reconversion in many countries after World War II was relatively painless because most factories had previously produced civilian goods. But today's military industries are largely unaccustomed to the nonmilitary market. Decades of serving military needs have let them, in the judgement of Seymour Melman, chair of the National Commission for Economic Conversion and Disarmament, develop a "trained incapacity" for civilian production.[19]

The experience of Boeing-Vertol, a military contracting firm in Philadelphia, illustrates the problems of attempts to enter civilian markets without proper retraining. Trolley cars manufactured for the Massachusetts Bay Transit Authority failed to meet such crucial civilian design criteria as simplicity and durability. They proved so unreliable and required such costly repairs and modifications that most were taken out of service after only a few years. Similar problems have been encountered by the Rohr Corporation and Westinghouse Military Electronics, which manufactured trains and electronic control systems for San Francisco's Bay Area Rapid Transit system, and by Grumman Corporation, which produced motor buses for New York City.[20]

No matter how well planned, conversion is unlikely to be entirely smooth. Some facilities (including ammunitions factories and missile-propulsion plants) are not readily convertible because there are no apparent civilian uses for them. And parts of the complex that designs, produces, assembles, and tests nuclear warheads may simply be too contaminated for alternative uses. These employees will need to find new jobs altogether. A special conversion fund could finance retraining programs and worker and community adjustment plans during a transition phase.[21]

BUILDING A CONVERSION COALITION

Although the political obstacles to conversion are formidable, a coalition for building a peaceful economy could change the equation. Conversion can be an instrument of social transformation and economic renewal, a theme around which various social change movements might coalesce. Dumas and Gordon point out that "the moment one advocates conversion, one must ask: 'Conversion to what?' This initiates a crucial discussion about what should be produced, what goods and services are needed, what technologies should be used in production of these goods, who should decide what to produce and be involved in planning, production, and distribution. Such steps help to empower people who have been taught that they are incapable of planning for their own future, and lead to their direct involvement in the activities and decision-making processes that govern their lives."[22]

Economic conversion is at the center of a number of acute concerns. By freeing up resources and providing a planning mechanism for their alternative use, conversion is an essential component of any attempt to reverse industrial decay and revitalize the civilian economy. But a conversion strategy must encompass more than just a shift away from all things military. An indiscriminate approach may simply worsen existing civilian overcapacities and deepen structural unemployment problems. The severely contracted civilian shipbuilding industry, for example, is most unlikely to offer a viable alternative to producing warships and attack submarines.

Furthermore, economic conversion is very much concerned with the *kind* of civilian economy to emerge. Indeed, for most conversion advocates the goal is the production of socially useful goods to satisfy unmet individual and collective needs in transportation, housing, health, energy, environmental protection, communications, and education. This implies a more differentiated approach than the open-ended pursuit of economic growth.[23]

Achieving more workplace democracy is an important motivation for unions to get involved in conversion initiatives. Plant-based alternative use committees would greatly augment worker involvement in decisions on investment and production, plant closings, automation, and layoffs. Several major trade unions in Western Europe and the United States have explicitly linked conversion planning proposals with ambitious demands for industrial democracy, perceiving conversion as a concept that could very well be applied to civilian industries in upheaval. Alternative use planning offers a chance to examine the direction of technological development, the scale and capital intensity of production technologies, and occupational health and safety.[24]

Economic conversion is very much concerned with the *kind* of civilian economy to emerge.

By releasing resources now absorbed by the military, conversion can make an important contribution to the social and economic regeneration of urban areas, particularly benefiting disadvantaged inner-city residents. In the United States, the ability of cities to address mounting social problems suffered a tremendous decline in the eighties as federal grant programs were cut sharply. A recent study for the U.S. Conference of Mayors examined the effects of transferring $30 billion annually over five years from the military to key urban programs, including housing, mass transit, education, health, and community services. Such a transfer would not only give these areas a tremendous boost, it would increase the gross national product (GNP), personal disposable income, fixed private investment, and number of jobs created.[25]

In an age of profound ecological crisis, the question 'conversion to what?' cannot avoid addressing the neglected relationship between economy and ecology. Producing more cars instead of tanks, for example, would simply exacerbate the environmental crisis. The environmental movement could considerably strengthen the alternative use concept by developing guidelines for conversion to a more sustainable economic system founded on environmentally sound products and production technologies.

The environmental community needs, in turn, its own conversion policy. Conversion could release desperately needed funds to clean up the planet. More important, the features embedded in the conversion concept can help smooth the transition to a sustainable society by addressing misunderstandings about tension between the goals of full employment and a healthy environment.

An ecologically inspired restructuring of the economy would involve moving away from activities that contribute most to global warming, ozone depletion, and other threats to human health and the environment. A sustainable economy relies more on renewable energy, emphasizes conservation and efficiency, minimizes waste and hazardous materials generation, maximizes recycling, and is generally oriented toward qualitative economic development. (See Chapter 10.) Oil and coal producers, auto manufacturers, and branches of the chemical industry (such as the plastic packaging industry) would unquestionably be among those most affected by such a transformation. The environmental movement has by and large given insufficient attention to the economic implications of sustainability for these workers.

The car industry, for instance, has long formed a cornerstone of economic development in industrial countries. Shifting to transportation that contributes less to urban smog, acid rain, and global warming would mean relying less on automobiles, affecting a large number of jobs. In the United States, the roughly 750,000 workers in the motor vehicle industry are matched by an equal number in road construction and maintenance. And more than 3 million people work in automotive sales and servicing. Altogether, this sector makes up approximately 5 percent of total U.S. employment, but that share is easily doubled if the secondary industries are included. Moving toward an ecologically sustainable society need not entail severe economic penalties if current employment is replaced with jobs in manufacturing, operating, and servicing mass transit systems.[26]

Unfortunately, little work has been done to assess either the impact of such fundamental economic shifts or the potential of "environmental" work opportunities. Currently, most environmental protection expenditures—and thus the jobs created—go to mitigate or repair ecological damage. Pollution control generates an estimated 1.4 million jobs in the United States and some 1.25 million in the European Community. There is added potential for alternative employment, however, if the focus shifts from "tail-pipe" technologies and remedial actions to developing new, environmentally more benign products and industrial processes. A key variable is capital-intensity: the number of jobs generated by each dollar spent. A sustainable society must wrestle with this difficult question if it is to provide sufficient employment.[27]

The promise of building a broad coalition around the conversion issue is matched by the difficulty of making it a reality. Conversion campaigns span a broad spectrum of efforts, ranging from specific local or regional issues to the need for national legislation, from advocates of conversion as a tool for social change to those favoring narrower approaches. Accordingly, they are often initiated by quite diverse groups with different, sometimes conflicting motivations, philosophies, strategies, and goals.

On occasion, this leads to friction. For example, the preoccupation of large segments of the peace movement with the moral dimensions of arms production—leading to demands for the closure of military factories, not conversion—has alienated parts of the labor movement. Plant-based conversion initiatives, on the other hand, are sometimes narrowly concerned with job security while eschewing broader social goals. These are often ad hoc campaigns, springing to life in response to the threat of layoffs, and sometimes ending abruptly when a new military contract is signed. Frequently, conversion is an option of last resort, considered only once the fight to cling to military production has been lost.

Weaving the fabric of a broad conversion coalition remains a challenging task. But the threads are gradually coming together. The conversion debate reached a new threshold as the military expansion of the eighties ended and as the needs of social and economic revitalization and environmental restoration became ever more imperative.

The Path Forged by China and the Soviet Union

Since the end of World War II, little systematic experience of planned conversion has been gained. Only one major power—the People's Republic of China—has undertaken a large-scale demobilization and conversion effort. Beijing's decision to rein in military spending and streamline the armed forces reflects both an evolution in security thinking, in particular the improved state of Sino-Soviet relations, and the fact that the economic dimensions of security are being taken more seriously. The military sector has been assigned the lowest priority of the "four moderni-

zations," behind agriculture, industry, and science and technology.[28]

Chinese armed forces have been reduced since 1981 by about 1.2 million, or almost one third, although it is unclear just how much advance planning has been done to keep economic dislocation to a minimum. After accounting for inflation, military spending has been cut by one sixth since 1979. Because the economy has at the same time grown quite remarkably, the military budget now absorbs only about 4 percent of GNP, down from an average of 13 percent during the seventies.[29]

By the late seventies, China faced a growing surplus of military production capacity. Resources devoted to a bloated but increasingly obsolete military sector drained the civilian economy, which suffered from a marked reduction in investment. While military factories were underutilized, the government was hard-pressed to satisfy people's growing thirst for consumer goods. So the share of military factories' output devoted to the civilian market was doubled between 1978 and 1983. By 1986, that share was at least 38 percent and possibly as high as 50. China's military industry now manufactures nearly 400 different consumer items, from bicycles and cars to television sets and washing machines.[30]

Although China's undertaking provides an interesting experiment, it does have its drawbacks. Rather than straight conversion away from military production, the aim has been dual-use capacity, with factories able to churn out war matériel when needed. Managers have been given greater responsibility for their operations, but little suggests any worker and community participation of significance in this endeavor. Instead, China's conversion effort appears to be a centrally directed, top-down one that could be reversed again if the leaders' priorities shift. It is too early to tell whether this will in fact occur in the wake of the violent crackdown on the prodemocracy movement.[31]

Military factories have also pursued a second, quite lucrative sideline business that may actually derail China's conversion. Dramatically increased arms exports account for an estimated 10 percent of military production and 6 percent of the country's total exports. From an average of 2 percent in the seventies, China has recently increased its share of the global arms export market to about 4 percent.[32]

Gorbachev's announcement of major unilateral troop and tank cuts in December 1988 further nourishes the need for conversion.

The Soviet Union, too, is beginning to tackle conversion in an effort to lighten the military burden on its economy and to have President Gorbachev's new foreign and military policies yield some economic dividends. The Intermediate-Range Nuclear Forces (INF) Treaty signed with the United States brought a net savings of 300 million rubles ($468 million, at mid-October 1989 official rate). This is small, but enough to build 30,000–40,000 apartments, for example, thus alleviating an acute housing shortage. Three factories that used to produce missiles now proscribed by the INF accord have shifted at least part of their capacity to washing machines and bicycles, and several missile design laboratories have been reoriented toward civilian work. In addition, in a venture with a U.S. company, the SS-20 missile factory will be converted to production of commercial satellite-launching rockets. President Gorbachev's announcement of major unilateral troop and tank cuts in December 1988 further nourishes the need for conversion.[33]

The stage is now set for an ambitious

Soviet conversion program, with some 200 factories slated for changeover. Overall arms production is to be slashed by one fifth and tank production even more—by 40 percent by the end of 1990. Military spending, frozen in 1987–88, is to be cut by 15 percent or 10 billion rubles ($15.6 billion) in 1990–91. Channeling these savings into civilian areas could work wonders: it would allow a 30-percent increase in housing construction or, if invested in environmental protection, increase current outlays more than fivefold. Over the next six years, Prime Minister Ryzhkov is hoping to trim defense expenditures further, by a third to half. Yet most of the expected savings would go toward balancing the Soviet budget.[34]

As in China, the Soviet Union is now trying to increase the share of civilian goods produced by military plants— from 40 percent currently to 50 percent in 1991 and around 60 percent by 1995. Their experience to date in this has not always gone smoothly: earlier attempts to convert some military facilities turned out washing machines that cost twice as much as ones of the same design produced under a civilian ministry. Still, nearly half the planned new equipment deliveries to the food industry under the investment plan for 1988–95 are to be provided by defense industry enterprises. The military has also been directed to produce 7–8 billion rubles ($10.9–12.5 billion) worth of goods for light industry, as well as to increase output of construction materials, medical equipment, and plumbing supplies.[35]

Resistance to conversion among military industry managers, who fear a loss of privilege and power, continues to be an obstacle. In an effort to increase the incentive to produce more civilian goods, the Council of Ministers decreed in September 1988 that defense factories could retain profits from above-plan production of consumer goods during 1989 and 1990.[36]

Previous Soviet conversion attempts also proved relatively unsuccessful because the Kremlin was much less willing then to pare back military demands on economic resources. Although current conversion plans are directed from the top, there is a growing public discussion and appreciation of the concept, and Soviet conversion scholars are seeking to share insights with their counterparts abroad. To go beyond ad hoc conversion, the government has begun to organize special working groups within the State Planning Commission and various military-related ministries to design a proper program. One particular obstacle is that even government agencies have little relevant data on hand—an ironic shortcoming in a centrally planned economy.[37]

The Chinese example, and probably soon the Soviet one too, demonstrates that conversion can succeed, even in a world that has barely begun to consider alternatives to the arms race seriously. But these two nations cannot be considered models for the West. Even though they are becoming less centralized and more democratic, their economies—indeed their societies—remain of a very different nature. For one, no private contractors trying to make a profit stand in the way of conversion. A strong government role virtually assures that conversion will be implemented once leaders are committed. By the same token, however, the lack of public accountability provides little guarantee that any change of course will be maintained.

GRASSROOTS INITIATIVES IN THE WEST

No western government has so far espoused an active conversion policy, but grassroots interest in the concept is growing. The discussions are propelled by a host of factors, among them the new East-West détente, the imminent threat of layoffs in the arms industries, the ethical implications of arms exports, the need to revitalize stagnant civilian economies, and the growing desire for socially useful and environmentally sound production.

Conversion is gaining new relevance as the United States adjusts to the priorities of a post–cold war world. The signs for the U.S. military industry are unmistakable: as defense spending has begun to decline in real terms, some 54,000 jobs have either been lost in the last couple of years or are currently imperiled. Congressional approval to shut down some 86 obsolete military bases by 1995 will eliminate another 21,000 jobs. Recent Defense Department proposals to cut spending, if implemented, will undoubtedly translate into even greater job loss. And if major arms control and disarmament treaties are concluded, as now seems a strong possibility, layoffs will hit the military industry like an avalanche.[38]

Opponents of conversion legislation argue that no new effort is needed because an Office of Economic Adjustment (OEA) has existed in the Department of Defense since 1961. Created to deal with the economic effects of military base closures, the OEA has no mandate to look at military factory shutdowns. Moreover, it has reacted to closures on a case-by-case basis rather than engaging in the comprehensive advance planning needed to deal with the economic consequences of disarmament. In fact, it opposes genuine conversion planning. Rather than weaning communities from dependence on military spending, during the eighties the OEA actually encouraged them to seek Pentagon procurement dollars.[39]

Conversion and diversification legislation has either passed or is pending in Colorado, Connecticut, Massachusetts, Minnesota, New Mexico, New York,

Pennsylvania, and Washington. In recent years, city councils and county supervisors have also become actively involved in the conversion issue. Regional and local studies on the economic impact of military spending have been prepared by official and private conversion task forces.[40]

For the popular conversion movement, the Center for Economic Conversion in California's Silicon Valley serves as a clearinghouse for U.S. initiatives and offers organizing assistance to local campaigns. In Washington, D.C., the National Commission for Economic Conversion and Disarmament (a private advocacy group with members of Congress, local elected officials, trade union leaders, and economists) was set up in 1988 to broaden understanding of political and economic alternatives to the arms race and to urge passage of comprehensive national conversion legislation.[41]

Conversion is gaining support among unions because the handwriting is on the wall: employment in arms industries is set to decline.

During the eighties, U.S. labor unions became increasingly active in the conversion debate, reawakening an interest from the late sixties and early seventies. In particular, the International Association of Machinists has a long record of active support for the concept. Conversion is gaining support among unions because the handwriting is on the wall: employment in arms industries is set to decline, perhaps drastically. And conversion is now being discussed in the context of civilian layoffs and plant closings.[42]

The Jobs With Peace campaign, founded in 1981, has been involved in local conversion efforts, but it has also approached the issue from a different vantage point: it seeks to organize and vocalize the discontent of people, mostly in the inner city, who fell victim to social program cuts that went hand in hand with the military buildup of the eighties. The campaign's major tool has been the use of local referenda in some 86 cities and towns (a number of which have been won) calling for an assessment of the military's impact on the local economy and for a transfer of funds to housing, health, education, and other social needs.[43]

The western nation with the second largest share of resources devoted to the military sector is the United Kingdom. Since the mid-eighties, U.K. military spending has been in decline, setting off a "contraction in the British defense industry on a scale unknown since the end of the Second World War," according to John Lovering of the University of Bristol. The squeeze on jobs is likely to intensify as the government moves ahead with costly nuclear programs that represent a large and growing share of its military budget but that provide relatively few jobs.[44]

The crisis of employment in the British defense industry has rekindled a debate first sparked by the Lucas Aerospace campaign well over a decade ago. Faced with a wave of layoffs in the mid-seventies, shop stewards at this major British military contractor formulated a detailed plan involving 150 goods that could be produced with existing skills and resources, such as medical equipment, alternative energy technologies, and remote control equipment for the handicapped. Although management ultimately rejected the plan, the publicity helped prevent most of the planned layoffs. More important, the Lucas campaign was a tremendous inspiration for many others in Britain and elsewhere.[45]

The drawn-out controversy at Lucas

Aerospace raised awareness and stimulated a reassessment of conversion-related issues within unions and the Labour Party. In 1983, the Trade and General Workers' Union (the largest single union) committed itself to supporting conversion planning; its widely quoted *A Better Future for Defence Jobs* proposed local alternative use committees and a national conversion fund. Similar proposals were made by the civil servants union. And the Trades Union Congress, the umbrella group for British unions, called for a large-scale public investment program to create a market for alternative products.[46]

In September 1984, a National Trades Union Defence Conversion Committee was set up. Meanwhile, committees involving a number of unions have sprung up at the regional level, most prominently in the northeast. The Labour Party National Executive Committee formally declared its support for conversion in 1986. And in a 1987 report Labour promised, if elected, to create a National Agency for Industrial Conversion and Recovery within the Department of Trade and Industry to administer a conversion fund, though they failed to be specific on how the program would work.[47]

According to Lovering, this "upswell of interest in conversion was of a 'top-down' nature, and it failed to feed through to substantial local activity." Labour's defeat in the 1987 general elections quelled hopes for a more constructive government stance toward conversion. Instead, confronting an intensifying employment crisis in military industry, new local, workplace-based campaigns such as those in Barrow, Llangenerch, and Bristol began to emerge. (See Table 9-2.)[48]

Meanwhile, an institutional structure for a conversion network is beginning to emerge in Britain. Conversion campaigns found support in many municipalities' struggles to formulate local strategies for economic development, although these initiatives face an uphill struggle, as the financial and legal powers of municipal governments were severely limited during the eighties. In 1985, an Arms Conversion Group was established at the School of Peace Studies at Bradford University to serve as a clearinghouse. And a number of small, independent agencies—including the Centre for Alternative Industrial and Technological Systems at the Polytechnic of North London and the Network of Product Development Agencies in Sheffield—offer practical assistance to local campaigns.[49]

For West Germany, adjusting the armaments industry to nonmilitary use should be easier than in the United States or the United Kingdom, for a number of reasons. First, the country devotes a considerably smaller share of financial and scientific resources to defense. Second, its vibrant civilian economy provides ample opportunities for alternative use of resources. And finally, the military industry is less specialized, and thus more integrated into the overall economy, than its U.S. or British counterparts are. For many West German companies involved in the military sector, arms sales are less significant than they are for similar firms in other countries.[50]

Still, the military industry faces tremendous overcapacities as the boom of the seventies—when contracts doubled in value—has given way to stagnation. Overcapacities are estimated at 30 percent in both the aerospace industry and the shipyards, and at 50 percent in tank production. (By contrast, the electronics industry expects continued growth in orders.) To escape the unpleasant dilemma of either lobbying for additional domestic and foreign military contracts or suffering possibly large-scale layoffs, alternative use working groups have

Table 9-2. United Kingdom: Recent Conversion Initiatives

Initiative, Location, and Year Started	Events and Reports
Vickers Shipyard, Barrow Alternative Employment Committee, 1984	1987 report "Oceans of Work" assesses alternatives to Trident submarine production in marine technology and renewable energy. Barrow City Council now involved in diversification efforts.
Greater London Conversion Council, London, 1985–86	Series of reports on impact of defense spending on London economy; sector studies of electronics and aerospace industries.
Royal Naval Stores Depot, Llangenerch, South Wales, 1986	Alternative product campaign "Alterplan" to save hundreds of jobs threatened by closure in 1990. Conversion to network of small businesses proposed. Some success in diversification efforts.
British Aerospace (BAe), Bristol, 1989	Study of economic impact of layoffs announced for 1990 and of alternative product potential. Bristol City Council set up a Local Defense Forum to pursue diversification efforts.
BAe, Warton, Northwest, 1989	Tornado project. Alternative use study proposed by shop stewards. Coordination with Tornado workers in other countries.

SOURCE: Worldwatch Institute, based on John Lovering, "Arms Conversion into the 1990s: The British Experience," paper presented at conversion conference in Cortona, Italy, April 22, 1989.

sprung up in numerous military contracting firms. By the mid-eighties, a dozen existed in the northern part of the country alone. Such initiatives became a driving force in conversion discussions. Many of their proposals focus on the need for production that is environmentally more benign.[51]

The oldest and most established West German alternative use working groups are found along the coast, which is not surprising as the region has a number of cities heavily dependent on arms production. In Kiel, one in every five industrial workers is employed in the military industry; in Emden, the figure is one in

six. Half the revenues of Kiel's shipyard and machine tool industries in 1984 came from military contracts. In Hamburg, one out of four shipyard workers and one tenth of those in the electronics industry are involved in military work. And in Bremen, one quarter of the workers in the electronics industry produce armaments.[52]

During the military industry's boom years of the early seventies, employment rose by 50 percent; job maintenance and alternative production were hardly topics on the union agenda. But since the beginning of the eighties, the discussion has intensified on all levels of labor, and conversion has been officially embraced. Indeed, confronted with continued mass unemployment, unions no longer see conversion as solely related to the military but seek to apply the concept to civilian sectors as well. Specifically, conversion is seen as a tool to put teeth into West Germany's "co-determination" laws, which rule worker participation in corporate decision making. In addition, it has been linked to ongoing demands for "humanization of work"—the improvement of working conditions.[53]

The metal workers' union (IG Metall) has been the most actively involved, calling for alternative use committees with equal representation by labor and management. By 1977, it had already formed a working group on Defense Technology and Employment to advance proposals for alternative production. The IG Metall has facilitated the coordination of various plant-based conversion initiatives, helped integrate them into the union organization, and broadened the conversion discussion within the union movement. In particular, its Innovation and Technology Advisory Bureau in Hamburg and the Cooperation Bureau in Bremen have lent important support to workers' alternative use committees.[54]

Among western countries, perhaps the government most sympathetic to conversion has been Sweden's. It is the only one that responded constructively to a 1982 U.N. General Assembly suggestion that member states do a detailed study of their military industries. The result was the landmark report *In Pursuit of Disarmament*, released in 1984. Known as the Thorsson report (Under-Secretary of State Inga Thorsson chaired the group that prepared the study), it analyzed the size, structure, and economic impact of the Swedish military industry. It found conditions quite favorable to conversion: only a few Swedish

firms depend on military orders for the majority of their production, and few communities depend heavily on revenues from military work.[55]

In Pursuit of Disarmament assessed the effect on the Swedish defense industry of halving military expenditures between 1990 and 2015. Even though 20,000 soldiers would be demobilized and some 14,000 military industry jobs would be lost, less than 1 percent of the Swedish labor force would be affected. The Thorsson report called for a Council for Disarmament and Conversion and a national conversion fund to be financed by a 5-percent levy on arms export sales. It further envisioned conversion funds at each defense facility, paid for by setting aside 1 percent of the value of domestic military contracts; to become eligible for such funds, contractors would have to match them with their own capital.[56]

The Thorsson report envisioned conversion funds at each defense facility, paid for by setting aside 1 percent of the value of domestic military contracts.

These recommendations remain proposals. A follow-up report, with some modifications to the original suggestions, was published in spring 1988. Despite continued interest in the Foreign Ministry, among some military firms, and within the trade union movement, the government eschews an active conversion policy. Yet even without it, the case for alternative use planning remains compelling. The Stockholm business magazine *Veckans Affärer* reports that Swedish arms manufacturers are struggling with plunging sales; over the next five years, some 4,000–5,000 jobs could be lost. Indeed, the prospects of lower defense budgets and a gradually shrinking military force ensure that conversion will increase in political importance.[57]

The litmus test for grassroots conversion initiatives in the West is frequently seen as preventing factory shutdowns and the associated job loss, and gaining management acceptance of alternative product proposals. Because local campaigns face an uphill struggle against defense departments and the management of military contractors, such expectations are unrealistic. A different yardstick should be applied. Indeed, conversion activists are conscious of the limits of defining success in this way. They seek primarily to generate greater awareness among both military industry employees and the general public and thus create a political opening for the conversion concept. Alternative use campaigns often (but not always) intend to dramatize the need for a public policy that establishes the conditions and framework for building a peaceful economy.

AN ALTERNATIVE AGENDA

Local conversion campaigns help raise awareness of the adverse impact of military spending. Ultimately, however, their success depends on the passage of comprehensive national legislation that provides a mandatory framework for the transfer of resources from military to civilian applications.

Currently, no such legislation is in force anywhere in the world. In China, conversion has been carried out by administrative decree rather than by public deliberation and legally codified measures. Legislative proposals have been made in a number of countries, however. In the Netherlands, a parliamentary effort in 1983 failed in the end due to lack of government enthusiasm. Conversion legislation has been introduced in On-

tario, Canada, and is awaiting further action. And in Italy, an arms export bill currently under consideration includes some relevant provisions. Pressure for more straightforward conversion legislation is mounting, as four bills have been introduced by Italian opposition parties.[58]

Three bills introduced in the U.S. Congress address the conversion issue. Only one—the Defense Economic Adjustment Act (H.R. 101) introduced by Representative Ted Weiss of New York—incorporates the features discussed in this chapter as crucial to the success of a conversion strategy. First offered in 1977, the Weiss bill is essentially a refined version of the National Economic Conversion Act introduced as early as 1963 by Senator George McGovern.[59]

The Weiss bill has become somewhat of a model for conversion proponents in other countries. It calls for a Defense Economic Adjustment Council charged, among other things, with preparing general guidelines for conversion and disseminating conversion-related information. The bill also requires every military base, laboratory, and production facility with more than 100 employees to establish an Alternative Use Committee as a prerequisite for future eligibility for military contracts. Management and labor would be evenly represented, and nonvoting representatives of the local community would be included. The committees would develop, and review every two years, a conversion blueprint. The Weiss bill facilitates conversion planning by requiring the Secretary of Defense to provide one year's advance notice of any changes in procurement contracts that lead to layoffs.[60]

To smooth adjustment, occupational retraining of military industry employees would be required. Workers displaced as a result of defense cutbacks would be eligible for temporary income maintenance and relocation allowances. A national employment network would be created to help them find new jobs. The legislation also specifies that communities "substantially or seriously affected" by reduced defense spending "shall be eligible for Federal assistance for planning for economic adjustment to avoid substantial dislocations."[61]

An "economic adjustment fund" would be created to finance these programs, funded by 10 percent of any projected savings in military spending. In addition, the Weiss bill would make the receipt of defense contracts contingent upon a contractor's payment into the fund of an amount equal to 1.25 percent per year of the company's gross revenue on its military sales.[62]

The success of any conversion program hinges on broader economic conditions. It is unlikely under the laissez-faire approach to macroeconomic policy that dominated in many western countries during the eighties, especially if conversion is to be geared toward socially useful and environmentally sound production and consumption. This prompts a question about the proper role of public policy: To what degree should government interfere in the market to bring about the rather fundamental redirection of priorities needed?

The answer, of course, varies from country to country. But the most effective approach might be for public policy to create the overall framework and provide the incentives that would stimulate demand for socially useful products for which there is no, or little, effective market demand. The centerpiece of such a policy would be an alternative research and capital investment agenda to create at least initial market demand. The program might include measures to support the development of nonpolluting, appropriate-scale production technologies, enhance renewable sources of energy, boost energy efficiency and

reforestation, strengthen public transportation, provide affordable housing and preventive health care, and improve educational services.

The Weiss bill has become somewhat of a model for conversion proponents in other countries.

In the United States, the Weiss bill addresses this need. It encourages the "preparation of concrete plans for . . . public projects addressing vital areas of national concern (such as transportation, housing, education, health care, environmental protection, and renewable energy resources) by the various civilian agencies of the Government, as well as by State and local governments."[63]

Too centralized an approach is objectionable not only because of the inherent danger of bureaucratization, but also because a decentralized effort is much more sensitive to the strengths and weaknesses of individual regions and communities. Enormous regional and local imbalances are created or exacerbated by military spending priorities, for example. To be effective, a conversion policy needs to be tailored to the specific needs of individual regions and to invest local governments with sufficient authority in the entire endeavor.

International coordination nevertheless has a role to play. The United Nations could become a forum where each nation's experiences with alternative use projects are made available. On the grassroots level, meanwhile, contacts among activists are growing. At Sweden's Uppsala University, a Center for Military Conversion is being set up to collect and distribute information on European activities and to establish an international conversion network.[64]

The lack of an active government conversion program in the West would seem to indicate that this concept has little hope of being translated into reality. At the moment, what prospects there are for such a policy appear brighter in Europe than in the United States. Nations there have a stronger tradition of public intervention in economic policy, and the labor movement is more influential. A realignment in the political geography—as appears in the making now in much of Western Europe, where Green movements are on the rise—provides hope that an ecologically inspired alternative use strategy will get a more serious hearing.

The gathering pressure for disarmament suggests strongly that conversion will be a topic of growing importance during the nineties. As the debate about the essence of security is broadened to embrace economic vitality, social justice, and ecological stability, there is a growing desire for ways to redirect resources to these areas of neglect. The challenge is to channel this desire into concrete action.

10

Picturing a Sustainable Society

Lester R. Brown, Christopher Flavin, and Sandra Postel

Societies everywhere are slowly coming to recognize that they are not only destroying their environments but undermining their futures. In response, governments, development agencies, and people the world over have begun to try to reverse obviously threatening trends. So far, this has resulted in a flurry of fragmented activity—a new pollution law here, a larger environment staff there—that lacks any coherent sense of what, ultimately, we wish to achieve.

Building a more environmentally stable future clearly requires some vision of it. If not fossil fuels to power society, then what? If forests are no longer to be cleared to grow food, then how is a larger population to be fed? If a throwaway culture leads inevitably to pollution and resource depletion, how can we satisfy our material needs? In sum, if the present path is so obviously unsound,

what picture of the future can we use to guide our actions toward a global community that can endure?

That, in essence, is the challenge taken up in this chapter—to draw the outlines of a sustainable society, to describe what it would look like, how it would function. It can, of course, only be a thumbnail sketch. Ideas and technologies yet unknown will fill in many of the gaps. But just as any technology of flight, however primitive or advanced, must abide by the basic principles of aerodynamics, so must a lasting society satisfy some immutable criteria. With that understanding and from accumulated experience to date, it is possible to create a vision of a society quite different, indeed preferable to today's.

A sustainable society is one that satisfies its needs without jeopardizing the prospects of future generations. Inherent in this definition is the responsibility of each generation to ensure that the next one inherits an undiminished natural and economic endowment. This concept of intergenerational equity, profoundly moral in character, is violated in numerous ways by our current society.[1]

Indeed, there are no existing models of sustainability. For the past several decades, most developing nations have aspired to the automobile-centered, fossil-fuel-driven economies of the industrial West. But from the localized problems of intractable air pollution to the global threat of climate change, it is now clear that these societies are far from durable; indeed they are rapidly bringing about their own demise.

Efforts to understand sustainability often focus on what it is not. Obviously, an economy that is rapidly changing the climate on which its food-producing capability depends is not sustainable. Neither is one that overcuts the forests that provide its fuel and timber. But this negative definition leads to a strictly reactive posture, one that has us constantly trying to repair the consequences of our destructive behavior.

The world economy of 2030 will not be powered by coal, oil, and natural gas.

The World Bank, for example, now tries piecemeal to assess the environmental side effects of projects it is considering funding. Not one of its member countries, however, has a coherent plan of action aimed at achieving sustainability, which logically should provide the basis for deciding what investments are needed in the first place. The United States has followed a similar tack for the last 20 years. Its National Environmental Policy Act requires that the environmen-

tal impacts of major proposed government actions be assessed. But this, too, is a defensive approach, one that attempts only to avert unwanted effects rather than working positively and consistently toward a sustainable economy.

In taking on the task of sketching an environmentally stable society, we have made several important assumptions. The first is that if the world is to achieve sustainability, it will need to do so within the next 40 years. If we have not succeeded by then, environmental deterioration and economic decline are likely to be feeding on each other, pulling us into a downward spiral of social disintegration. Our vision of the future therefore looks to the year 2030.

Second, new technologies will of course be developed. Forty years ago, for example, some renewable energy technologies now on the market did not even exist. Under the pressure of finding a means to slow global warming, researchers are likely to develop a range of new energy technologies, some of which may be difficult to imagine at the moment. In the interest of being conservative, however, the future we sketch here is based only on existing technologies and foreseeable improvements in them.

Our third assumption is that the world economy of 2030 will not be powered by coal, oil, and natural gas. It is now well accepted that continuing heavy reliance on fossil fuels will cause catastrophic changes in climate. The most recent scientific evidence suggests that stabilizing the climate depends on eventually cutting annual global carbon emissions to some 2 billion tons per year, about one third the current level. Taking population growth into account, the world in 2030 will therefore have per capita carbon emissions that are one eighth the level in Western Europe today.[2]

The choice then becomes whether to make solar or nuclear power the centerpiece of energy systems. We believe societies will reject nuclear power because of its long list of economic, social, and environmental liabilities. Though favored by many political leaders during the sixties and seventies, the nuclear industry has been in decline for over a decade. Only 94 plants remain under construction, and most will be completed in the next few years. Meanwhile, worldwide orders for new plants have slowed to a trickle. The accidents at Three Mile Island and Chernobyl and the failure to develop a safe way to store nuclear waste permanently have turned governments and citizens alike away from nuclear power.[3]

It is of course possible that scientists could develop new nuclear technologies that are more economical and less accident-prone. Yet this would not solve the waste problem. Nor would it alleviate growing concern about the use of nuclear energy as a stepping stone to developing nuclear weapons. Trying to prevent this in a plutonium-based economy with thousands of operating plants would require a degree of control that is probably incompatible with democratic political systems. Societies are likely to opt instead for diverse, solar-based systems.[4]

The fourth major assumption is about population size. Current U.N. projections have the world headed for nearly 9 billion people by 2030. This figure implies a doubling or tripling of the populations of Ethiopia, India, Nigeria, and scores of other countries where human numbers are already overtaxing natural support systems. Either these societies will move quickly to encourage smaller families and bring birth rates down, or rising death rates from hunger and malnutrition will check population growth.[5]

The humane path to sustainability by the year 2030 therefore requires a dramatic drop in birth rates. More countries will do as China has done, and as Thailand is doing: cut their population growth rates in half in a matter of years. As of 1990, 13 European countries have stable or declining populations; by 2030, most countries are likely to be in that category. For the world as a whole, human numbers will total well below 9 billion. We assume a population of at most 8 billion that will either be essentially stable or declining slowly—toward a number the earth can support comfortably and indefinitely.[6]

The last assumption we make is that the world in 2030 will have achieved a more equitable and secure economy. Unless Third World debt can be reduced to the point where the net flow of capital from industrial to developing countries is restored, the resources and incentives to invest in sustainability will simply be inadequate. In the final section, we briefly discuss issues related to jobs, economic growth, and social priorities, recognizing that major changes in other areas will be needed as well.

In the end, individual values are what drive social change. Progress toward sustainability thus hinges on a collective deepening of our sense of responsibility to the earth and to future generations. Without a re-evaluation of our personal aspirations and motivations, we will never achieve an environmentally sound global community.

POWERED BY THE SUN

During the seventies and eighties, policymakers assumed that changes in the world energy system would be driven by depletion of the world's fossil fuel resources: as we gradually ran out of oil, coal, and natural gas, we would be forced to develop alternatives. Such a transition would have been comfortably gradual, extending over more than a century. But now the world faces a new set of limits. Long before fossil fuels are exhausted, rising global temperatures from their use could spell an end to civilization as we know it.

The world energy system in the year 2030 is thus likely to bear little resemblance to today's. No longer dominated by fossil fuels, it will be run by solar resources daily replenished by incoming sunlight and by geothermal energy. And it will be far more energy-efficient.

In many ways, the solar age today is where the coal age was when the steam engine was invented in the eighteenth century. At that time, coal was used to heat homes and smelt iron ore, but the notion of using coal-fired steam engines to power factories or transportation systems was just emerging. Yet only a short time later the first railroad started running, and fossil fuels began to transform the world economy.

Direct conversion of solar energy will be the cornerstone of a sustainable world energy system.

The late twentieth century, then, is the dawn of the solar age. Many technologies have been developed that allow us to harness the energy of the sun effectively, but these devices are not yet in widespread use, and their potential impact is barely imagined. When it comes to solar technologies, today's political leaders, still captivated by coal and nuclear power, are akin to the steam engine's eighteenth century skeptics.

Countering such skepticism are glimmerings of the new age. Some nations—Norway and Brazil, for example—already obtain over half their energy from renewables. And these resources are available in immense quantity. The U.S. Department of Energy estimates that the annual influx of currently accessible re-

newable resources in the United States is 250 times the country's annual use of energy.[7]

Solar power is by nature diverse; the mix of energy sources will reflect the climate and natural resources of each particular region. Northern Europe, for example, is likely to rely heavily on wind and hydropower. The economies of northern Africa and the Middle East may instead use direct sunlight. Japan and the Philippines will tap their abundant geothermal energy. And Southeast Asian economies will be powered largely by wood and agricultural wastes, along with sunshine.

Although some countries are likely to import renewable energy, the enormous oil-related bills that characterize modern trade relationships will dwindle. And renewable energy sources are to a large extent inflation-proof: solar, wind, and geothermal power plants require no fuel and so are not vulnerable to fuel price increases.

Due to the abundance of sunlight, direct conversion of solar energy will be the cornerstone of a sustainable world energy system. Not only is sunshine available in great quantity, but it is more widely distributed than any other energy source, renewable or fossil fuel. By 2030, solar panels will heat most residential water around the world, following the model of Japan and Israel, which already use them extensively. A typical urban landscape will have thousands of collectors sprouting from rooftops, much as television antennas do today. And passive solar architecture may by then cut artificial heating and cooling needs to virtually zero in millions of buildings.[8]

Solar thermal power is one means of harnessing sunlight. It uses mirrored troughs to focus sunlight onto oil-filled tubes that convey heat to a turbine and generator that then produce electricity. An 80-megawatt solar thermal plant built in the desert east of Los Angeles in 1989 converts an extraordinary 22 percent of the incoming sunlight into electricity. It does so at a cost of 8¢ per kilowatt-hour—a third less than the 12¢ per kilowatt-hour cost of power from new nuclear plants.[9]

Forty years from now, solar thermal plants may stretch across the deserts of the United States, North Africa, and central Asia. As the technology becomes widespread, these regions may become large exporters of electricity. They may also become suppliers of hydrogen fuel, which can be manufactured in desert solar plants and shipped by pipeline to run automobiles in distant cities.

Photovoltaic solar cells are a semiconductor technology that converts sunlight directly into electricity without using the mechanical processes involved in solar thermal conversion. Currently, photovoltaic systems are less efficient and four times as expensive as solar thermal power is. But they are already used in many remote locations, and the cost is likely to decline rapidly. Systems with efficiencies approaching 20 percent may soon be available, and scientists hope to bring costs down to less than 10¢ per kilowatt-hour by the end of this decade.[10]

The key advantage of photovoltaics is their versatility. They can be used not only in large electricity plants but to power small water pumps and rural communications systems. As they become economical, the completion of the solar revolution will be possible: all Third World villages can be electrified with this technology. Unlike communities today, these villages will not have to depend on extended power lines connected to centralized plants. Rather, they will contain their own power sources—photovoltaic electric systems with battery storage for lighting and other uses in the dark evening hours.[11]

Using this technology, homeowners throughout the world may become producers as well as consumers of electricity. Indeed, photovoltaic shingles have already been developed that allow roofing material itself to become a power source. As costs continue to decline, many homes are apt to get much of their electricity from photovoltaics, and in sunny regions residents will sell any surplus to the utility company for use by others.[12]

Wind power is an indirect form of solar energy, generated by the sun's differential heating of the earth's atmosphere. The cost of wind energy has already fallen by about 70 percent during the eighties, to 6–8¢ per kilowatt-hour, making it close to competitive with new coal-fired power plants.[13]

Engineers are confident that they will soon have improved wind turbines that are economical not just in California's blustery mountain passes, where some wind farmers are already generating $30,000 worth of electricity per hectare annually, but in vast stretches of the U.S. northern plains, the U.K. North Sea Coast, and central Europe. Among the regions where major wind farms are now on the drawing boards are West Germany and the state of Gujarat in India.[14]

The United States could be deriving 10–20 percent of its electricity from the wind by 2030. Wind power is likely then to rival hydro as a low-cost form of energy, and so attract new industries into windswept regions. These businesses will be set up to take advantage of the wind whenever it is available, such as in the middle of the night, when electricity demand is normally quite low.[15]

The wind resources of the U.S. Great Plains—which have long pumped water for millions of cattle—may one day be used to generate vast amounts of electricity for sale to Denver, Kansas City, and other major cities. On the better sites, income from electricity sales could dwarf the $30 or so per hectare from cattle grazing, tempting many cattle ranchers to become wind farmers as well.[16]

For hydropower, which now supplies nearly one fifth of the world's electricity, prospects for future growth are most promising in the Third World, where the undeveloped potential is still large. Small-scale projects are likely to be more appealing than the massive projects favored by governments and international lending agencies in past decades. In deciding which hydropower resources to develop, environmental issues such as land flooding and siltation will play an important role. These considerations will likely keep most nations from developing their total potential, though hydro will still be an important energy source.[17]

One promising approach is to grow energy crops on marginal lands not currently used for food.

Living green plants provide another means of capturing solar energy. Through photosynthesis, they convert sunlight into biomass that can be burned. Until the industrial revolution, wood supplied most of the world's energy. Today, it still provides 12 percent of the total, chiefly in the form of firewood and charcoal in developing countries.[18]

The use of bioenergy will surely expand during the next 40 years, but its growth will be constrained. With many forests and croplands already overstressed, and with food needs pressing against agricultural resources, it is unrealistic to think that ethanol derived from corn or sugarcane, for instance, can supply more than a tiny fraction of the world's liquid fuels.

One promising approach is to grow energy crops on marginal lands not currently used for food. Land that is too steeply sloping or not sufficiently fertile or well watered for crops might support trees that are periodically harvested. The wood could then be burned directly in a wood-fired power plant or converted to ethanol; the Solar Energy Research Institute has already developed a process that brings the cost of wood-derived ethanol down to $1.35 per gallon.[19]

In the United States, for example, the 13 million hectares of marginal cropland retired under the Conservation Reserve Program could be planted in trees that would yield as much as 265 million barrels of ethanol each year. Equivalent to 10 percent of current U.S. gasoline consumption, this amount will be a far larger share of supplies in the energy-efficient world of 2030. Since in any given year most of the land would be covered by trees, there would be a carbon storage benefit as well.[20]

Biomass energy thus has a major role to play, although resource constraints must be kept in mind. The conversion efficiency of the photosynthetic process is only a fraction of that of direct solar technologies. And there are substantial pressures on biological systems already, as well as new stresses likely to be exerted by global warming. Projects will have to be carefully chosen and properly managed.

Geothermal energy employs the huge reservoir of heat that lies beneath the earth's surface, making it the only renewable source that does not rely on sunlight. For this reason, geothermal resources must be tapped slowly enough so as not to deplete the accessible reservoir of heat, and thus be truly renewable. Continuing advances will allow engineers to use previously unexploitable, lower-temperature reservoirs that are hundreds of times as abundant as those in use today. In the future, many countries may find themselves where Kenya, Nicaragua, and the Philippines are today—getting much of their electricity from geothermal resources.[21]

Virtually all Pacific Rim countries, as well as those along East Africa's Great Rift and the Mediterranean Sea, are well endowed with geothermal energy. Iceland, Indonesia, and Japan are among the nations with the greatest potential. Geothermal energy can provide not only electricity that is transmitted long distances, but also direct heat for industries that locate near accessible heat reservoirs.[22]

Land use will inevitably be shaped by the development of economic systems based on renewable energy. Solar technologies are land-intensive, but no more so than those based on fossil fuels. In fact, if all the land devoted to mining coal is included, many solar technologies actually require less area than coal power does.[23]

A 1,000-megawatt solar thermal facility, for example, would require 24 square kilometers of land. Meeting all U.S. electricity needs with this technology would thus require about 29,000 square kilometers, an area one tenth the size of Arizona. Although wind farms that produce an equivalent amount of power would cover a wider area, they would not really occupy the land in the same way; indeed, cattle-grazing and other activities could go on as before. Moreover, solar, wind, and geothermal systems will use land that may not be much in demand today—windy mountain passes or stretches of dry desert. Land values, traditionally determined largely by agriculture, ranching, and forestry prospects, will be influenced by energy production potential as well.[24]

Nations now constituting what is called the Third World face the immense challenge of continuing to develop their economies without massive use of fossil fuels. Yet they have an advantage in being less hooked on oil, coal, and gas, and so their transition to the solar age will be easier. One option is to rely on biomass energy in current quantities but to step up replanting efforts and to burn the biomass much more efficiently, using gasifiers and other devices. Another is to turn directly to the sun, which most of the Third World has in abundance— solar ovens for cooking, solar collectors for hot water, and photovoltaics for electricity.

In both industrial and developing nations, energy production inevitably will be much more decentralized, breaking up the huge industries that have been a dominant part of the economic scene in the late twentieth century. Indeed, a world energy system based on the highly efficient use of renewable resources is liable to be not only more decentralized but also less vulnerable to disruption and more conducive to market principles and democratic political systems. In ecological terms, it is the only kind of world likely to endure for more than a few decades.

EFFICIENT IN ALL SENSES

Getting total global carbon emissions down to 2 billion tons a year requires vast improvements in energy efficiency. Fortunately, many of the technologies to accomplish such reductions are already at hand and cost-effective. No technical breakthroughs are needed, for example, to double automobile fuel economy, triple the efficiency of lighting systems, or cut typical heating requirements by 75 percent. Technologies developed in the decades ahead will undoubtedly allow even greater gains.[25]

Automobiles in 2030 are apt to get at least 100 miles per gallon of fuel, four times the current average for new cars. A hint of what such vehicles may be like is given in a recently developed prototype, the Volvo LCP 2000. It is an aerodynamic four-passenger car that weighs just half as much as today's models due to the use of lightweight synthetic materials. Moreover, it has a highly efficient and clean-burning diesel engine. With the addition of a continuously variable transmission and a flywheel energy storage device, this vehicle could get 90 miles to the gallon.[26]

Forty years from now, Thomas Edison's revolutionary incandescent light bulbs may be found only in museums—replaced by a variety of new lighting systems, including halogen and sodium lights. The most important new light source may be compact fluorescent bulbs that, for example, use 18 watts rather than 75 to produce the same amount of light. The new bulbs, already available today, not only reduce consumers' electricity bills, they last over seven times as long.[27]

In 2030, homes are likely to be weather-tight and highly insulated, greatly reducing the need for both heating and cooling. Superinsulated homes in the Canadian province of Saskatchewan are already so tightly built that it does not pay to install a furnace; a small electric baseboard heater is more than adequate. Such homes use one third as much energy as modern Swedish homes do, or one tenth the U.S. average. They have double the normal insulation and airtight liners in the walls. Mechanical ventilation systems keep the air fresh. The most welcome change for residents may be the almost complete elimination of utility bills.[28]

Inside these homes, people will have appliances that are on average three to four times as efficient as those in use today. Probably the greatest savings will come in refrigeration. Commercial models now on the market can reduce electricity use from 1,500 kilowatt-hours per year to 750; other models under development would bring that figure down to 240 kilowatt-hours. Gains nearly as great are possible in air-conditioners, water heaters, and clothes dryers.[29]

Industry will also be shaped by the need to improve energy efficiency. Steelmaking is likely to rely heavily on efficient electric arc furnaces that require half the energy of the open hearth ones of today. Some energy-intensive materials, such as aluminum, may be used only in select applications, replaced in large measure by less energy-intensive synthetics. Vast improvements in the design and maintenance of electric motors could by themselves eliminate the need for hundreds of large power plants around the world.[30]

Cogeneration (the combined production of heat and power) will also spread widely. Many factories will generate their own power with biomass, using the waste heat for industrial processes as well as heating and cooling. Such systems are in wide use in some parts of the world already, and can raise total plant efficiency from 50–70 percent to as high as 90 percent. Excess power can be transferred to the electric grid and used by other consumers.[31]

Improving energy efficiency will not noticeably change life-styles or economic systems. A highly efficient refrigerator or light bulb provides the same service as an inefficient one—just more economically. Gains in energy efficiency alone, however, will not reduce fossil fuel carbon emissions by the needed amount. Additional steps to limit the use of fossil fuels are likely to reshape cities, transportation systems, and industrial patterns, fostering a society that is more efficient in all senses.

By the year 2030, a much more diverse set of transportation options will exist. The typical European or Japanese city today has already taken one step toward this future. Highly developed rail and bus systems move people efficiently between home and work: in Tokyo only 15 percent of commuters drive cars to the office. The world of 2030 is apt to rely particularly heavily on light rail—systems built at street level, relatively inexpensively, that allow people to move quickly between neighborhoods.[32]

Automobiles will undoubtedly still be in use four decades from now, but their numbers will be fewer and their role smaller. Within cities, only electric or clean hydrogen-powered vehicles are likely to be permitted, and most of these will be highly efficient "city cars." The energy to run them may well come from solar power plants. Families might rent efficient larger vehicles for vacations.

The bicycle will also play a major role, as it already does in much of Asia as well as in some industrial-country towns and cities. In Amsterdam and many other communities such as Davis, California, bike-path networks have been developed that encourage widespread use of this form of personal transport. Also likely to develop rapidly is the concept of bike-and-ride, using the bicycle to reach rail systems that then move commuters into the center city. There are already twice as many bikes as cars worldwide. In the bicycle-centered transport system of 2030, the ratio could easily be 10 to 1.[33]

Forty years from now, people will live closer to their jobs, and much socializing and shopping will be done by bike rather than in a one-ton automobile. Computerized delivery services may allow people to shop from home—consuming less time as well as less energy. In addition, a world that allows only 2 billion tons of carbon emissions cannot be trucking vast quantities of food and other items thousands of kilometers.

Telecommunications will substitute for travel as well. Many people may work at home or in special satellite offices, connected to colleagues and supervisors by electronic lines rather than crowded highways. Daily trips to the office could be replaced by occasional visits. The saved time and frustration will raise both worker productivity and the quality of life. The current need of many businesspeople to jet frequently from place to place may be substituted by ever more creative uses of electronic communications. And the current transition from overnight delivery services to computerized facsimile transmissions will be nearly complete.

Many people may work at home or in special satellite offices, connected by electronic lines rather than crowded highways.

The automobile-based modern world is now only about 40 years old, but it is already apparent to many that, with its damaging air pollution and traffic congestion, it does not represent the pinnacle in human social evolution. Although a world where cars play a minor role may be difficult for some to imagine, it is worth remembering that our grandparents would have had a hard time visualizing today's world of traffic jams and smog-filled cities. Ultimately, a more efficient world is likely to be less congested and less polluted.[34]

REUSING AND RECYCLING MATERIALS

In the sustainable, efficient economy of 2030, waste reduction and recycling industries will have largely replaced the garbage collection and disposal companies of today. The throwaway society that has emerged during the late twentieth century uses so much energy, emits so much carbon, and generates so much air pollution, acid rain, water pollution, toxic waste, and rubbish that it is strangling itself. Rooted in the concept of planned obsolescence and appeals to convenience, it will be seen by historians as an aberration.

Most materials used today are discarded after one use—roughly two thirds of all aluminum, three fourths of all steel and paper, and an even higher share of plastic. Society will become dramatically less energy-intensive and less polluting only if the throwaway mentality is replaced by a recycling ethic. Just 5 percent as much energy is needed to recycle aluminum as to produce it from bauxite, the original raw material. For steel produced entirely from scrap, the saving amounts to roughly two thirds. Newsprint from recycled paper takes 25-60 percent less energy to make than that from wood pulp. And recycling glass saves up to a third of the energy embodied in the original product.[35]

Recycling is also a key to getting land, air, and water pollution down to acceptable levels. For example, steel produced from scrap reduces air pollution by 85 percent, cuts water pollution by 76 percent, and eliminates mining wastes altogether. Paper from recycled material reduces pollutants entering the air by 74 percent and the water by 35 percent, as well as reducing pressures on forests in direct proportion to the amount recycled.[36]

A hierarchy of options can guide materials policy: The first priority, of course, is to avoid using any nonessential item. Second is to directly reuse a product—for example, refilling a glass beverage container. The third is to recycle the material to form a new product. Fourth, the material can be burned to extract whatever energy it contains, as long as this can be done safely. And finally, the option of last resort is disposal in a landfill.

The first check on the worldwide movement toward a throwaway society came during the seventies as oil prices and environmental consciousness climbed. Rising energy costs made recycling more attractive, reversing the trend toward tossing out more metal, glass, and paper. The second boost came during the eighties as many urban landfill sites filled, forcing municipal governments to ship their garbage to faraway places for disposal. For many U.S. cities, garbage disposal costs during the last decade increased severalfold, making it cost-effective for them to help establish recycling industries.[37]

During the nineties, this trend will be reinforced by the need to reduce carbon emissions, air pollution, acid rain, and toxic waste. In the early stages, countries will move toward comprehensive, systematic recycling of metal, glass, paper, and other materials, beginning with source separation at the consumer level. Many communities in Europe, Japan, and, more recently, the United States have already taken steps in this direction.

Steady advances in technologies are speeding the transition. The electric arc furnace, as mentioned earlier, produces high-quality steel from scrap metal using far less energy than a traditional open-hearth furnace does. In the United States, a leader in this technology, roughly a third of all steel is already produced from scrap in such furnaces.[38]

Historically, the steel industry has been concentrated near areas with coal and iron ore, such as Wales in the United Kingdom or western Pennsylvania in the United States. By 2030, the industry will be widely dispersed. Electric arc furnaces can operate wherever there is electricity and a supply of scrap metal, and they can be built on a scale adapted to the volume of locally available scrap.

The steel mills of the future will feed heavily on worn-out automobiles, household appliances, and industrial equipment. Further, they will provide local jobs and revenue, while eliminating a source of waste.

In the sustainable economy of 2030, the principal source of materials for industry will be recycled goods. Most of the raw material for the aluminum mill will come from the local scrap collection center, not from the bauxite mine. Paper and paper products will be produced at recycling mills, with recycled paper moving through a hierarchy of uses, from high-quality bond to newsprint and, eventually, into cardboard boxes. When, after several rounds of recycling, the fibers are no longer reusable, they can be burned as fuel in a cogenerating plant. In a paper products industry that continually uses recycled materials, wood pulp will play a minor role. Industries will feed largely on what is already within the system, turning to virgin raw materials only to replace any losses in use and recycling.

Although early moves away from the throwaway society are concentrating on recycling, sustainability over the long term depends more on eliminating waste flows. One of the most obvious places to reduce the volume of waste generated is in industry, where a restructuring of manufacturing processes can easily slash wastes by a third or more. The 3M Company halved its hazardous waste flows within a decade of launching a corporation-wide program. A pioneer in waste reduction, 3M also boosted its profits in the process.[39]

Another major potential source of waste reduction lies in the simplification of food packaging. In the United States, consumers spent more on food packaging in 1986 than American farmers earned selling their crops. In the interest of attracting customers, items are sometimes buried in three or four layers of packaging. For the final trip from supermarket to home, yet another set of materials is used in the form of paper or plastic bags, also typically discarded after one use. Forty years from now, government regulation is likely to have eliminated excessive packaging. Throwaway grocery bags will have been replaced by durable, reusable bags of canvas or other material.[40]

In the sustainable economy of 2030, the principal source of materials for industry will be recycled goods.

Societies in 2030 may also have decided to replace multisized and shaped beverage containers with a set of standardized ones made of durable glass that can be reused many times. These could be used for most, if not all, beverages, such as fruit juices, beer, milk, and soda pop. Bottlers will simply clean the container, steam off the old label, and add a new one. Containers returned to the supermarket or other outlet might become part of an urban or regional computerized inventory, which would permit their efficient movement from supermarkets or other collection centers to local dairies, breweries, and soda bottling plants as needed. Such a system will save an enormous amount of energy and materials.

In addition to recycling and reusing metal, glass, and paper, a sustainable society also recycles nutrients. In nature, one organism's waste is another's sustenance; in urban societies, however, human sewage has become a troublesome source of pollutants in rivers, lakes, and coastal waters. The nutrients in human wastes can be reused safely as long as the process includes measures to prevent the spread of disease.

Fortunately, cities in Japan, South Korea, and China already provide some examples of this kind of nutrient recycling. In these countries, human waste is systematically returned to the land in vegetable-growing greenbelts around cities. Intensively farmed cropland surrounding some cities there produces vegetables year-round using greenhouses or plastic covering during the winter to extend the growing season. Perhaps the best model is Shanghai: after modestly expanding its urban political boundaries to facilitate sewage recycling, the city now produces an exportable surplus of vegetables.[41]

Some cities will probably find it more efficient to use treated human sewage to fertilize aquacultural operations. A steady flow of nutrients from human waste into ponds can supply food for a vigorously growing population of algae that in turn are consumed by fish. In Calcutta, a sewage-fed aquaculture system now provides 20,000 kilograms of fresh fish each day for sale in the city. In a society with a scarcity of protein, such an approach, modeled after nature's nutrient recycling, can both eliminate a troublesome waste problem and generate a valuable food resource.[42]

As recycling reaches full potential over the next 40 years, households will begin to compost yard wastes rather than put them out for curbside garbage pickup. A lost art in many communities, composting not only reduces garbage flows, it provides a rich source of humus for gardening, lessening the need to buy chemical fertilizers to maintain lawn and garden fertility.

By systematically reducing the flow of waste and reusing or recycling most remaining materials, the basic needs of the planet's growing number of human residents can be satisfied without destroying our very life-support systems. Moving in this direction will not only create a far more livable environment with less air and water pollution, it will also reduce the unsightly litter that blights the landscape in many industrial societies today.

WITH A RESTORED BIOLOGICAL BASE

Imagine trying to meet the food, fuel, and timber needs of some 8 billion people—nearly 3 billion more than the current population—with 960 billion fewer tons of topsoil (more than twice the amount on all U.S. cropland) and 440 million fewer hectares of trees (an area more than half the size of the continental United States). That, in a nutshell, will be the predicament of society in 2030 if current rates of soil erosion and deforestation continue unaltered for the next 40 years.[43]

Fortunately, or unfortunately, that predicament will not arise. If the earth's croplands, forests, and rangelands are not soon restored and stabilized, world population will never grow that large; human numbers will drop because of malnutrition, famine, and rising death rates, as stated earlier. If, on the other hand, the population of 2030 is adequately being supported, it will be because our use of the earth's biological resources was put on a sound footing long before then, and because population growth was slowed before it completely overwhelmed life-support systems.

Of necessity, societies in 2030 will be using the land intensively; the needs of a population more than half again as large as today's cannot be met otherwise. But unlike the present, land use patterns would be abiding by basic principles of biological stability: nutrient retention, carbon balance, soil protection, water conservation, and preservation of species diversity. Rather than the earth's photosynthetic productivity being eaten away, as is the case now, it will be safeguarded or even enhanced. Harvests will rarely exceed sustainable yields.

Meeting food needs will pose monumental challenges, as some simple numbers illustrate. Currently, 0.28 hectares of cropland is available for every man, woman, and child. By 2030, assuming cropland area expands by 5 percent between now and then and that population grows to 8 billion, cropland per person will have dropped to 0.19 hectares—a third less than we have in today's inadequately fed world. Virtually all of Asia, and especially China, will be struggling to feed its people from a far more meager cropland area per person.[44]

In light of these constraints, the rural landscapes of 2030 are likely to exhibit greater diversity than they do now. Variations in soils, slope, climate, and water availability will require different patterns and strains of crops grown in different ways so as to maximize sustainable output. For example, farmers may adopt numerous forms of agroforestry—the combined production of crops and trees—to provide food, biomass, and fodder, while also adding nutrients to soils and controlling water runoff. Many such systems already are successfully used.[45]

Efforts to arrest desertification, now claiming 6 million hectares each year, may by then have transformed the gullied highlands of Ethiopia, China's Loess Plateau, and other degraded areas into productive, income-generating terrain. A mobilization of villagers, scientists, and development workers will have spread widely the successes in land restoration evident by the late eighties. Much sloping land now losing topsoil rapidly will be terraced, enhanced by shrubs or nitrogen-fixing trees planted along the contour. With improved crop varieties and planting patterns that conserve soil and water, many of these areas

are likely to be largely self-sufficient and economically more diverse by 2030.[46]

Halting desertification also depends on eliminating overgrazing. The global livestock herd in 2030 is thus likely to be much smaller than today's 3 billion. Since open grazing is likely to diminish, more farmers will integrate livestock into their diverse farming systems, using for fodder the leaves from trees in their agroforestry systems or the cover crop in their rotational cropping patterns. It seems inevitable that adequately nourishing a world population 60 percent larger than today's will preclude feeding one third of the global grain harvest to livestock and poultry, as is currently the case. As meat becomes more scarce and expensive, the diets of the affluent will move down the food chain.[47]

If the population of 2030 is adequately being supported, it will be because our use of the earth's biological resources was put on a sound footing.

It is difficult to foresee whether the vast areas of monoculture wheat, rice, and corn so crucial to the global food supply in the late twentieth century will play as great a role in 2030. In the Corn Belt of the United States, farmers will undoubtedly be rotating crops much more extensively to help curb soil erosion, conserve moisture, and reduce pesticide and fertilizer use, a trend already under way.

Farmers in some parts of the world may opt to grow seed-bearing perennial grasses, creating a cropping pattern that resembles the native prairies that much of modern agriculture replaced. Such "polycultures," like their monoculture counterparts, would yield edible grains, oils, and other commodities. They take advantage of the prairie's natural diversity, drought resistance, and soil-renewing capacity, and thus greatly reduce the need for chemicals, irrigation water, and other intensive inputs. The Land Institute in Salina, Kansas, is pioneering such work, which could result in an altogether different, inherently more enduring form of crop production.[48]

Another option some farmers may choose is that of a mixed enterprise of food and energy production. They might grow a winter grain, such as wheat, followed by sweet sorghum as a summer energy crop, which would be used to make ethanol. Besides increasing the amount of sunlight converted into biochemical energy, this cropping pattern would help check soil erosion since the land would be covered year-round.

The rate of deforestation will have had to slow dramatically by the end of this decade, and come to a halt soon thereafter.

Successfully adapting to changed climatic regimes resulting from greenhouse warming, as well as to water scarcity and other resource constraints, may lead scientists to draw on a much broader base of crop varieties. A greater area will be devoted, for example, to crops that are salt-tolerant and drought-resistant, whether they are new varieties of familiar crops such as wheat, or less familiar crops such as amaranth, a grain native to the Andean highlands of South America. The extensive gene pool needed to revamp agricultural systems depends on the collection of wild plants from their native areas and storage of their genetic material in international gene banks. So far this effort has focused on the world's 30 major food crops, but

less on unfamiliar plants that could become an important component of diets 40 years from now.[49]

Forests and woodlands will be valued more highly and for many more reasons in 2030 than is the case today. The planet's mantle of trees, already a third less than in preagricultural times and shrinking by more than 11 million hectares per year now, will be stable or expanding as a result of serious efforts to slow deforestation and to replant vast areas.[50]

Long before 2030, the clearing of most tropical forests will have ceased. Since the nutrients in these ecosystems are held in the leaves and biomass of the vegetation rather than in the soil, only activities that preserve the forest canopy are sustainable. Those forestlands that can support crop production will already have been identified and converted during the intervening 40 years. While it is impossible to say how much virgin tropical forest would remain in 2030 if sustainability is achieved, certainly the rate of deforestation will have had to slow dramatically by the end of this decade, and come to a halt soon thereafter.[51]

Efforts to identify and protect unique parcels of forest will probably have led to a widely dispersed network of preserves. But a large portion of tropical forests still standing in 2030 will be exploited in a variety of ways by people living in and around them. Hundreds of "extractive reserves" will exist, areas in which local people harvest rubber, resins, nuts, fruits, medicinal substances, and other nontimber forest products for domestic use or export. Long-term benefits—both economic and ecological—from a tropical forest area managed in this way are greater than those from burning off an equivalent area and planting it in crops or pasture. Although the latter yields greater monetary returns in the initial few years after clearing, income drops to zero when the land's productivity is gone, usually within a decade or so, and virtually all ecological benefits are lost.[52]

By definition, a sustainable society will not be overcutting or degrading its forests for lumber or other wood products. Today's logging operations often damage or destroy large areas, and regulations governing timber practices—including reforestation—routinely are ignored. Needed efforts to understand how to exploit a natural forest for timber without diminishing its productivity, species diversity, and overall health are still in their infancy.[53]

Since societies will be recycling most of their paper products, demand for pulping wood per person may be less in 2030 than today. Still, large areas of partially desertified land, degraded watersheds, railroad and highway borders, and open countryside will need to be in trees. Serious efforts to alleviate the fuelwood crisis in developing countries, to reduce flooding and landslides in hilly regions, to meet industrial wood needs sustainably, and to slow the buildup of carbon dioxide may have spurred the planting of an additional 200 million hectares or so of trees.[54]

Many of these plantings will be on private farms as part of agroforestry systems. But plantations may also have an expanded role. Cities and villages will turn to managed woodlands on their outskirts to contribute fuel for heating, cooking, and electricity. Wood from these plantations will substitute for some portion of coal and oil use, and, since harvested on a sustained-yield basis, will make no net contribution of carbon dioxide to the atmosphere.

Restoring and stabilizing the biological resource base by 2030 depends on a pattern of land ownership and use far more equitable than today's. Much of the degradation now occurring stems from the heavily skewed distribution of land that, along with population growth, pushes poor people into ever more marginal environments. Good stewardship requires that people have plots large enough to sustain their families without abusing the land, access to the technological means of using their land productively, and the right to pass it on to their children.[55]

This inevitably will require large landholdings in densely populated agrarian societies to be broken up and redistributed to the poorer majority who lack viable livelihoods. Similarly, much government-owned common land, such as forests and pastures, likely will revert to communities and villages that have a stake in optimizing the productivity of these lands and managing them sustainably.[56]

No matter what technologies come along, including unforeseen advances in biotechnology, the biochemical process of photosynthesis, carried out by green plants, will remain the basis for meeting human needs. Given that humanity already appropriates an estimated 40 percent of the earth's annual photosynthetic product on land, leaving only 60 percent for the millions of other species and for protecting basic ecosystem functions, the urgency of slowing the growth in human numbers and demands is obvious. The sooner societies stabilize their populations, the greater will be their opportunities for achieving equitable and stable patterns of land use that can meet their needs indefinitely.[57]

WITH A NEW SET OF VALUES

The fundamental changes we have outlined in energy, forestry, agriculture, and other physical systems cannot occur without corresponding shifts in the social, economic, and moral character of human societies. During the transition to sustainability, political leaders and citizens alike will be forced to reevaluate their goals and aspirations, to redefine their measures of success, and to adjust work and leisure to a new set of principles that have at their core the welfare of future generations.

Given the enormity of the tasks involved, many people may assume that moving in this direction will be painful and limiting, and thus something to resist. But given the choice of repairing your house or having it collapse around you, you would not question whether to undertake the project.

Global military expenditures will be cut heavily as countries recognize that environmental threats to security have supplanted traditional military ones.

Shifts in employment will be among the most visible as the transition gets under way. Moving from fossil fuels to a diverse set of renewable energy sources, extracting fewer materials from the earth and recycling more, and revamping farming and forestry practices will greatly expand opportunities in new areas even as the number of some traditional jobs contracts.

Losses in coal mining, auto production, road construction, and metals prospecting will be offset by gains in the manufacture and sale of photovoltaic solar cells, wind turbines, bicycles, mass transit equipment, and a host of materials recycling technologies. In land-rich countries and those with an abundance of agricultural wastes, alcohol-fuel plants will replace oil refineries. Since planned obsolescence will itself be obsolete in a sustainable society, a far greater share of workers will be employed in repair, maintenance, and recycling activi-

ties than in the extraction of virgin materials and production of new goods.

Wind prospectors, energy efficiency auditors, and solar architects will be among the booming professions stemming from the shift to a highly efficient, renewable energy economy. Numbering in the hundreds of thousands today, jobs in these fields may collectively total in the millions worldwide within a few decades. Opportunities in forestry will expand markedly with the need to design and extend highly productive agroforestry systems, to better manage natural forests, and to plant and maintain vast areas of trees. Similarly, as new cropping patterns are devised and toxic chemical use reduced or eliminated, agronomists will be in far higher demand, as will specialists in biological methods of pest control.

Many people will find their skills valued in new or expanded lines of work. Petroleum geologists may be retrained as geothermal geologists, for example, while traditional midwives continue to broaden their roles to include the spectrum of family planning needs.

Long before 2030, the trend toward ever larger cities and an increasing ratio of urban-to-rural dwellers is likely to have reversed. Each person who moves from countryside to city requires that a corresponding surplus of food be produced in some rural part of the world. Getting more food from less labor requires the use of more energy, as does processing the food and transporting it to the distant city center. An inexorable trend over the last several decades, the increasing energy intensity of food production and distribution cannot continue indefinitely.

Smaller human settlements will also be favored by the shift to renewable energy sources. In contrast to the large centralized fossil-fuel and nuclear plants that dominate energy systems today, power from renewable technologies—

whether photovoltaic cells, wood-fired plants, or wind generators—can be developed cost-effectively across a range of scales. They also allow local areas to capitalize on their natural endowments, whether that be strong winds, bright sunshine, abundant woodlands, or proximity to geothermal reservoirs. In so doing, they foster greater local self-reliance.

As the transition to a more environmentally benign economy progresses, sustainability will gradually eclipse growth as the focus of economic policymaking. Over the next few decades, government policies will encourage investments that promote stability and endurance at the expense of those that simply expand short-term production. Companies would thus devote a greater share of their investment capital, for example, to converting to renewable energy sources and to installing recycling technologies—and proportionately less to expanding the output of goods.

As a yardstick of progress, the gross national product (GNP) will be seen as a bankrupt indicator. By measuring flows of goods and services, GNP undervalues qualities a sustainable society strives for, such as durability and resource protection, and overvalues many it does not, such as planned obsolescence and waste.[58]

Shoddy appliances that need frequent repair and fast replacement, for instance, raise the GNP more than a well-crafted product that lasts, even though the latter is really more valuable. Similarly, the pollution caused by a coal-burning power plant raises GNP by requiring expenditures on lung disease treatment and the purchase of a scrubber to control emissions. Yet society would be far better off if power were generated in ways that did not pollute the air in the first place.

In 2030, planners will measure economic and social advances by sustaina-bility criteria rather than simply by growth in short-term output. As economist Herman Daly observes, a new direction of technical progress is needed, "one that squeezes more service per unit of resource, rather than one that just runs more resources through the system." Furthermore, he points out, "as long as the GNP is thought to measure human well-being, the obstacles to change are enormous. . . . The market sees only efficiency—it has no organs for hearing, feeling, or smelling either justice or sustainability."[59]

National military budgets in a sustainable world will be a small fraction of what they are today. Now totaling some $1 trillion per year, global military expenditures will be cut heavily as countries recognize that environmental threats to security have supplanted traditional military ones. Moreover, sustainability cannot be achieved without a massive shift of resources from military endeavors into energy efficiency, soil conservation, tree planting, family planning, and other needed development activities. Rather than maintaining their own large defense establishments, governments may come to rely on a greatly expanded and strengthened U.N peacekeeping force, one that would have the power and authority to defend any member country against an aggressor. This might allow more countries to follow Costa Rica's example, and eliminate their armies altogether.[60]

Nations will undoubtedly be cooperating in numerous other ways as well. Careful tracking of changes in atmospheric chemistry, forest cover, land productivity, and ocean resources will be among the many efforts handled by multinational teams of scientists and government workers. It may be among the ironies of future political development that even as individual nations move to decentralize power and decisionmaking within their own borders, they simultaneously establish a degree of cooperation and coordination at the international level that goes well beyond anything witnessed to date.

Materialism simply cannot survive the transition to a sustainable world.

Movement toward a lasting society cannot occur without a transformation of individual priorities and values. Throughout the ages, philosophers and religious leaders have denounced materialism as a viable path to human fulfillment. Yet societies across the ideological spectrum have persisted in equating quality of life with increased consumption. Personal self-worth typically is measured by possessions, just as social progress is judged by GNP growth.

Because of the strain on resources it creates, materialism simply cannot survive the transition to a sustainable world. As public understanding of the need to adopt simpler and less consumptive lifestyles spreads, it will become unfashionable to own fancy new cars and clothes. This shift, however, will be among the hardest to make, since consumerism so deeply permeates societies of all political stripes. Yet the potential benefits of unleashing the tremendous quantities of human energy now devoted to designing, producing, advertising, buying, consuming, and discarding material goods are enormous. Much undoubtedly would be channeled into forming richer human relationships, stronger communities, and greater outlets for cultural diversity, music, and the arts.

As the amassing of personal and national wealth becomes less of a goal, the gap between haves and have-nots will gradually close, eliminating many societal tensions. Ideological differences may fade as well, as nations adopt sustainability as common cause, and as they come to recognize that achieving it requires a shared set of values that includes democratic principles, freedom to innovate, respect for human rights, and acceptance of diversity. With the cooperative tasks involved in repairing the earth so many and so large, the idea of waging war could become an anachronism.

Fortunately, there is as much a pull as a push toward an enduring society. As economist Herman Daly and theologian John Cobb write, "People can be *attracted* by new ways of ordering their lives, as well as *driven* by the recognition of what will happen if they do not change."[61]

The opportunity to build a lasting foundation will pass us by if we do not seize it soon. To get under way, we need only stop resisting the push, and embrace the pull, of building a sustainable society.

Notes

Chapter 1. The Illusion of Progress

1. United Nations (UN), Department of International Economic and Social Affairs (DIESA), *World Demographic Estimates and Projections, 1950–2025* (New York: 1988); world economic output from Herbert R. Block, *The Planetary Product in 1980: A Creative Pause* (Washington, D.C.: U.S. Department of State, 1981), and from International Monetary Fund (IMF), *International Financial Statistics Yearbook* (Washington, D.C.: 1989).

2. United States Department of Agriculture (USDA), Economic Research Service (ERS), *World Grain Harvested Area, Production, and Yield 1950–88* (unpublished printout) (Washington, D.C.: 1989).

3. Worldwide loss of topsoil is author's estimate; forest loss is author's estimate based on U.N. Food and Agriculture Organization (FAO) data; loss of species from E.O. Wilson, ed., *Biodiversity* (Washington, D.C.: National Academy Press, 1988).

4. Carbon dioxide levels from A. Neftel et al., "Evidence in Polar Ice Cores for the Increase in Atmospheric CO_2 in the Last Two Centuries," *Nature*, May 2, 1985, and Charles D. Keeling, Scripps Institution of Oceanography, private communication, July 31, 1989; ozone levels from National Aeronautics and Space Administration, "Executive Summary of the Ozone Trends Panel," Washington, D.C., March 15, 1988; see Chapter 6 for a discussion of air pollution's effects on lakes and forests.

5. Annual loss of topsoil is author's estimate, based on Lester R. Brown and Edward C. Wolf, *Soil Erosion: Quiet Crisis in the World Economy*, Worldwatch Paper 60 (Washington, D.C.: Worldwatch Institute, September 1984); damage to crops by air pollution from James J. MacKenzie and Mohamed T. El-Ashry, *Ill Winds: Airborne Pollution's Toll on Trees and Crops* (Washington, D.C.: World Resources Institute, 1988).

6. Grain production data from USDA, ERS, *World Grain*, except 1989 from USDA, World Agricultural Outlook Board (WAOB), *World Agriculture Supply and Demand Estimates*, Washington, D.C., October 12, 1989; per capita production derived using population figures in Francis Urban and Philip Rose, *World Population by Country and Region, 1950–86, and Projections to 2050* (Washington, D.C.: USDA, ERS, 1988), which are based on data from U.S. Bureau of the Census; amount of decreased production offset by stock depletion based on grain consumption and reserves data in USDA, Foreign Agricultural Service (FAS), *World Grain Situation and Outlook*, Washington, D.C., September 1989.

7. Per capita grain production from USDA, ERS, *World Grain*; African grain imports from USDA, FAS, *World Rice Reference Tables* (unpublished printout)(Washington, D.C.: July 1988) and USDA, FAS, *World Wheat and Coarse Grains Reference Tables* (unpublished printout)(Washington, D.C.: August 1988); hunger in Africa from World Bank, *Report of the Task Force on Food Security in Africa* (Washington, D.C.: 1988); quote from World

Bank, *Sub-Saharan Africa: From Crisis to Sustainable Growth—A Long-Term Perspective Study* (Washington, D.C.: 1989); the "nightmare scenario" was first developed by the U.N. Economic Commission for Africa in 1983.

8. Food consumption per person in Africa and Latin America derived from USDA, ERS, *World Grain*, from USDA, FAS, *World Wheat and Coarse Grains*, from USDA, FAS, *World Rice*, and from Urban and Rose, *World Population;* infant mortality in Brazil, Ghana, Mexico and Uruguay from Per Pinstrup-Andersen, "Food Security and Structural Adjustment," in Colleen Roberts, ed., *Trade, Aid and Policy Reform: Proceedings of the Eighth Agriculture Symposium* (Washington, D.C.: World Bank, 1988); in Zambia from United Nations Children's Fund (UNICEF), *State of the World's Children 1989* (New York: Oxford University Press, 1989); in El Salvador from Lindsey Grusen, "Salvador's Poverty Is Called Worst in Century," *New York Times*, October 16, 1988; in Madagascar from Michael Griffin, *Madagascar* (Mauritius: UNICEF, undated); in Dominican Republic from Charles F. McCoy, "Debt Crisis is Inflicting a Heavy Toll in Dominican Republic," *Wall Street Journal*, August 20, 1987; in Peru from Michael Reid, "Rebels Gain As Economy Worsens," *Christian Science Monitor*, August 2, 1989.

9. Environmental politics in Western Europe from Serge Schmemann, "Environmentalists and Socialists Gain in European Vote," *New York Times*, June 19, 1989; in Poland from "Report of the Round Table Subunit on Ecology," Warsaw, March 1989, submitted as supplementary material to the Congressional Hearings on East-West Environmental Cooperation, Commission on Security and Cooperation in Europe, September 28, 1989; in Soviet Union from Alexei Yablokov, Deputy, Congress of People's Deputies, Union of Soviet Socialist Republics, private communication, Washington, D.C., October 30, 1989; in Japan from Masao Kunihiro, Member, Japanese Diet, private communication, Washington, D.C., September ber 13, 1989; in Australia from Michael Malik, " Greens Set the Pace," *Far Eastern Economic Review*, July 13, 1989.

10. UN, DIESA, *World Population Prospects 1988* (New York: 1989).

11. 1987 carbon emissions increase derived from Gregg Marland et al., *Estimates of CO_2 Emissions from Fossil Fuel Burning and Cement Manufacturing, Based on the United Nations Energy Statistics and the U.S. Bureau of Mines Cement Manufacturing Data* (Oak Ridge, Tenn.: Oak Ridge National Laboratory, 1989), and Gregg Marland, Oak Ridge National Laboratory, private communication and printout, July 6, 1989; 1988 total emissions and increase from 1987 are Worldwatch Institute estimates based on above and British Petroleum Company, *BP Statistical Review of World Energy* (London: 1989).

12. Peter M. Vitousek et al., "Human Appropriation of the Products of Photosynthesis," *BioScience*, June 1986.

13. Share of earth's surface that is land from ibid.; land in crops, pasture, forest, and other uses from FAO, *1987 Production Yearbook* (Rome: 1988).

14. Cropland, grassland, and forest area trends from FAO, *Production Yearbook*, various issues, from Francis Urban, "Agricultural Resources Availability," *World Agriculture Situation and Outlook Report*, USDA, ERS, Washington, D.C., June 1989, and from USDA, ERS, *World Grain*; historical forest cover trends from Sandra Postel and Lori Heise, *Reforesting the Earth*, Worldwatch Paper 83 (Washington, D.C.: Worldwatch Institute, April 1988).

15. MacKenzie and El-Ashry, *Ill Winds*; Siberian forest decline from George Woodwell, Director, Woods Hole Research Center, private communication, September 16, 1989.

16. MacKenzie and El-Ashry, *Ill Winds;* Woodwell, private communication.

17. Edward C. Wolf, "Managing Rangelands," in Lester R. Brown et al., *State of the*

World 1986 (New York: W.W. Norton & Co., 1986); African livestock data for 1950 from FAO, *Production Yearbook 1954* (Rome: 1955) and for 1987 from *FAO Monthly Bulletin of Statistics*, Vol. 1, No. 3, 1988; African population from Urban and Rose, *World Population*.

18. *FAO Monthly Bulletin of Statistics*, Vol. 1, No. 3, 1988; Southern African Development Coordination Conference, *SADCC Agriculture: Toward 2000* (Rome: FAO, 1984).

19. Wolf, "Managing Rangelands"; Government of India, "Strategies, Structures, Policies: National Wastelands Development Board," New Delhi, mimeographed, February 6, 1986; M.V. Desai, "Gujarat: Lending Helping Hand," *Hindustan Times* (New Delhi), October 8, 1988.

20. Loss of cropland from Lester R. Brown, *The Worldwide Loss of Cropland*, Worldwatch Paper 24 (Washington, D.C.: Worldwatch Institute, October 1978), and from FAO, *Protect and Produce: Soil Conservation for Development* (Rome: 1983); growth in world automobile fleet from Motor Vehicle Manufacturers Association, *Facts and Figures '89* (Detroit, Mich.: 1989).

21. Vitousek et al., "Human Appropriation."

22. Global economic output from IMF, *International Financial Statistics Yearbook*; job creation from *Economic Report of the President* (Washington, D.C.: U.S. Government Printing Office, 1989).

23. Robert Repetto et al., *Wasting Assets: Natural Resources in the National Income Accounts* (Washington, D.C.: World Resources Institute, 1989).

24. Ibid. Nations may derive more from leaving forests standing than from logging them: one economist has calculated that for Japan's forests the annual economic benefits of water resources conservation, soil erosion prevention, landslide prevention, and recreation are greater than the value of the annual timber harvest; Katsuya Fukuoka, *The Economics of Coexistence Between Man and Nature* (Tokyo: Rissho University Institute for Economic Research, 1985).

25. Repetto et al., *Wasting Assets.*

26. See Chapter 4 for more information on the unsustainable production of grain.

27. Area of rapid topsoil loss (now included in U.S. Conservation Reserve Program) from Brad Karmen, ERS, USDA, Washington, D.C., private communication, October 25, 1989; Great Plains wheat yield estimate based on USDA, ERS, *World Grain*.

28. U.S. irrigated area from Clifford Dickason, ERS, USDA, Washington, D.C., private communication, October 19, 1989; area with falling water tables from Clifford Dickason, "Improved Estimates of Groundwater Mining Acreage," *Journal of Soil and Water Conservation*, May/June 1988; the greatest excess world production of grain during the eighties was 67.2 million tons in 1985, according to USDA, FAS, *World Grain Situation and Outlook*.

29. Alex Echols, Senior Legislative Assistant to Senator Robert W. Kasten, Jr., Washington, D.C., private communication, November 1, 1989; U.S. Public Law 101–45, June 30, 1989. At a meeting in Caracas, Venezuela, the South Commission, a group of political leaders from developing countries, called for the adjustment of GNP statistics to reflect a variety of other indicators, including pollution and depletion of natural resources; Hazel Henderson, "National Economic Indexes Aren't Enough," *Christian Science Monitor*, August 17, 1989.

30. Population data from Urban and Rose, *World Population*.

31. See Chapter 4 for further discussion of food production and food security; grain production from USDA, ERS, *World Grain*; grain prices from IMF, *International Financial Statistics*, Washington, D.C., monthly, various issues.

32. World grain production from USDA, ERS, *World Grain*, except 1989 from USDA, WAOB, *Supply and Demand Estimates*.

33. World grain production from USDA, ERS, *World Grain*, except 1989 from USDA, WAOB, *Supply and Demand Estimates*; grain prices from IMF, *International Financial Statistics*; U.S. idled cropland from Karmen, private communication.

34. Fertilizer consumption 1984–87 from FAO, *FAO Quarterly Bulletin of Statistics*, Vol. 1, No. 2, 1988; 1988 from Fertilizer Institute, *Fertilizer Facts and Figures 1989* (Washington, D.C.: 1989); 1989 is author's estimate.

35. World grain consumption was an estimated 322 kilograms per capita in 1989, so the additional 88 million people that year would need 28 million tons of grain; figure derived from consumption estimate in USDA, WAOB, *Supply and Demand Estimates*, and from population estimate in Urban and Rose, *World Population*.

36. Malnutrition information from UNICEF, *State of the World's Children 1989*; declining food aid from USDA, ERS, *World Food Needs and Availabilities, 1988/89: Fall* (Washington, D.C.: 1988), and from Ray Nightingale et al., "Higher Prices Strain Food Aid Budgets," *Agricultural Outlook*, USDA, ERS, Washington, D.C., September 1989.

37. Grain stocks and production from USDA, FAS, *World Grain Situation and Outlook*; countries that import U.S. grain from USDA, ERS, *World Agricultural Trends and Indicators, 1970–88* (Washington, D.C.: 1989).

38. Grain fed to livestock from USDA, FAS, *World Wheat and Coarse Grains Reference Tables*, and from Gerald Ostrowski, USDA, FAS, Washington, D.C., private communication, September 1989.

39. U.N. World Food Council (WFC), "The Global State of Hunger and Malnutrition: 1988 Report," 14th Ministerial Session, Nicosia, Cyprus, March 24, 1988.

40. John Hunt, "Conference Puts Risk to Ozone on World Map," *Financial Times* (London), March 8, 1989.

41. Centre for Our Common Future, "Background Information on the Hague Declaration," press release, Geneva, undated (conference held March 10–11, 1989).

42. Peter Calamai, "Earth Council Next Step," in Southam Newspapers (Canada), *Our Fragile Future* (special supplement on the environment), October 7, 1989; Crispin Tickell, Permanent Representative of the United Kingdom to the United Nations, "Global Climate Change," Statement to the Economic and Social Council, New York, May 8, 1989; Eduard A. Shevardnadze, Minister for Foreign Affairs of the Union of Soviet Socialist Republics, Statement to the U.N. General Assembly, New York, September 27, 1988; Geoffrey Palmer, Prime Minister of New Zealand, General Debate Statement, U.N. General Assembly, New York, October 2, 1989.

43. Edward Cody, "Pact Seeks to Shield Third World States," *Washington Post*, March 23, 1989.

44. Ibid.

45. "Seven Nations' Leaders Endorse Action to Protect Earth's 'Ecological Balance'," *International Environment Reporter*, August 1989; Summit of the Arch, "Economic Declaration," Paris, July 16, 1989.

46. "Is Europe Turning Green?" (editorial), *New York Times*, June 22, 1989; Schmemann, "Environmentalists and Socialists Gain in European Vote"; Sheila Rule, "Battling for England's Green and Pleasant Land," *New York Times*, July 10, 1989.

47. "Environmental Party Gains 20 Seats, Gets 5.4 Percent of Vote in Election," *International Environment Reporter*, October 1988; number of seats in Riksdag from John Haggert, Embassy of Sweden, Washington, D.C., private communication, October 25, 1989;

Greens in Italian House of Deputies from press office, Embassy of Italy, Washington, D.C., private communication, November 2, 1989; Malik, "Greens Set the Pace"; Mark Hunter, "Greens Join Mainstream in France," *Washington Post*, March 22, 1989; "Greens Capture Europe's Imagination" (editorial), *Nature*, June 22, 1989.

48. Poland from "Report of the Round Table Subunit on Ecology"; Soviet Union from Yablokov, private communication; Japan from Kunihiro, private communication.

49. Government of the Netherlands, Ministry of Housing, Physical Planning, and the Environment, "Netherlands National Environmental Policy Plan: Extra Expenditure of 6.7 Billion Guilders in 1994," press release, May 25, 1989; Government of the Netherlands, Ministry of Housing, Physical Planning, and the Environment, "Netherlands National Environmental Policy Plan: Drastic Decisions Inevitable to Improve Environment," press release, May 25, 1989.

50. Ministry of Environment, *Environment and Development: Programme for Norway's Follow-Up of the Report of the World Commission on Environment and Development*, Report to the Storting No. 46, 1988–89 (Oslo: Government of Norway, 1989).

51. Robert J.L. Hawke, Prime Minister of Australia, "Speech by the Prime Minister: Launch of Statement on the Environment," Wentworth, N.S.W., Australia, July 20, 1989.

52. "Score One for the Trees," *Economist*, January 14, 1989; James Brooke, "Rain and Fines, But Mostly Rain, Slow Burning of Amazon Forest in Brazil," *New York Times*, September 17, 1989.

53. "New Jersey Residents Required to Sort Trash," *Washington Post*, April 21, 1987; Timothy Egan, "Curbside Pickup and Sludge Forests: Some Cities Make Recycling Work," *New York Times*, October 24, 1988; Joan Mulhern, Vermont Public Interest Research

Group, Warren, Vt., private communication, November 24, 1989; Madeleine M. Kunin, Governor, State of Vermont, "A Legacy for the Next Generation: the Vermont Third Century Trust," remarks at 25th anniversary celebration of the Vermont Natural Resources Council, Montpelier, Vt., September 16, 1989.

54. Alan B. Durning, *Action at the Grassroots: Fighting Poverty and Environmental Decline*, Worldwatch Paper 88 (Washington, DC: Worldwatch Institute, January 1989).

55. "U.N.-Commissioned Harris Poll Shows Environmental Concerns," *Multinational Environmental Outlook*, May 16, 1989; the countries surveyed were Argentina, China, Hungary, India, Jamaica, Japan, Kenya, Mexico, Nigeria, Norway, Saudi Arabia, Senegal, the Federal Republic of Germany, and Zimbabwe; other polls have found strong and rising concern over environmental issues; see "Environmental Worry Up," *New York Times*, July 2, 1989; Richard Morin, "Polls Show Public Wants Cleanup, But Will It Pay?" *Washington Post*, June 18, 1989.

56. Environmental and defense spending from Office of Management and Budget, *Historical Tables, Budget of the United States Government, Fiscal Year 1990* (Washington, D.C.: U.S. Government Printing Office, 1989); figures do not completely reflect the relative size of military and environmental spending: for example, the military spending figure does not include military aid to foreign governments or interest on previous military expenditures financed by government borrowing, and the natural resources and environment figure includes expenditures (e.g., on water projects) that are not for environmental protection.

57. Countries that have signed or agreed to sign the Montreal Protocol from Hunt, "Conference Puts Risk to Ozone on World Map"; as of October 1989, only 48 nations had actually ratified the Montreal Protocol—other nations' ratifications are still pending—

according to Liz Cook, Friends of the Earth, Washington, D.C., private communication, October 31, 1989; Swedish CFC phaseout from Government of Sweden, "Environmental Policy for the 1990s," Environmental Bill, March 4, 1988, and from Sverker Hogberg, scientific counselor, Embassy of Sweden, Washington, D.C., private communication, October 13, 1988; Kara Swisher, "4 Big Firms Pledge to Cut Use of Ozone-Sapping CFCs," *Washington Post*, August 8, 1989.

58. Annual addition to world population from UN, DIESA, *World Population Prospects 1988*; carbon dioxide emissions from Marland et al., *Estimates of CO$_2$ Emissions*, and from Marland, private communication and printout; soil erosion reduction derived from information in USDA, ERS, *Agricultural Resources: Cropland, Water, and Conservation Situation and Outlook Report*, Washington, D.C., September 1989, and from USDA, Soil Conservation Service, *1982 National Resources Inventory* (Washington, D.C.: 1984).

Chapter 2. Slowing Global Warming

1. "UNEP/WMO Panel from 30 Countries to Work Toward Global Warming Treaty," *International Environment Reporter*, December 1988.

2. U.S. Environmental Protection Agency (EPA), *Policy Options for Stabilizing Global Climate* (draft) (Washington, D.C.: 1989).

3. James E. Hansen, NASA Goddard Institute for Space Studies, "The Greenhouse Effect: Impacts on Current Global Temperature and Regional Heat Waves," Testimony before the Committee on Energy and Natural Resources, U.S. Senate, Washington, D.C., June 23, 1988; Stephen Schneider, *Global Warming: Are We Entering the Greenhouse Century?* (San Francisco: Sierra Club Books, 1989); James E. Hansen et al., "Global Climate Changes as Forecast by the GISS 3-D Model," *Journal of Geophysical Research*, August 20, 1988.

4. Worldwatch Institute estimates based on Gregg Marland et al., *Estimates of CO$_2$ Emissions from Fossil Fuel Burning and Cement Manufacturing, Based on the United Nations Energy Statistics and the U.S. Bureau of Mines Cement Manufacturing Data* (Oak Ridge, Tenn.: Oak Ridge National Laboratory, 1989); Gregg Marland, private communication and printout, Oak Ridge National Laboratory, Oak Ridge, Tenn., July 6, 1989; British Petroleum (BP), *BP Statistical Review of World Energy* (London: 1989); R.A. Houghton et al., "The Flux of Carbon from Terrestrial Ecosystems to the Atmosphere in 1980 Due to Changes in Land Use: Geographic Distribution of the Global Flux," *Tellus*, February-April 1987; values in this chapter generally will be in terms of carbon. To derive carbon dioxide weight from carbon multiply by 3.67 (carbon has a molecular weight of 12, oxygen of 16; therefore, one molecule of carbon dioxide (CO$_2$) contains one part carbon (12) to two parts oxygen (2 times 16), which equals 44, a factor of 3.67 above the simple carbon value).

5. Marland, private communication; BP, *BP Statistical Review*.

6. Marland et al., *Estimates of CO$_2$ Emissions*; Marland, private communication; Worldwatch estimates for 1988 carbon emissions derived from above and from BP, *BP Statistical Review*.

7. Worldwatch Institute estimates based on Marland et al., *Estimates of CO$_2$ Emissions*; pre-1973 growth rate was 4.5 percent annually; since 1973, the average annual growth rate has been 3.5 percent.

8. International Energy Agency (IEA), "Energy Policies and Programmes of IEA Countries," *General Report* (Paris: Organisation for Economic Co-operation and Development (OECD), in press); Worldwatch Institute estimates based on Marland et al., *Estimates of CO$_2$ Emissions*, and BP, *BP Statistical Review*.

9. Lee Schipper and Ruth Caron Cooper, "Energy Conservation in the USSR: Realistic or Mirage?" presented to the Meeting on Soviet Energy Options and Global Environmental Issues, Georgetown University, Washington, D.C., June 29, 1989.

10. Marland, private communication; Sandra Postel and Lori Heise, *Reforesting the Earth*, Worldwatch Paper 83 (Washington, D.C.: Worldwatch Institute, April 1988).

11. Christopher Flavin, *Electricity for a Developing World*, Worldwatch Paper 70 (Washington, D.C.: Worldwatch Institute, June 1986); Marland et al., *Estimates of CO₂ Emissions*.

12. Alberto Setzer et al., "Relatório de Atividades do Projeto IBDF-INPE 'SEQE'— Ano 1987," Instituto de Pesquisas Espaciais, São José dos Campos, Brazil, May 1988.

13. EPA, *Policy Options*; "Conference Statement," The Changing Atmosphere: Implications for Global Security, Toronto, June 27–30, 1988.

14. Worldwatch estimate based on Marland et al., *Estimates of CO₂ Emissions*.

15. Ibid.; Worldwatch estimates also based on Bert Bolin et al., eds., *The Greenhouse Effect, Climate Change and Ecosystems* (Chichester, U.K.: John Wiley and Sons, 1986).

16. Centre for Our Common Future, "Background Information on the Hague Declaration," press release, Geneva, undated (conference held March 10–11, 1989); Summit of the Arch, "Economic Declaration," Paris, July 16, 1989.

17. Population Reference Bureau, *1989 World Population Data Sheet*, 1989; Francis Urban and Philip Rose, *World Population by Country and Region, 1950–1986, and Projections to 2050* (Washington, D.C.: U.S. Department of Agriculture (USDA), Economic Research Service (ERS), 1988).

18. Fossil fuel contribution and breakdown are Worldwatch Institute estimates based on D.O. Hall et al., *Biomass for Energy in Developing Countries* (Elmsford, N.Y.: Pergamon Press, 1982), on Daniel Deudney and Christopher Flavin, *Renewable Energy: The Power to Choose* (New York: W.W. Norton & Co., 1983), and on BP, *BP Statistical Review*.

19. Gregg Marland, "Carbon Dioxide Emission Rates for Conventional and Synthetic Fuels," *Energy*, Vol. 8, No. 12, 1983; Dean Abrahamson, University of Minnesota, "Relative Greenhouse Heating from the Use of Fuel Oil and Natural Gas," unpublished, May 20, 1989.

20. Marland et al., *Estimates of CO₂ Emissions*; George Borris, Electric Power Research Institute (EPRI), Palo Alto, Calif., private communication, July 18, 1989; M. Steinberg et al., *A Systems Study for the Removal, Recovery, and Disposal of Carbon Dioxide from Fossil Fuel Power Plants in the U.S.* (Springfield, Va.: National Technical Information Service, 1984).

21. Equipment costs from EPA, *Policy Options*.

22. IEA, *Energy Conservation in IEA Countries* (Paris: OECD, 1987); William U. Chandler et al., *Energy Efficiency: A New Agenda* (Washington, D.C.: American Council for an Energy-Efficient Economy, 1988); José Goldemberg et al., *Energy for a Sustainable World* (Washington, D.C.: World Resources Institute, 1987).

23. Lighting electricity figures are Worldwatch estimates based on U.S. figures and on United Nations, *1985 Energy Statistics Yearbook* (New York: 1987); projected growth is based on an assumed 3-percent annual growth rate, with carbon emissions based on replacing coal-fired generation through efficiency improvements; Arthur H. Rosenfeld and David Hafemeister, "Energy-Efficient Buildings," *Scientific American*, April 1988.

24. Michael Renner, *Rethinking the Role of the Automobile*, Worldwatch Paper 84 (Washington, D.C.: Worldwatch Institute, June 1988); a 3-percent annual growth in vehicle numbers, or a 75-percent increase by 2010, is

consistent with recent trends; carbon emission figures are Worldwatch estimates based on Gregg Marland, "The Impact of Synthetic Fuels on Global Carbon Dioxide," in W.C. Clark, ed., *Carbon Dioxide Review 1982* (New York: Oxford University Press, 1982); Deborah Bleviss, *The New Oil Crisis and Fuel Economy Technologies: Preparing the Light Transportation Industry for the 1990's* (New York: Quorum Press, 1988).

25. Worldwatch Institute estimate based on Marland, private communication and on BP, *BP Statistical Review*.

26. Carbon estimates were derived by determining the additional generation of nuclear power and hydroelectric power from 1973 through 1988 in BP, *BP Statistical Review of World Energy* (London: various years).

27. Christopher Flavin, Testimony in Support of Statement of Case of Greenpeace, U.K., Hinkley Inquiry, April 1989; Louis Harris, "Sentiment Against Nuclear Power Plants Reaches Record High," *The Harris Poll*, Creators Syndicate, Inc., Los Angeles, Calif., January 15, 1989; Christopher Flavin, *Reassessing Nuclear Power: The Fallout From Chernobyl*, Worldwatch Paper 75 (Washington, D.C.: Worldwatch Institute, March 1987); number of plants under construction is Worldwatch estimate based on "World List of Nuclear Power Plants," *Nuclear News*, August 1989; International Atomic Energy Agency, *Annual Report* (Vienna: 1974).

28. Ridley quote from "Nuking the Greenhouse," *New Scientist*, November 5, 1988; Flavin, *Reassessing Nuclear Power*; Irvin C. Bupp and Jean-Claude Derian, *The Failed Promise of Nuclear Power* (New York: Basic Books, 1981).

29. John J. Taylor, "Improved and Safer Nuclear Power," *Science*, April 21, 1989; U.K. study from "Outlook on Advanced Reactors," *Nucleonics Week*, March 30, 1989.

30. Worldwatch Institute estimate based on current global nuclear generating capacity of 312,000 megawatts and global carbon emissions of 7.3 billion tons.

31. Mark Newham, "West Germany to Build 150 MWe of Wind Farms by Mid-1990s," *International Solar Energy Intelligence Report*, November, 22, 1988; Department of Non-Conventional Energy Sources, Ministry of Energy, *Annual Report 1987–88* (New Delhi, India: undated); Paul Gipe, "Wind Energy Comes of Age in California," Paul Gipe and Associates, Tehachapi, Calif., July 1989; Robert R. Lynette, "Wind Energy Systems," presented to the Forum on Renewable Energy and Climate Change, Washington, D.C., June 14–15, 1989; Christopher Flavin, *Wind Power: A Turning Point*, Worldwatch Paper 45 (Washington, D.C.: Worldwatch Institute, July 1981); the 10-percent projection is a Worldwatch estimate.

32. Susan Williams and Kevin Porter, *Power Plays* (Washington, D.C.: Investor Responsibility Research Center, 1989); Paul Kruger, "1987 EPRI Survey of Geothermal Electric Utilities," Electric Power Research Institute, Palo Alto, Calif., undated; Ronald DiPippo, "International Developments in Geothermal Power Development," *Geothermal Resources Council Bulletin*, May 1988.

33. Williams and Porter, *Power Plays*; Michael Lotker, Vice President, Luz International, "Solar Electric Generating System (SEGS)," presented to SOLTECH '89, March 7, 1989; Paul Savoldelli, Luz International, Los Angeles, Calif., private communication and printout, July 11, 1989.

34. Solar Energy Research Institute (SERI), "Photovoltaics: Electricity from Sunshine," Golden, Colo., unpublished, 1989; Christopher Flavin, *Electricity From Sunlight: The Emergence of Photovoltaics* (Washington, D.C.: U.S. Government Printing Office, 1984); H.M. Hubbard, "Photovoltaics Today and Tomorrow," *Science*, April 21, 1989; Maheshwar Dayal, Director, Department of Non-Conventional Energy Sources, New Delhi, India, private communication, July 13, 1989.

35. Hubbard, "Photovoltaics Today and Tomorrow"; SERI, "Photovoltaics: Electricity from Sunshine."

36. Hall et al., *Biomass for Energy in Developing Countries*; World Commission on Environment and Development, *Our Common Future* (New York: Oxford University Press, 1987); Deudney and Flavin, *Renewable Energy*; Brazil example is a Worldwatch Institute estimate based on Gregg Marland and Tom Boden, Oak Ridge National Laboratory, Testimony before Committee on Energy and Natural Resources, U.S. Senate, Washington, D.C., July 26, 1989, on Philip M. Fearnside, National Institute for Research in the Amazon, "The Charcoal of Carajas: Pig-Iron Smelting Threatens the Forests of Brazil's Eastern Amazon Region," unpublished paper, April 28, 1988, and on SUMEI, Companhia Vale do Rio Doce (CVRD), Brazil, "Internal Memorandum Re: Steel Industry Along the Carajas Railway—The Korf Metallurgy Technology Ltd. Viability Study," April 7, 1987.

37. Williams and Porter, *Power Plays*; Peter Gleick et al., "Greenhouse-Gas Emissions from the Operation of Energy Facilities," prepared for the Independent Energy Producers Association, Sacramento, Calif., July 22, 1989.

38. Patrick Knight, "Brazil's Alcohol Programme in Crisis: Imports Probably Needed in 1990," *Energy Economist*, May 1989; Sally M. Kane and John M. Reilly, "Economics of Ethanol Production in the United States," USDA, ERS, March 1989; Migdon Segal, "Ethanol Fuel and Global Warming," Congressional Research Service, Washington, D.C., March 6, 1989.

39. U.S. Department of Energy (DOE), Energy Information Administration (EIA), *Historical Plant Cost and Annual Production Expenses for Selected Electric Plants 1987* (Washington, D.C.: 1989); pollution costs are Worldwatch Institute estimates based on Olav Hohmeyer, *Social Costs of Energy Consumption: External Effects of Electricity Generation in the Federal Republic of Germany* (Berlin: Springer-Verlag, 1988).

40. Electricity costs based on Howard Geller, American Council for an Energy Efficient Economy, Washington, D.C., private communication, August 15, 1989; Tom Gray, American Wind Energy Association, Norwich, Vt., private communication, March 21, 1989; Shepard Buchanan, Bonneville Power Administration, Portland, Ore., private communication and printout, July 28, 1989; Richard Nishkian, California Energy Company, San Francisco, Calif., private communication, August 22, 1989; Charles Komanoff, Komanoff Energy Associates, New York, N.Y., private communication and printout, February 10, 1989; Savoldelli, private communication and printout, July 9, 1989; David Rinebolt, Director of Research, National Wood Energy Association, Arlington, Va., private communication and printout, August 14, 1989; DOE, EIA, *Historical Plant Cost*. Carbon emissions based on Gleick et al., "Greenhouse-Gas Emissions"; Robert San Martin, DOE, "Environmental Emissions from Energy Technology Systems: The Total Fuel Cycle," Spring 1989. Ethanol costs based on Lynn Wright, Oak Ridge National Laboratory, private communication, August 25, 1989; Norman Hinman, SERI, Boulder, Colo., private communication, August 25, 1989. Pollution costs based on Hohmeyer, *Social Costs of Energy*. Carbon and pollution avoidance costs, Worldwatch Institute estimates.

41. Worldwatch Institute estimates based on Gray, private communication; Nishkian, private communication; Savoldelli, private communication; Rinebolt, private communication; Geller, private communication.

42. "Canadian Energy Ministers Fail to Agree on Reduction in Carbon Dioxide," *International Environment Reporter*, September 1989.

43. IEA, *Emission Controls in Electricity Generation and Industry* (Paris: OECD, 1988); Ken

Maize, "DOE Clean Coal Program Takes Greenhouse Effect, Environment Into Account," *Energy Daily*, March 17, 1989; "Researchers Say Methanol May Not Fulfill Clean Air Hopes," *New York Times*, August 1, 1989; David E. Gushee, "Carbon Dioxide Emissions from Methanol as a Vehicle Fuel," Congressional Research Service, Washington, D.C., June 3, 1988.

44. Robert Reinhold, "Southern California Takes Steps to Curb its Urban Air Pollution," *New York Times*, March 18, 1989; Zorik Pirveysian, South Coast Air Quality Management District, private communication, September 12, 1989.

45. A carbon tax would be proportional to the carbon content of various fuels; for example, coal averages 24.12 kilograms carbon per gigajoule, oil averages 19.94 kilograms (82 percent of coal), and natural gas, 13.78 (57 percent of coal), from Marland, "Carbon Dioxide Emission Rates"; Nicholas Wood, "Ridley Backs 'Carbon Tax' to Force Polluters to Pay," *The Times* (London), June 6, 1989.

46. Schneider, *Global Warming*.

47. "West German Social Democrats Outline Possible Energy Changes," *European Energy Review*, July 28, 1989.

48. Worldwatch Institute estimate for carbon tax revenues based on Marland et al., *Estimates of CO$_2$ Emissions*; DOE, EIA, *Monthly Energy Review*, March 1989. The proposed carbon tax of 5.65¢ per kilogram of carbon would increase conventional coal-generated electricity by 1.72¢ per kilowatt-hour in the United States.

49. DOE, *Budget Highlights, FY 1989* (Washington, D.C.: 1988); Christopher Flavin and Alan Durning, *Building on Success: The Age of Energy Efficiency*, Worldwatch Paper 82 (Washington, D.C.: Worldwatch Institute, March 1988).

50. Flavin and Durning, *Building on Success*.

51. Ibid.; Roger W. Sant et al., *Creating Abundance: America's Least Cost Energy Strategy* (New York: McGraw-Hill, 1984).

52. David H. Moskowitz, "Cutting the Nation's Electric Bill," *Issues in Science and Technology*, Spring 1989; News Release, SESCO, Inc., West Springfield, Mass., February 14, 1989.

53. Howard S. Geller, "Energy and Economic Saving from National Appliance Efficiency Standards," American Council for an Energy-Efficient Economy, Washington, D.C., 1987; Flavin and Durning, *Building on Success*.

54. Bolin et al., *The Greenhouse Effect*; Houghton et al., "The Flux of Carbon."

55. Lester R. Brown et al., "Outlining a Global Action Plan," in Lester R. Brown et al., *State of the World 1989* (New York: W.W. Norton & Co., 1989).

56. Mark C. Trexler et al., "Forestry as a Response to Global Warming: An Analysis of the Guatemala Agroforestry and Carbon Sequestration Project," World Resources Institute, June 1989.

57. Ibid.; Worldwatch Institute estimate for levelized cost of offsetting carbon emissions based on Shepard Buchanan, "Costs of Mitigating Greenhouse Effect for Generic Coal-, Oil-, and Gas-Fired Plants" (draft), Bonneville Power Administration, Portland, Ore., January 1989.

58. Trexler et al., "Forestry as a Response"; Postel and Heise, *Reforesting the Earth*; Setzer et al., "Relatório de Atividades."

59. Brown et al., "Global Action Plan"; Postel and Heise, *Reforesting the Earth*.

60. World Bank, "Bank Policy and Operational Options with Regard to Greenhouse Gases and Global Warming," Washington, D.C., 1989.

61. Postel and Heise, *Reforesting the Earth*; Houghton et al., "The Flux of Carbon"; John Lancaster, "House Votes to Limit Logging in Tongass Forest," *Washington Post*, July 14, 1989; Barry Johnstone and Mike Gismondi, "A Forestry Boom in Alberta?" *Probe Post*, Spring 1989; Robert J.L. Hawke, Prime Minister of Australia, "Speech by the Prime Minister: Launch of Statement on the Environment," Wentworth, N.S.W., Australia, July 20, 1989.

62. Worldwatch Institute estimates based on Brad Karmen, USDA, private communication, October 25, 1989; on Brown et al., "Global Action Plan"; and on Marland, private communication.

63. R. Neil Sampson, "ReLeaf for Global Warming," *American Forests*, November/December 1988; Gregory Byrne, "Let 100 Million Trees Bloom," *Science*, October 21, 1988.

64. Robert Reinhold, "Los Angeles Takes Step in Environmental Battle," *New York Times*, January 12, 1989; Dona Chambers, Trees for Houston, private communication, September 7, 1989; Karen Fedor, Global ReLeaf, American Forestry Association, Washington, D.C., private communication, August 8, 1989.

65. Country-specific CFCs' contribution to global greenhouse effect is a Worldwatch estimate based on James E. Hansen et al., "Greenhouse Effect of Chlorofluorocarbons and Other Trace Gases," *Journal of Geophysical Research*, in press; on "The Ozone Treaty: A Triumph for All," *Update from State*, U.S. Department of State, May/June 1988; on Marland et al., *Estimates of CO$_2$ Emissions*; on Cynthia Pollock Shea, *Protecting Life on Earth: Steps to Save the Ozone Layer*, Worldwatch Paper 87 (Washington, D.C.: Worldwatch Institute, December 1988); on The Ad Hoc Group on Global Environmental Problems, "Japan's Activities to Cope with Global Environmental Problems," Environment Agency,

Japan, June 1988; and on Houghton et al., "The Flux of Carbon"; EPA *Policy Options*.

66. Douglas Cogan, *Stones in a Glass House: CFCs and Ozone Depletion* (Washington, D.C.: Investor Responsibility Research Center, 1988); Pollock Shea, *Protecting Life on Earth*; NASA, "Executive Summary of the Ozone Trends Panel," Washington, D.C., March 15, 1988.

67. Cogan, *Stones in a Glass House*.

68. Kara Swisher, "4 Big Firms Pledge to Cut Use of Ozone-Sapping CFCs," *Washington Post*, August 8, 1989; John Maggs, "Nissan to Stop Using Chlorofluorocarbons," *Journal of Commerce*, August 3, 1989; Ministry of Environment, press release, Oslo, Norway, April 28, 1989; Anders Boeryd, National Energy Administration of Sweden, "Options for Constraining the Emissions of CO$_2$ in Sweden, Including the Role of Forestry Biomass," undated; Ed H.T.M. Nijpels, Minister of Housing, Physical Planning and Environment of the Netherlands, Address at the International Conference on the Changing Atmosphere, Toronto, June 30, 1988; Robert Reinhold, "Frustrated by Global Ozone Fight, California City Offers Own Plan," *New York Times*, July 19, 1989; "Helsinki Declaration on the Protection of the Ozone Layer," *International Environment Reporter*, May 1989; Craig R. Whitney, "80 Nations Favor Ban to Help Ozone," *New York Times*, May 3, 1989; United Nations Environment Programme (UNEP), "Decisions Adopted by the Governing Council at Its Fifteenth Session," Nairobi, May 25, 1989.

69. "More than 130 Bills on Climate Change Introduced in 1989 in State Legislatures," *World Climate Change Report*, July 1989; Ministry of Housing, Physical Planning and Environment, "Highlights of the National Environmental Policy Plan," The Netherlands, undated; "National Environmental Policy Plan Issued, Deep Pollution Cuts Sought from Industries," *International Environment Reporter*, June 1989; Pier Vellinga,

Ministry of Housing, Physical Planning and Environment, The Netherlands, private communication, August 28, 1989; Ministry of Environment, press release, Norway; Magnar Norderhaug, Worldwatch Institute—Norden, private communication, September 1, 1989; Boeryd, "Options for Constraining Emissions in Sweden"; The Energy Committee, House of Commons, *Energy Implication of the Greenhouse Effect* (London: Her Majesty's Stationery Office, July 1989); "Unilateral Action Urged," *Nature*, July 20, 1989; United States, congressional legislation, S. 201, S. 324, S. 491, HR 1078; West Germany, "Background Information," First Meeting, Response Strategies Working Group, Intergovernmental Panel on Climate Change (IPCC), Washington, D.C., January 30-February 1, 1989; Tracy Bischoff, Institute for European Environmental Politics, Bonn, private communication, August 24, 1989; California Energy Commission, *The Impacts of Global Warming on California*, Interim Report (Sacramento, Calif.: 1989); Senate Bill 576, Enrolled, 65th Oregon Legislative Assembly—1989 Regular Session; Margie Durery, Governor's Office, Salem, Ore., private communication, September 1, 1989; George Hamilton, Governor's Office, Montpelier, Vt., private communication, September 6, 1989.

70. "National Energy Policy Act of 1989," S. 324, U.S. Senate, introduced by Senator Timothy Wirth, March 14, 1989; "Global Warming Prevention Act," HR 1078, U.S. House of Representatives, introduced by Representative Claudine Schneider, February 22, 1989.

71. The Netherlands, Ministry of Housing, "Highlights of the National Environmental Policy Plan"; Ministry of Environment, press release, Norway; Norderhaug, private communication; Boeryd, "Options for Constraining Emissions in Sweden"; "Swedish CO_2 Ruling Begins to Bite," *European Energy Review*, June 16, 1989; Christine McGourty, "Thatcher Hosts Symposium," *Nature*, May 4, 1989; Energy Committee, *Energy Policy Implications*; "West German Social Democrats Outline Possible Energy Changes"; "Background Information," First Meeting, Response Strategies Working Group; Bischoff, private communication; Serge Schmemann, "Environmentalists and Socialists Gain in European Vote," *New York Times*, June 19, 1989; "Coalition of Greens, Other Leftists Gains Slim Majority in EC Parliament," *International Environment Reporter*, July 1989.

72. U.S. Department of State, Unclassified Cable, Summary of the IPCC Response Strategies Working Group Sub-Groups (May 8–9) and Steering Committee (May 10–12) in Geneva, May 16, 1989; Department of State, Unclassified Cable, Summary of the IPCC Meeting in Nairobi (June 28–30), July 6, 1989; "Japan Reluctant on Early Control of Greenhouse Gas Emissions, Paper Reveals," *International Environment Reporter*, September 1989; "Canadian Energy Ministers Fail to Agree on Reduction in Carbon Dioxide"; Maumoon Abdul Gayoom, speech before the Forty-second session of the U.N. General Assembly, New York, October, 19, 1987; Jane Rosen, "Suddenly, Everyone's Talking About the Weather," *The Interdependent*, Fall 1988.

73. Summit of the Arch, "Economic Declaration."

74. "UNEP/WMO Panel from 30 Countries"; World Meteorological Organization (WMO), WMO/UNEP Intergovernmental Panel on Climate Change, *Report of the First Session* (Geneva: World Climate Programme Publications Series, 1988).

75. "UNEP/WMO Panel from 30 Countries"; WMO, *Report of the First Session*.

76. Worldwatch Institute targets based on Marland, private communication, and on Urban and Rose, *World Population*.

77. Worldwatch Institute estimates based on Marland, private communication; on Houghton et al., "The Flux of Carbon"; on

Urban and Rose, *World Population*; and on BP, *BP Statistical Review*.

78. Worldwatch Institute projections based on Marland, private communication, and on Houghton et al., "The Flux of Carbon"; "China's Problem: Economic Balance," *Oil and Gas Journal*, August 22, 1988.

79. Daniel Lashof, Natural Resources Defense Council, Washington, D.C., private communication, September 11, 1989; United Nations Secretariat, "Long-Range Global Population Projections as Assessed in 1980," *Population Bulletin of the United Nations* (New York: 1983).

80. Centre for Our Common Future, "Background Information."

81. Julie VanDomelen, *Power to Spare: The World Bank and Electricity Conservation* (Washington, D.C.: Osborn Center, 1989); Peter Miller, Environmental Defense Fund, Washington, D.C., private communication, September 6, 1989; Barber Conable, President, World Bank, "Development and the Environment: A Global Balance," presented to Conference on Global Environment and Human Response Toward Sustainable Development, Tokyo, September 11, 1989.

Chapter 3. Saving Water for Agriculture

1. Sandra Postel, *Water: Rethinking Management in an Age of Scarcity*, Worldwatch Paper 62 (Washington, D.C.: Worldwatch Institute, December 1984); W. Robert Rangeley, "Irrigation and Drainage in the World," in Wayne R. Jordan, ed., *Water and Water Policy in World Food Supplies* (College Station, Tex.: Texas A&M University Press, 1987).

2. Rangeley, "Irrigation and Drainage in the World"; U.N. Food and Agriculture Organization (FAO), *Production Yearbook 1987* (Rome: 1988).

3. FAO, *Production Yearbook 1988* (unpublished printout) (Rome: 1989).

4. Ibid.; United Nations, Department of International Economic and Social Affairs, *World Population Prospects 1988* (New York: 1989).

5. G. Levine et al., "Irrigation in Asia and the Near East in the 1990s: Problems and Prospects," prepared for the Irrigation Support Project for Asia and the Near East at the request of the Asia/Neareast Bureau, U.S. Agency for International Development (AID), August 1988.

6. India figures from Mark Svendsen, "Sources of Future Growth in Indian Irrigated Agriculture," presented to the Planning Workshop on Policy Related Issues in Indian Irrigation, Ootacamund, Tamil Nadu, India, April 26–28, 1988; China estimate from Daniel Gunaratnam, China Agriculture Operations Division, World Bank, Washington, D.C., private communication, June 20, 1989; supporting figures and Mexico estimate from Robert Repetto, *Skimming the Water: Rent-Seeking and the Performance of Public Irrigation Systems* (Washington, D.C.: World Resources Institute, 1986); Brazil figure from Jean-Louis Ginnsz, Brazil Agriculture Operations Division, World Bank, Washington, D.C., private communication, June 7, 1989; see also Rangeley, "Irrigation and Drainage in the World"; Montague Yudelman, "Sustainable and Equitable Development in Irrigated Environments," in H. Jeffrey Leonard et al., *Environment and the Poor: Development Strategies for a Common Agenda* (New Brunswick, N.J.: Transaction Books, 1989).

7. Rangeley, "Irrigation and Drainage in the World"; Thayer Scudder, "Conservation Vs. Development: River Basin Projects in Africa," *Environment*, March 1989; FAO, *Consultation on Irrigation in Africa* (Rome: 1987).

8. Hubertus von Pogrell, Latin American Agriculture Operations Division, World Bank, Washington, D.C., private communication, May 19, 1989; Ginnsz, private commu-

nication; Yudelman, "Sustainable and Equitable Development."

9. Possibility of rising prices and dwindling stocks from Lester R. Brown, *The Changing World Food Prospect: The Nineties and Beyond*, Worldwatch Paper 85 (Washington, D.C.: Worldwatch Institute, October 1988).

10. Rangeley, "Irrigation and Drainage in the World."

11. Li Zhuoyan and Gao Jin'an, "Neglect of Water Projects Hurts Farmland," *China Daily*, March 10, 1989; U.S. Department of Agriculture (USDA), Economic Research Service (ERS), "USSR: Agriculture and Trade Report," Washington, D.C., 1989.

12. 150 million figure from M.E. Jensen et al., "Irrigation Trends in World Agriculture," in B.A. Stewart and D.R. Nielsen, eds., *Irrigation of Agricultural Crops* (Madison, Wisc.: American Society of Agronomy, in press).

13. Ibid.

14. Manzur Ahmad, "Water as a Constraint to World Food Supplies," in Jordan, *Water and Water Policy*.

15. Ashok K. Mitra, "Underutilisation Revisited: Surface Irrigation in Drought Prone Areas of Western Maharashtra," *Economic and Political Weekly*, April 26, 1986; see also Robert Chambers, "Food and Water as if Poor People Mattered: A Professional Revolution," in Jordan, ed., *Water and Water Policy*.

16. Irrigation estimate made by multiplying net irrigated area (from FAO) of 227 million hectares by average per-hectare water use of 11,400 cubic meters, and multiplying that total by 1.3 to account for conveyance and storage losses. This yields 3,364 cubic kilometers.

17. V.A. Kovda, "Loss of Productive Land Due to Salinization," *Ambio*, Vol. 12, No. 2, 1983.

18. Yudelman, "Sustainable and Equitable Development."

19. Tom Harris, "A Valley Filled with Selenium," *Sacramento Bee*, July 16, 1989; The Wilderness Society, "Ten Most Endangered National Wildlife Refuges," Washington, D.C., October 1988; see also Eliot Marshall, "High Selenium Levels Confirmed in Six States," *Science*, January 10, 1986; and Cass Peterson, "Toxic Time Bomb Ticks in San Joaquin Valley," *Washington Post*, March 19, 1989.

20. Harris, "A Valley Filled with Selenium"; National Research Council, *Irrigation-Induced Water Quality Problems: What Can Be Learned from the San Joaquin Valley Experience* (Washington, D.C.: National Academy Press, 1989); San Joaquin Valley Drainage Program, *Preliminary Planning Alternatives for Solving Agricultural Drainage and Drainage-Related Problems in the San Joaquin Valley* (Sacramento, Calif.: 1989).

21. Clifford Dickason, "Improved Estimates of Groundwater Mining Acreage," *Journal of Soil and Water Conservation*, May-June 1988; Clifford Dickason, USDA, ERS, Washington, D.C., private communication, October 19, 1989; Comer Tuck, Texas Water Development Board, Austin, Tex., private communication, September 14, 1989.

22. "Advance Census Reports Show Irrigation Rebound," *Agricultural Outlook*, May 1989.

23. James Nickum and John Dixon, "Environmental Problems and Economic Modernization," in Charles E. Morrison and Robert F. Dernberger, *Focus: China in the Reform Era*, Asia-Pacific Report 1989 (Honolulu: East-West Center, 1989); reference to Tamil Nadu in Carl Widstrand, ed., *Water Conflicts and Research Priorities* (Elmsford, N.Y.: Pergamon Press, 1980); Raj Chengappa, "India's Water Crisis," *India Today*, May 31, 1986, excerpted in *World Press Review*, August 1986.

24. Philip P. Micklin, "The Water Management Crisis in Soviet Central Asia," final report to the National Council for Soviet and

East European Research, Washington, D.C., February 1989.

25. Micklin, "The Water Management Crisis in Soviet Central Asia"; data for figure also from I.B. Vol'ftsun et al., "Concerning the Changes in the Structure of the Use of River Flow in the Irrigated Zone of the Amu Dar'ya and Syr Dar'ya Basins," *Vodnyye Resursy*, No. 3, 1988; 1987 estimate from Philip P. Micklin, Western Michigan University, Kalamazoo, Mich., private communication, June 6, 1989.

26. Micklin, "The Water Management Crisis in Soviet Central Asia"; Philip P. Micklin, "Desiccation of the Aral Sea: A Water Management Disaster in the Soviet Union," *Science*, September 2, 1988; Micklin, private communication, October 23, 1989.

27. Micklin, "The Water Management Crisis in Soviet Central Asia"; cost estimate from Micklin, private communication, October 23, 1989.

28. Judith Perera, "Kremlin Moves to Save the Aral Sea," *New Scientist*, November 26, 1988; Micklin, "The Water Management Crisis in Soviet Central Asia"; cost estimate from Micklin, private communication, October 23, 1989.

29. Barbara Crossette, "Water, Water Everywhere? Many Now Say 'No!'," *The New York Times*, October 7, 1989; Lori Udall, Environmental Defense Fund (EDF), "The Environmental and Social Impacts of the World Bank Financed Sardar Sarovar Dam in India," Testimony before the Subcommittee on Natural Resources, Agricultural Research and Environment, Committee on Science, Space and Technology, U.S. House of Representatives, Washington, D.C., October 24, 1989; Omar Sattaur, "India's Troubled Waters," *New Scientist*, May 27, 1989; Baba Amte, *Cry, the Beloved Narmada* (Chandrapur, Maharashtra, India: Maharogi Sewa Samiti, 1989).

30. Scudder, "River Basin Projects in Africa."

31. Council on Environmental Quality, Executive Office of the President, "Findings and Recommendations on a Referral from the Environmental Protection Agency Concerning the Proposal by the Department of the Interior's Bureau of Reclamation to Renew Long-Term Water Contracts for the Orange Cove and Other Friant Unit Irrigation Districts of the Central Valley Project in California," Washington, D.C., June 29, 1989; Natural Resources Defense Council, "Environmentalists and Fishing Groups Announce Lawsuit to Block Federal Water Contracts," press release, December 21, 1988; "Federal Irrigation Contracts Debated," *U.S. Water News*, August 1989; "Calif. Contract Renewal Sets Precedent for Federal Leases," *U.S. Water News*, May 1989.

32. The six East African countries referred to are Burundi, Ethiopia, Kenya, Rwanda, Somalia, and Tanzania; Malin Falkenmark, "The Massive Water Scarcity Now Threatening Africa—Why Isn't it Being Addressed?" *Ambio*, Vol. 18, No. 2, 1989; grain imports from USDA, Foreign Agricultural Service (FAS), *World Rice Reference Tables* (unpublished printout) (Washington, D.C.: July 1988); USDA, FAS, *World Wheat and Coarse Grains Reference Tables* (unpublished printout) (Washington, D.C.: August 1988).

33. Joyce R. Starr and Daniel C. Stoll, *U.S. Foreign Policy on Water Resources in the Middle East* (Washington, D.C.: Center for Strategic & International Studies, 1987); Joyce R. Starr and Daniel C. Stoll, eds., *The Politics of Scarcity: Water in the Middle East* (Boulder, Colo.: Westview Press, 1988); population figure calculated from Population Reference Bureau, *1989 World Population Data Sheet* (Washington, D.C.: 1989).

34. U.S. AID, Office of Irrigation and Land Development, "Irrigation Briefing Paper," Cairo, April 1987; U.S. AID, Agricultural Resources Directorate, "Agricultural Briefing Paper," Cairo, December 1988; Joyce R. Starr, "Global Water Scarcity: The Middle East as a Case Study," Testimony

before Subcommittee on Natural Resources, Agricultural Research and Environment, Committee on Science, Space and Technology, U.S. House of Representatives, February 28, 1989.

35. Starr and Stoll, *U.S. Foreign Policy on Water Resources*; Starr and Stoll, *Politics of Scarcity*; Alan Cowell, "Next Flashpoint in Middle East: Water," *New York Times*, April 16, 1989; World Bank prediction in Starr, "Global Water Scarcity."

36. James E. Nickum, "Beijing's Rural Water Use," prepared for East-West Center North China Project, Honolulu, March 1987; The Chinese Research Team for Water Resources Policy and Management in Beijing-Tianjin Region of China, *Report on Water Resources Policy and Management for the Beijing-Tianjin Region of China* (Beijing: Sino-US Cooperative Research Project on Water Resources Policy and Management, 1987); "Water Rules Tightened; Fines Levied," *China Daily*, May 18, 1989; Li Hong, "Beijing Set To Tackle Water Thirst," *China Daily*, October 17, 1989; North China Plain grain output from Frederick W. Crook, *Agricultural Statistics of the People's Republic of China, 1949–86* (Washington, D.C.: USDA, ERS, 1988).

37. Chinese Research Team for Water Resources Policy and Management in Beijing-Tianjin Region, *Report on Water Resources Policy*; Nickum and Dixon, "Environmental Problems and Economic Modernization"; Nickum, "Beijing's Rural Water Use."

38. For a good overview of markets, see the interview with Steven J. Shupe by Nancy Zeilig, "Face to Face—Water Marketing: An Overview," *Journal of the American Water Works Association*, March 1988.

39. "Conservation and Drought Strategies," *Water Market Update*, December 1988.

40. Elizabeth Checchio, *Water Farming: The Promise and Problems of Water Transfers in Arizona* (Tucson: University of Arizona, 1988);

Arizona Cooperative Extension Service, "Extension Project Plan: Reclamation of Retired Farmland," Tucson, Ariz., November 1986; Gary C. Woodard et al., "The Water Transfer Process in Arizona: Analysis of Impacts and Legislative Options," College of Business and Public Administration, University of Arizona, Tucson, April 1988.

41. Chinese Research Team on Water Resources Policy and Management in Beijing-Tianjin Region, *Report on Water Resources Policy*.

42. William L. Jackson et al., "An Interdisciplinary Process for Protecting Instream Flows," *Journal of Soil and Water Conservation*, March/April 1989; "N.M. Still Has No Instream Law," *U.S. Water News*, May 1989; James Huffman, "Instream Water Use: Public and Private Alternatives," in Terry L. Anderson, ed., *Water Rights: Scarce Resource Allocation, Bureaucracy, and the Environment* (Cambridge, Mass.: Ballinger Publishing Co., 1983); Philip C. Metzger and Jennifer A. Haverkamp, "Instream Flow Protection: Adaptation to Intensifying Demands," Conservation Foundation, Washington, D.C., June 1984.

43. Ralph W. Johnson, "Public Trust Doctrine Challenges Appropriation," *U.S. Water News*, July 1988.

44. For general background on global warming, see Stephen H. Schneider, *Global Warming: Are We Entering the Greenhouse Century?* (San Francisco: Sierra Club Books, 1989).

45. Marshall E. Moss and Harry F. Lins, "Water Resources in the Twenty-First Century—A Study of the Implications of Climate Uncertainty," U.S. Geological Survey Circular No. 1030, 1989; Norman J. Rosenberg et al., "Climate Change, CO_2 Enrichment and Evapotranspiration," in American Association for the Advancement of Science (AAAS), *Climatic Variability, Climate Change, and the Planning and Management of U.S. Water Resources* (New York: John Wiley & Sons, in press).

46. Peter H. Gleick, "Climate Change and California: Past, Present, and Future Vulnerabilities," in M.H. Glantz, ed., *Societal Responses to Regional Climate Change: Forecasting By Analogy* (Boulder, Colo.: Westview Press, 1988); Peter H. Gleick, "Regional Hydrologic Consequences of Increases in Atmospheric CO_2 and Other Trace Gases," *Climatic Change*, Vol. 10, 1987; John C. Schaake, "From Climate to Flow," in AAAS, *Climatic Variability*.

47. Gleick, "Regional Hydrologic Consequences"; Schaake, "From Climate to Flow."

48. Schaake, "From Climate to Flow."

49. The work of Stockton and Boggess is summarized in Roger R. Revelle and Paul E. Waggoner, "Effects of a Carbon Dioxide-Induced Climatic Change on Water Supplies in the Western United States," in National Research Council, *Changing Climate* (Washington, D.C.: National Academy Press, 1983). Irrigated area calculation assumes an annual consumptive demand of 5,500 cubic meters per hectare; for additional explanation of this analysis see, Sandra Postel, *Altering the Earth's Chemistry: Assessing the Risks*, Worldwatch Paper 71 (Washington D.C.: Worldwatch Institute, July 1986).

50. Dean F. Peterson and Andrew A. Keller, "Irrigated Agriculture," in AAAS, *Climatic Variability*.

51. Ibid.

52. Ibid. According to FAO, there were 1,474 million hectares of cropland worldwide in 1986. Subtracting from this figure the estimated 250 million hectares of irrigated land gives a total of 1,224 million hectares of rainfed cropland; 5 percent of this is 61.2 million hectares.

53. James E. Hansen, NASA Goddard Institute for Space Studies, "Modeling Greenhouse Climate Effects," testimony before Subcommittee on Science, Technology, and Space, Committee on Commerce, Science and Transportation, U.S. Senate, Washington, D.C., May 8, 1989.

54. Repetto, *Skimming the Water*.

55. E. Phillip LeVeen and Laura B. King, *Turning off the Tap on Federal Water Subsidies*, Vol. 1 (San Francisco: Natural Resources Defense Council and California Rural Legal Assistance Foundation, 1985); U.S. Department of the Interior, Bureau of Reclamation, *1987 Summary Statistics*, Vol. 1, Water, Land, and Related Data (Denver: 1988); Michael R. Moore and Catherine A. McGuckin, "Program Crop Production and Federal Irrigation Water," in USDA, ERS, *Agricultural Resources: Cropland, Water, and Conservation Situation and Outlook Report*, Washington, D.C., September 1988.

56. "Texas Irrigation Use Declines," *U.S. Water News*, April 1989; Soviet figure from Micklin, private communication, October 13, 1989; for descriptions of Israeli water-saving measures, see "Israel's Water Policy: A National Commitment," in Office of Technology Assessment, *Water-Related Technologies for Sustainable Agriculture in Arid/Semiarid Lands: Selected Foreign Experience* (Washington, D.C.: U.S. Government Printing Office, May 1983).

57. Romana P. de los Reyes and Sylvia Ma. G. Jopillo, *An Evaluation of the Philippine Participatory Communal Irrigation Program* (Quezon City: Institute of Philippine Culture, Ateneo de Manila University, 1986).

58. Repetto, *Skimming the Water*; Robert Wade, "The Management of Irrigation Systems: How to Evoke Trust and Avoid Dilemma," *World Development*, Vol. 16, No. 4, 1988.

59. See Jael Silliman and Roberto Lenton, "Irrigation and the Land-Poor," in Jordan, *Water and Water Policy*; FAO, *Consultation on Irrigation in Africa*.

60. Sandra Postel, *Conserving Water: The Untapped Alternative*, Worldwatch Paper 67 (Washington, D.C.: Worldwatch Institute,

September 1985); World Bank, *Vetiver Grass (Vetiveria zizanioides): A Method of Vegetative Soil and Moisture Conservation* (New Delhi: 1987); John C. Greenfield, seminar on the vetiver system presented at the World Bank, Washington, D.C., August 4, 1988.

61. "Watershed Management in State Impressive," *Deccan Herald*, January 6, 1989.

62. Analysis by Valmik Ahuja, EDF, in Peter Miller, EDF, "An Alternative Development Strategy to the Sardar Sarovar Dam," Testimony before the Subcommittee on Natural Resources, Agricultural Research and Environment, Committee on Science, Space and Technology, U.S. House of Representatives, Washington, D.C., October 24, 1989; see also "Sardar Sarovar Project: An Economic, Environmental and Human Disaster," Narmada Bachao Andolan, Bombay, 1989.

63. "Irrigation is Key to Bangladesh Food Needs," *World Water*, May 1989; IFAD quotation from "Asia's Green Revolution: Pause or Setback," *Hunger Notes*, May 1989.

64. FAO, *Consultation on Irrigation in Africa*.

65. Charles F. Hutchinson, "Will Climate Change Complicate African Famine?" *Resources*, Spring 1989.

66. Scudder, "River Basin Projects in Africa."

67. See Sandra Postel and Lori Heise, *Reforesting the Earth*, Worldwatch Paper 83 (Washington, D.C.: Worldwatch Institute, April 1988).

68. Hillel I. Shuval, "The Development of Water Reuse in Israel," *Ambio*, Vol. 16, No. 4, 1987.

69. Anne Charnock, "Plants with a Taste for Salt," *New Scientist*, December 3, 1988; Brian Forster, "Wheat Can Take on More than a Pinch of Salt," *New Scientist*, December 3, 1988.

70. "Farm Grows Salt-Tolerant Crops," *U.S. Water News*, September 1988; Charnock, "Plants with a Taste for Salt."

Chapter 4. Feeding the World in the Nineties

1. U.N. World Food Council (WFC), "The Global State of Hunger and Malnutrition: 1988 Report," 14th Ministerial Session, Nicosia, Cyprus, March 24, 1988.

2. Grain stocks from U.S. Department of Agriculture (USDA), Foreign Agricultural Service (FAS), *World Grain Situation and Outlook*, Washington, D.C., October 1989; grain price changes calculated in constant dollars, using prices in International Monetary Fund (IMF), *International Financial Statistics*, Washington, D.C., various issues, and deflators in U.S. Department of Commerce, Bureau of Economic Analysis, *Survey of Current Business*, Washington, D.C., various issues.

3. USDA, FAS, *World Grain Situation and Outlook*, October 1989.

4. Francis Urban and Philip Rose, *World Population by Country and Region, 1950–86, and Projections to 2050* (Washington, D.C.: USDA, Economic Research Service (ERS), 1988).

5. Estimate of cropland affected by soil erosion from Lester R. Brown and Edward C. Wolf, *Soil Erosion: Quiet Crisis in the World Economy*, Worldwatch Paper 60 (Washington, D.C.: Worldwatch Institute, September 1984).

6. Robert J.L. Hawke, Prime Minister of Australia, "Speech by the Prime Minister: Launch of Statement on the Environment," Wentworth, N.S.W., Australia, July 20, 1989; *Pravda* cited in Yuri Markish, "Soviet Environmental Problems Mount," *CPE Agriculture Report* (Washington, D.C.: USDA, ERS), May/June 1989; Government of India, "Strategies, Structures, Policies: National Wastelands Development Board," New Delhi, mimeographed, February 6, 1986.

7. Annual soil erosion figure is authors' estimate, based on Brown and Wolf, *Soil Erosion*; assuming an average topsoil depth of 18 centimeters, 156 tons of soil per hectare-centimeter, and 24 million hectares each in corn, wheat, and soybean land, there are 202 billion tons of soil on such land in the United States; area figures are typical for the eighties and are from USDA, ERS, *World Grain Harvested Area, Production, and Yield, 1950–88* (unpublished printout) (Washington, D.C., 1989) (for wheat and corn), and from USDA, FAS, *World Oilseed Situation and Market Highlights*, September 1989 (for soybeans).

8. U.N. Food and Agriculture Organization (FAO), *Protect and Produce: Soil Conservation for Development* (Rome: 1983).

9. Gbdebo Jonathan Osemeobo, "The Human Causes of Forest Depletion in Nigeria," *Environmental Conservation*, Spring 1988.

10. Leon Lyles, "Possible Effects of Wind Erosion on Soil Productivity," *Journal of Soil and Water Conservation*, November/December 1975.

11. Figures are rounded; calculations based on unrounded figures.

12. Irrigated area planted in grain is authors' estimate; yield per hectare of irrigated grain land is authors' estimate, based on data in USDA, ERS, *World Grain*.

13. Richard M. Weintraub, "Bangladesh Braces for More Flooding as Casualties Climb," *Washington Post*, September 7, 1988.

14. Current Madhya Pradesh population is authors' estimate based on 1981 population and growth rate in Frank B. Hobbs, *Demographic Estimates, Projections, and Selected Social Characteristics of the Population of India*, CIR Staff Paper No. 21 (Washington, D.C.: U.S. Bureau of the Census, Center for International Research, 1986); Madhya Pradesh fuel consumption from Centre for Science and Environment, *The State of India's Environment, 1984–85* (New Delhi: 1985); the spiraling process of fuelwood consumption, deforestation, and loss of soil nutrients is described in Kenneth Newcombe, *An Economic Justification for Rural Afforestation: The Case of Ethiopia*, Energy Department Paper No. 16 (Washington, D.C.: World Bank, 1984), and in Kenneth Newcombe, "Household Energy Supply: The Energy Crisis That Is Here To Stay!" presented to the World Bank Senior Policy Seminar—Energy, Gaborone, Botswana, March 18–22, 1985.

15. The results of the U.S. National Crop Loss Assessment Network study are discussed in James J. MacKenzie and Mohamed T. El-Ashry, *Ill Winds: Airborne Pollution's Toll on Trees and Crops* (Washington, D.C.: World Resources Institute, 1988).

16. Ibid.

17. Ibid.

18. 1987 grain production data from USDA, ERS, *World Grain*; 1987 was the most recent year for which data are available in which none of the regions or countries listed experienced drought-reduced harvests.

19. Worldwide ozone depletion from National Aeronautics and Space Administration, "Executive Summary of the Ozone Trends Panel," Washington, D.C., March 15, 1988; relationship between ozone levels and ultraviolet radiation from William Booth, "Severe Ozone Depletion Likely Again This Year," *Washington Post*, October 6, 1989; effect of increased ultraviolet radiation on soybean yields in Alan H. Teramura and N. S. Murali, "Intraspecific Differences in Growth and Yield of Soybeans Exposed to Ultraviolet-B Radiation Under Greenhouse and Field Conditions," *Environmental and Experimental Botany*, Vol. 26, No. 1, 1986.

20. USDA, FAS, *World Grain Situation and Outlook*, October 1989.

21. James E. Hansen et al., "Global Climate Changes as Forecast by the GISS 3-D

Model," *Journal of Geophysical Research*, August 20, 1988.

22. For the effects of high temperatures on corn pollination, see R.H. Shaw, "Estimates of Yield Reductions in Corn Caused by Water and Temperature Stress," in C.D. Raper and P.J. Kramer, eds., *Crop Reactions to Water and Temperature Stresses in Humid, Temperate Climates* (Boulder, Colo.: Westview Press, 1983) and R.F. Dale, "Temperature Perturbations in the Midwestern and Southeastern United States Important for Corn Production," in ibid.; for effects on rice, see Alan H. Teramura, Testimony before the Committee on Agriculture, Nutrition, and Forestry and the Subcommittee on Foreign Operations of the Committee on Appropriations, U.S. Senate, Washington, D.C., May 10, 1989.

23. Annual addition to irrigated area is authors' estimate based on recent growth rates derived from data in FAO, *1988 Production Yearbook* (unpublished printout) (Rome: 1989); annual growth in fertilizer use is authors' estimate based on recent trends in FAO data.

24. Historical data on cultivated area from USDA, ERS, *World Grain*; idled U.S. cropland from Brad Karmen, USDA, Washington, D.C., private communication, October 25, 1989.

25. FAO, *Protect and Produce*; cropland area from USDA, ERS, *World Grain*.

26. USDA, ERS, *World Grain*; area enrolled in Conservation Reserve from Karmen, private communication; reduced Conservation Reserve funding from Thomas E. Kuhnle, Natural Resources Defense Council, Washington, D.C., private communication, October 3, 1989.

27. 1980 grain area per capita derived from harvested area in USDA, ERS, *World Grain* and from population in Urban and Rose, *World Population*; 1990 is authors' estimate, based on above sources, assuming a return to production of all U.S. grain land

currently enrolled in supply management programs (not including Conservation Reserve land).

28. Fertilizer consumption in 1950 from Paul Andrilenas, U.S. Department of Agriculture, Washington, D.C., private communication, May 9, 1986; 1989 is authors' estimate.

29. Data on adoption of high-yielding wheats in India and Mexico from Dana G. Dalrymple, *Development and Spread of High-Yielding Wheat Varieties in Developing Countries* (Washington, D.C.: U.S. Agency for International Development, 1986); corn in U.S. from Lester R. Brown, *Increasing World Food Output*, Foreign Agriculture Economic Report No. 25 (Washington, D.C.: USDA, ERS, 1965); rices in Indonesia from Dana G. Dalrymple, *Development and Spread of High-Yielding Rice Varieties in Developing Countries* (Washington, D.C.: U.S. Agency for International Development, 1986).

30. U.S. fertilizer consumption 1950–81 from USDA, *Agricultural Statistics* (Washington, D.C.: Government Printing Office, annual, 1967 and 1987 issues); U.S. fertilizer consumption since 1981 from Fertilizer Institute, *Fertilizer Facts and Figures 1989* (Washington, D.C.: 1989); 1989 is authors' estimate based on ibid.; Edward C. Cook, "First Quarter Economic Results and Agriculture," *CPE Agriculture Report*, Washington, D.C., USDA, ERS, May/June 1989.

31. In 1986 (the most recent year for which data are available in which none of the three countries experienced a major crop failure), Chinese and Indian farmers produced 17.8 and 16.1 kilograms of grain, respectively, for every kilogram of fertilizer they used, while U.S. farmers produced 17.6 kilograms of grain for the same amount of fertilizer; Worldwatch estimates, based on USDA, ERS, *World Grain*, and on FAO, *FAO Quarterly Bulletin of Statistics*, Vol. 1, No. 3, 1988.

32. Jerome Mintz, in *Venture* magazine, quoted in Jack Ralph Kloppenburg, *First the*

Seed: The Political Economy of Plant Biotechnology, 1492–2000 (New York: Cambridge University Press, 1988); USDA, FAS, *World Grain Situation and Outlook*, Washington, D.C., July 1989.

33. Kloppenburg, *First the Seed*.

34. Jane E. Brody, "Quest for Lean Meat Prompts New Approach," *New York Times*, April 12, 1988; Debora MacKenzie, "Science Milked for All It's Worth," *New Scientist*, March 24, 1988; Charles C. Muscoplat, "Commercialization and Research Perspectives for Vaccines," in Joel I. Cohen, ed., *Strengthening Collaboration in Biotechnology: International Agricultural Research and the Private Sector* (Washington, D.C.: U.S. Agency for International Development, 1989); George E. Seidel, Jr., "Geneticists in the Pasture," *Technology Review*, April 1989.

35. Joseph J. Molnar and Henry Kinnucan, "Introduction: The Biotechnology Revolution," in Joseph J. Molnar and Henry Kinnucan, eds., *Biotechnology and the New Agricultural Revolution*, AAAS Selected Symposium 108 (Boulder, Colo.: Westview Press, 1989); Inter-American Development Bank, *Economic and Social Progress in Latin America: 1988 Report* (Washington, D.C.: 1988).

36. Livestock, poultry, and fish and other aquatic animals feed originally upon plants; Kloppenburg, *First the Seed*.

37. Norman E. Borlaug, "Conference Comments: Biotechnology and World Food Demand," in Cohen, *Strengthening Collaboration in Biotechnology*.

38. Kloppenburg, *First the Seed*; Jean L. Marx, "Foreign Gene Transferred into Maize," *Science*, April 8, 1988.

39. Kloppenburg, *First the Seed*; Michael Lipton with Richard Longhurst, *New Seeds and Poor People* (Baltimore, Md.: Johns Hopkins University Press, 1989).

40. Nagesh Kumar, "Biotechnology Revolution and the Third World: An Overview," in Research and Information System for the Non-Aligned and Other Developing Countries, *Biotechnology Revolution and the Third World* (New Delhi: 1988); M.S. Swaminathan, "Biotechnology and Sustainable Agriculture," in ibid.; Nyle C. Brady, "Application of Biotechnology and Other New Technologies in Developing-Country Agriculture," in Molnar and Kinnucan, *Biotechnology and the New Agricultural Revolution*; Peter R. Day, "The Impact of Biotechnology on Agricultural Research," in Martin Gibbs and Carla Carlson, eds., *Crop Productivity: Research Imperatives Revisited*, proceedings of conference held at Boyne Highlands Inn, October 13–18, 1985, and at Airlie House, December 11–13, 1985.

41. U.S. biotechnology research spending estimates from John Chirichiello, National Science Foundation, private communication, October 6, 1989; for discussion of the direction of federal biotechnology research funding, see Kloppenburg, *First the Seed*; industry supports an estimated 16–24 percent of U.S. university biotechnology research, according to David Blumenthal et al., "University-Industry Research Relationships in Biotechnology: Implications for the University," *Science*, June 13, 1986 (cited in Kloppenburg, *First the Seed*); for further discussion of industry funding of university research, see Martin Kenney, *Biotechnology: The University-Industrial Complex* (New Haven, Conn.: Yale University Press, 1986).

42. CGIAR budget from Henni Deboeck, CGIAR Secretariat, Washington, D.C., private communication, October 6, 1989; CGIAR biotechnology spending estimate from Kerri Wright, CGIAR Secretariat, Washington, D.C., private communication, November 16, 1989.

43. Kloppenburg, *First the Seed*; Calestous Juma, *The Gene Hunters: Biotechnology and the Scramble for Seeds* (Princeton, N.J.: Princeton University Press, 1989).

44. Jack Doyle, *Altered Harvest: Agriculture, Genetics, and the Fate of the World's Food Supply*

(New York: Viking Penguin, 1985); Kloppenburg, *First the Seed*; Lipton, *New Seeds and Poor People*.

45. Sandra Postel, *Defusing the Toxics Threat: Controlling Pesticides and Industrial Wastes*, Worldwatch Paper 79 (Washington, D.C.: Worldwatch Institute, September 1987); Doyle, *Altered Harvest*.

46. Lipton, *New Seeds and Poor People*.

47. Increased milk production of BGH-treated cows from MacKenzie, "Science Milked for All It's Worth"; BGH cost estimate from Richard Fallert, USDA, ERS, October 26, 1989.

48. Joel I. Cohen, U.S. Agency for International Development, Washington, D.C., private communication, September 22, 1989; Roger N. Beachy, "Genetic Transformation for Virus Resistance: Needs and Opportunities in Developing Countries," in Cohen, *Strengthening Collaboration in Biotechnology*; Roger N. Beachy and Claude Fauquet, "Cassava Viruses and Genetic Engineering," International Cassava-Trans Project, Washington University, St. Louis, Mo., undated.

49. Kloppenburg, *First the Seed*.

50. Kenney, *Biotechnology: The University-Industrial Complex*.

51. Donald L. Plucknett et al., "Future Role of the IARCs in the Application of Biotechnology in Developing Countries," in G.J. Persley, ed., *Agricultural Biotechnology: Opportunities for International Development* (Wallingford, U.K.: Commonwealth Agricultural Bureau International, in press).

52. Ibid.

53. Jack R. Kloppenburg, Jr., ed., *Seeds and Sovereignty: The Use and Control of Plant Genetic Resources* (Durham, N.C.: Duke University Press, 1988); Juma, *The Gene Hunters*; Cary Fowler et al., "The Laws of Life: Another Development and the New Biotechnologies," *Development Dialogue*, Issue 1–2, 1988.

54. Chinese fertilizer consumption from *FAO Fertilizer Yearbook*, various years; grain production from USDA, ERS, *World Grain*.

55. USDA, ERS, *USSR Agriculture and Trade Report*, Washington, D.C., May 1989.

56. Canadian and Soviet population from Population Reference Bureau (PRB), *1989 World Population Data Sheet* (Washington, D.C.: 1989).

57. USDA, ERS, *World Grain*.

58. Philip P. Micklin, "The Water Management Crisis in Soviet Central Asia," final report to the National Council for Soviet and East European Research, Washington, D.C., February 1989.

59. A. Kashtanov, cited in Markish, "Soviet Environmental Problems Mount."

60. Ibid.; Soviet grain output from USDA, ERS, *World Grain*; fertilizer use from FAO, *FAO Fertilizer Yearbook*, various years, and *FAO Quarterly Bulletin of Statistics*, Vol. 1, No. 2, 1988.

61. Roughly 25 percent of all Soviet investment goes to agriculture, whereas 10 percent or less is typical of most other nations, according to Edward C. Cook, USDA, ERS, Washington, D.C., private communication, October 25, 1989; Soviet output gains from USDA, ERS, *World Grain*.

62. PRB, *1989 World Population Data Sheet*.

63. USDA, ERS, *World Grain*.

64. Ibid.; yield growth in this decade is authors' estimate, based on ibid. and estimate for 1989 in USDA, World Agricultural Outlook Board (WAOB), *World Agriculture Supply and Demand Estimates*, October 12, 1989.

65. USDA, ERS, *World Grain*.

66. Ibid.

67. Historical data on Japanese grain yields from Brown, *Increasing World Food Output*; 1989 Chinese rice yield is authors' esti-

mate derived from production estimate in USDA, WAOB, *World Agriculture*, and from 1988 area figure in USDA, ERS, *World Grain*; 1989 U.S. corn yield from USDA, WAOB, *World Agriculture*; wheat yield for Western Europe is authors' estimate based on 1989 European Community production estimate in USDA, WAOB, *World Agriculture,* and from 1988 production and area figures in USDA, ERS, *World Grain*; 1989 Indian rice yield is authors' estimate based on production estimate in USDA, FAS, *World Grain Situation and Outlook*, September 1989, and on 1988 area figure in USDA, ERS, *World Grain*; all other yield figures from USDA, ERS, *World Grain*.

68. World population growth projection from United Nations, Department of International Economic and Social Affairs, *World Population Prospects 1988* (New York: 1989).

69. William T. Coyle, "Government Assistance in Japanese Agriculture," *East Asia and Oceania Situation and Outlook Report*, USDA, ERS, Washington, D.C., May 1987.

70. Information on Africa from World Bank, *Report of the Task Force on Food Security in Africa* (Washington, D.C.: 1988); see Chapter 8 for information on poverty in Latin America; food production per person from USDA, ERS, *World Grain*; food prices from IMF, *International Financial Statistics*, various issues; infant mortality from United Nations Children's Fund, *State of the World's Children 1989* (New York: Oxford University Press, 1989).

71. Carl Haub, Population Reference Bureau, Washington, D.C., private communication, October 17, 1988; see also Lester R. Brown et al., "Outlining a Global Action Plan," in Lester R. Brown et al., *State of the World 1989* (New York: W.W. Norton & Co., 1989).

Chapter 5. Holding Back the Sea

1. James E. Hansen et al., "Global Climate Changes as Forecast by the GISS 3-D Model," *Journal of Geophysical Research*, August 20, 1988.

2. United Nations Environment Programme (UNEP), *Criteria for Assessing Vulnerability to Sea-Level Rise: A Global Inventory to High Risk Areas* (Delft, Netherlands: Delft Hydraulics Laboratory, 1989).

3. John S. Hoffman et al., *Projecting Future Sea Level Rise* (Washington, D.C.: U.S. Environmental Protection Agency (EPA), 1983); J.S. Hoffman et al., "Future Global Warming and Sea Level Rise," in Per Brun, ed., *Iceland Symposium '85* (Reykjavik: National Energy Authority, 1986); G.P. Hekstra, "Global Warming and Rising Sea Levels: The Policy Implications," *The Ecologist*, January/February 1989.

4. John D. Milliman, "Sea Levels: Past, Present, and Future," *Oceanus*, Summer 1989; James G. Titus, EPA, "Causes and Effects of Sea Level Rise," presented to the First North American Conference on Preparing for Climate Change: A Cooperative Approach, Washington, D.C., October 27–29, 1987; James Titus, "Sea Level Rise," in *Potential Effects of Global Climate Change on the United States*, Vol. II (draft) (Washington, D.C.: EPA, 1989); The Oceanography Report, "Changes in Relative Mean Sea Level," *EOS*, November 5, 1985.

5. Milliman, "Sea Levels: Past, Present, and Future."

6. Ibid.; Paolo Antonio Pirazzoli, "Sea Level Change," *Nature and Resources*, October/December 1985.

7. Titus, "Causes and Effects"; T.P. Barnett, "Recent Changes in Sea Level and Their Possible Causes," *Climatic Change*, No. 5, 1983.

8. Milliman, "Sea Levels: Past, Present, and Future."

9. I.M. Salinas et al., "Changes Occurring Along A Rapidly Submerging Coastal Area: Louisiana, USA," *Journal of Coastal Resources*, Summer 1986; John D. Milliman et al., "Environmental and Economic Implications of

Rising Sea Level and Subsiding Deltas: The Nile and Bengal Examples," *Ambio,* Vol. 18, No. 6, 1989.

10. G. Sestini et al., *Implications of Expected Climate Changes in the Mediterranean Region: An Overview*, MAP Technical Reports Series, No. 27 (Athens: UNEP, 1989); Titus, "Sea Level Rise."

11. Titus, "Sea Level Rise."

12. Gordon de Q. Robin, "Projecting the Rise in Sea Level Caused by Warming of the Atmosphere," in Bert Bolin et al., eds., *The Greenhouse Effect, Climate Change and Ecosystems* (Chichester, U.K.: John Wiley and Sons, 1986); Titus, "Sea Level Rise."

13. Ann Henderson-Sellers and Kendall McGuffie, "The Threat from Melting Ice Caps," *New Scientist*, June 12, 1986.

14. Milliman, "Sea Levels"; Titus, "Causes and Effects."

15. Eric Golanty, "The Global Experiment," *Oceans*, January 1984.

16. Salinas et al., "Changes Occurring Along Submerging Coastal Area."

17. Peter Hawxhurst, "Louisiana's Responses to Irreversible Environmental Change; Strategies for Mitigating Impacts from Coastal Land Loss," in Mark Meo, ed., *Proceedings of Symposium on Climate Change in the Southern United States* (Washington, D.C.: EPA, 1987).

18. Salinas et al., "Changes Occurring Along Submerging Coastal Area"; Hawxhurst, "Louisiana's Responses."

19. James G. Titus, ed., *Greenhouse Effect, Sea Level Rise and Coastal Wetlands* (Washington, D.C.: EPA, 1987).

20. Ibid.

21. L.A. Boorman et al., *Climatic Change, Rising Sea Level and the British Coast*, Institute of Terrestrial Ecology Report No. 1 (London: Her Majesty's Stationary Office, 1989).

22. Eric C.F. Bird, "Potential Effects of Sea Level Rise on the Coasts of Australia, Africa, and Asia," in J.G. Titus, ed., *Effects of Changes in Stratospheric Ozone and Global Climate, Vol IV: Sea Level Rise* (Washington D.C.: EPA, 1986).

23. Ibid.

24. James Broadus et al., "Rising Sea Level and Damming of Rivers: Possible Effects in Egypt and Bangladesh," in Titus, *Effects of Changes*; Bird, "Potential Effects."

25. E.C.F. Bird, "The Modern Prevalence of Beach Erosion," *Marine Pollution Bulletin,* Vol. 18, No. 4, 1987; Eric C.F. Bird, "Coastal Erosion and a Rising Sea Level," *Coastal Subsidence: Problems and Strategies* (Chichester, U.K.: John Wiley and Sons, in press); Cory Dean, "As Beach Erosion Accelerates, Remedies are Costly and Few," *New York Times*, August 1, 1989; John Gribbin, "The World's Beaches Are Vanishing," *New Scientist,* May 10, 1984.

26. Titus, "Causes and Effects."

27. T.S. Murty et al., "Storm Surges in the Bay of Bengal," *IOC/UNESCO Workshop on Regional Cooperation in Marine Science in the Central Indian Ocean and Adjacent Seas and Gulfs* (Colombo, July 8–13, 1985) Workshop Report No. 37-Supplement (Paris: Intergovernmental Oceanographic Commission, UNESCO, 1988); Titus, "Causes and Effects."

28. Murty et al., "Storm Surges."

29. Ibid.

30. Ibid.

31. Hoffman et al., "Future Global Warming."

32. Ted R. Miller et al., *Impact of Global Climate Change on Urban Infrastructure* (draft) (Washington, D.C.: The Urban Institute, 1988).

33. Titus, "Causes and Effects."

34. Sestini et al., *Implications of Expected Climate Changes*.

35. John D. Milliman, "Rising Sea Level and Changing Sediment Influxes: Real and Future Problems for Indian Ocean Coastal Nations," in *IOC/UNESCO Workshop on Regional Cooperation*.

36. Ibid.; R. J. N. Devoy, "Sea Level Applications and Management" *Progress in Oceanography*, Vol. 18, 1987; Broadus et al., "Rising Sea Level."

37. UNEP, *Criteria for Assessing Vulnerability*.

38. Ibid.

39. Milliman et al., "Environmental and Economic Implications."

40. Milliman, "Rising Sea Level and Changing Sediment"; Broadus et al., "Rising Sea Level."

41. Milliman et al., "Environmental and Economic Implications"; for further discussion of environmental refugees, see Jodi L. Jacobson, *Environmental Refugees: a Yardstick of Habitability*, Worldwatch Paper No. 86 (Washington, D.C.: Worldwatch Institute, November 1988).

42. UNEP, *Criteria for Assessing Vulnerability*.

43. Milliman et al., "Environmental and Economic Implications."

44. Ibid.

45. Broadus et al., "Rising Sea Level."

46. Ibid.

47. Milliman et al., "Environmental and Economic Implications."

48. Broadus et al., "Rising Sea Level."

49. Milliman, "Rising Sea Level and Changing Sediment."

50. Hekstra, "Global Warming."

51. Broadus et al., "Rising Sea Level."

52. "National Strategy for Beach Preservation," Conference Summary for Second Skidway Institute of Oceanography Conference on America's Shoreline, Savannah, Ga., June 1985.

53. Tom Goemans and Pier Vellinga, "Low Countries and High Seas," presented to the First North American Conference on Preparing for Climate Change: A Cooperative Approach, Washington, D.C., October 27–29, 1987; UNEP and the Government of the Netherlands, *Impact of Sea Level Rise on Society: A Case Study for the Netherlands* (Delft, Netherlands: Delft Hydraulics Laboratory, 1988).

54. Goemans and Vellinga, "Low Countries"; UNEP and Government of Netherlands, *Impact of Sea Level Rise on Society*.

55. *The Times Atlas of the World* (New York: Times Books Ltd., 1985).

56. Titus, "Sea Level Rise."

57. Hekstra, "Global Warming"; Bird, "Potential Effects."

58. John Yoo, "Bad Trip: As Highways Decay, Their State Becomes Drag on Economy," *Wall Street Journal*, August 30, 1989.

59. Orrin H. Pilkey and Howard L. Wright III, "Seawalls Versus Beaches," *Journal of Coastal Research*, Autumn 1988; Testimony of Orrin H. Pilkey, Jr. before the Environment, Energy, and Natural Resources Subcommittee, Committee on Government Operations, U.S. House of Representatives, Washington, D.C, April 28, 1989; Robert G. Dean, "Managing Sand and Preserving Shorelines," *Oceanus*, Fall 1988.

60. Titus, "Causes and Effects."

61. Bruce Smith, "South Carolina on Long Road to Recovery After Devastating Storm," Associated Press, October 21, 1989; Donna Griffiths, Public Information Officer, South Carolina Coastal Council, Charleston, S.C., personal communication, October 26, 1989; Titus, "Sea Level Rise."

62. Titus, "Causes and Effects."

63. Gary W. Yohe, "The Cost of Not Holding Back the Sea—Economic Vulnerability" (draft), Department of Economics, Wesleyan University, Middletown, Conn., 1989.

64. Titus, "Causes and Effects."

65. Graham S. Giese and David G. Aubrey, "Losing Coastal Upland to Relative Sea-Level Rise: 3 Scenarios for Massachusetts," *Oceanus*, Fall 1987; Graham S. Giese and David G. Aubrey, "The Relationship Between Relative Sea-level Rise and Coastal Upland Retreat in New England," in J.C. Topping, ed., *Coping with Climate Change* (Washington, D.C.: Climate Institute, 1989).

66. Giese and Aubrey, "Losing Coastal Upland"; Giese and Aubrey, "The Relationship Between Sea Level Rise and Coastal Upland Retreat."

67. Giese and Aubrey,"Losing Coastal Upland"; Giese and Aubrey, "The Relationship Between Relative Sea-level Rise and Coastal Upland Retreat."

Chapter 6. Clearing the Air

1. U.S. number from American Lung Association, update of "Breath in Danger," New York, January 1989; "University Study Correlating Pollution, Mortality Prompts Complaints in Athens," *International Environment Reporter*, August 10, 1988; figure on Hungary cited in Don Hinrichsen, "The Danube Blue No Longer," *Earthwatch* (International Planned Parenthood Federation), No. 36, 1989; India figure from "Vehicular Pollution Makes Breathing Dangerous," *Indian Post* (Bombay), February 11, 1989; Larry Rohter, "Mexico City's Filthy Air," *New York Times*, April 12, 1989; brochure about Ashoka Fellow Luis Manuel Guerra, Ashoka, Arlington, Va., undated.

2. U.S. Environmental Protection Agency (EPA), *The Toxics-Release Inventory: A National Perspective* (Washington, D.C.: U.S. Government Printing Office, 1989).

3. For historical background on the London fog and similar incidents, see Erik P. Eckholm, *Down to Earth* (New York: W. W. Norton & Co., 1982); for a review of scientific studies related to London fog incident, see American Lung Association, *Health Effects of Air Pollution* (New York: 1989).

4. National Clean Air Coalition, *The Clean Air Act* (Washington, D.C.: 1985); see also Dr. Morton Lippman, "Health Benefits from Controlling Exposure to Criteria Air Pollutants," in John Blodgett, ed., *Health Benefits of Air Pollution Control: A Discussion* (Washington, D.C.: Congressional Research Service, 1989).

5. U.S. figures from EPA, *National Air Pollution Emission Estimates, 1940–1987* (Research Triangle Park, N.C.: 1989); Japan and Europe information from United Nations Environment Programme (UNEP) and World Health Organization (WHO), *Assessment of Urban Air Quality* (Nairobi: Global Environment Monitoring System, 1988); Claire Holman, *Particulate Pollution from Diesel Vehicles* (London: Friends of the Earth Ltd., 1989).

6. For information on environmental problems in Eastern Europe, see Hilary F. French, "Industrial Wasteland," *World Watch*, November/December 1988; for problems in the Soviet Union, see Hilary F. French, "The Greening of the Soviet Union," *World Watch*, May/June 1989.

7. Sandra Postel, *Air Pollution, Acid Rain, and the Future of Forests*, Worldwatch Paper 58 (Washington, D.C.: Worldwatch Institute, March 1984); projections for the year 2000 from Lori Heise, "Air Pollution Attacks China's Forests," *World Watch*, January/February 1988; information on India from C.K. Varshney and J.K. Garg, "A Quantitative Assessment of Sulfur Dioxide Emission from Fossil Fuels in India," *Journal of the Air Pollu-*

tion Control Association, November 1978, cited in Postel, *Air Pollution*.

8. UNEP and WHO, *Assessment of Urban Air Quality*.

9. Ibid.

10. Lippman, "Health Benefits from Controlling Exposure"; Michael Weisskopf, "Acid Aerosols and a Link to Childhood Bronchitis," *Washington Post*, June 6, 1989; U.S. Office of Technology Assessment (OTA), *Acid Rain and Transported Air Pollutants: Implications for Public Policy* (Washington, D.C.: U.S. Government Printing Office, 1984).

11. OTA, *Acid Rain and Transported Air Pollutants*; Sandra Postel, *Altering the Earth's Chemistry: Assessing the Risks*, Worldwatch Paper 71 (Washington, D.C.: Worldwatch Institute, July 1986); Daniel an Atta, "Thousands of Swedish Wells Acidified," *Acid*, March 6, 1988.

12. Philip A. Bromberg, M.D., University of North Carolina School of Medicine, Testimony, and Morton Lippmann, Ph.D., New York University Medical Center, Testimony, Hearings on Health Effects of Air Pollution, Subcommittee on Environmental Protection, Committee on Environment and Public Works, U.S. Senate, Washington, D.C., April 18, 1989; David Stipp, "Regular Ozone Levels in Cities May Hurt Lungs," *Wall Street Journal*, September 18, 1989.

13. EPA, "EPA Lists Places Failing to Meet Ozone or Carbon Monoxide Standards," press release, Washington, D.C., July 27, 1989; Don Theiler, State and Territorial Air Pollution Program Administrators, Testimony, Hearings on the Health Effects of Air Pollution, Subcommittee on Health and Environment, Committee on Energy and Commerce, U.S. House of Representatives, Washington, D.C., February 28, 1989; information on Washington and New York from Richard E. Ayres, Testimony, ibid.; Los Angeles number from "Smog Season One of Worst in California," *Washington Post*, December 3, 1988; national ozone nonattainment figures from American Lung Association, "Breath in Danger."

14. Friends of the Earth Ltd., "Clean Air Week Off to a Bad Start as WHO Ozone Levels are Exceeded Several Times," press release, London, May 26, 1989; P. Grennfelt et al., "Regional Ozone Concentrations in Europe," in P. Mathy, ed., *Air Pollution and Ecosystems: Proceedings of an International Symposium held in Grenoble, France, 18–22 May 1987* (Dordrecht: D. Reidel, 1988).

15. William Branigin, "Bracing for Pollution Disaster," *Washington Post*, November 28, 1988; Michael P. Walsh, transportation consultant, private communication, October 12, 1989.

16. In addition, Brazil, Taiwan, Mexico, and the European Community are planning to implement U.S.-equivalent standards soon; the European plans are discussed later in this chapter. Michael Renner, *Rethinking the Role of the Automobile*, Worldwatch Paper 84 (Washington, D.C.: Worldwatch Institute, June 1988); Michael P. Walsh, "Global Trends in Motor Vehicle Pollution Control—A 1988 Perspective," Society of Automotive Engineers Technical Paper Series, Schenectady, N.Y., 1989; UNEP and WHO, *Assessment of Urban Air Quality*.

17. EPA, *National Air Pollution Emission Estimates*; Michael Weisskopf, "Leaded Gasoline Cut Improves Air Greatly," *Washington Post*, April 15, 1987; on lead level in blood see Barry Commoner, "A Reporter at Large: The Environment," *The New Yorker*, June 15, 1987.

18. WHO and UNEP, *Assessment of Urban Air Quality*; Guillermo X. Garcia, "High Lead Levels Threatening Mexico's Next Generation," *Times Advocate* (Escondido, Calif.), April 14, 1988.

19. EPA, *Toxics Release Inventory*; Michael Weisskopf, "U.S. Air Pollution Exceeds Estimates," *Washington Post*, March 23, 1989.

20. EPA, *Unfinished Business: A Comparative Assessment of Environmental Problems, Appendix 1. Report of the Cancer Risk Work Group* (Washington, D.C.: Office of Policy Analysis, 1987); Alexandra Allen, "Poisoned Air," *Environmental Action*, January/February 1988.

21. Stanley Kabala, "Poland: Facing the Hidden Costs of Development," *Environment*, November 1985; Michael Kenney, "Poland's Environmental Crisis," *Journal of Commerce*, August 23, 1988.

22. N. Namiki, "International Redeployment of Pollution-Intensive Industries and the Role of Multinational Corporations," report prepared for the World Commission on Environment and Development, Geneva, 1986; information on chemical industry in India from "Environmental Pollution Control in Relation to Development," World Health Organization Technical Report Series, Geneva, 1985.

23. Michael Weisskopf, "'Toxic Clouds' Can Carry Pollutants Far and Wide," *Washington Post*, March 16, 1988; "Airborne Pollution May be Hurting Eskimo Health," *Multinational Environmental Outlook*, January 5, 1989; see also Jost Heintzenberg, "Arctic Haze: Air Pollution in Polar Regions," *Ambio*, Vol. 18, No. 1, 1989.

24. Crocker study described in James S. Cannon, *The Health Costs of Air Pollution* (New York: American Lung Association, 1985); Jay Mathews, "Smog-Control Study Targets Medical Costs," *Washington Post*, July 11, 1989.

25. For a description of the early interactions between American and Scandinavian scientists see Gene E. Likens, "Some Aspects of Air Pollutant Effects on Terrestrial Ecosystems and Prospects for the Future," *Ambio*, Vol. 18, No. 3, 1989.

26. David W. Schindler, "Effects of Acid Rain on Freshwater Ecosystems," *Science*, January 8, 1988; see also Henning Rodhe

and Rafael Herrera, ed., *Acidification of Tropical Countries Scope 36* (New York: John Wiley & Sons, 1988).

27. David W. Schindler, "Acid Precipitation and Global Change," speech delivered to the U.S. National Academy of Sciences Forum on Global Change and Our Common Future, Washington, D.C., May 2–3, 1989.

28. Diane Fisher et al., *Polluted Coastal Waters: The Role of Acid Rain* (New York: Environmental Defense Fund, April 1988); Nordic Council of Ministers, *Europe's Air—Europe's Environment* (Stockholm: International Conference on Transboundary Air Pollution, 1986).

29. 1982 figure from Postel, *Air Pollution*; other years from Ministry of Food, Agriculture, and Forestry, "1988 Forest Damage Survey," Bonn, West Germany, November 1988. It should be noted that some portion of the dramatic increases between 1982, 1983, and 1984 may have been due to a change in the surveying method. Also, there is an emerging consensus that "Class 1" (slight) damage, which accounts for much of the total in West Germany as well as elsewhere in Europe, serves as an early warning sign but does not necessarily indicate permanent damage.

30. James J. MacKenzie and Mohamed T. El-Ashry, *Ill Winds: Airborne Pollution's Toll on Trees and Crops* (Washington, D.C.: World Resources Institute, 1988).

31. Jon Luoma, "Acid Murder No Longer a Mystery," *Audubon*, November 1988; James J. MacKenzie and Mohamed T. El-Ashry, "Ill Winds: Air Pollution's Toll on Trees and Crops," *Technology Review*, April 1989; Bruck quoted in Philip Shabecoff, "Deadly Combination Felling Trees in East," *New York Times*, July 24, 1988.

32. Don Hinrichsen, "Poland's Chemical Cauldron," *The Amicus Journal*, Spring 1988; Schulz and Wicke cited in Per Elvingson, "Air Pollution Costing West Germany Bil-

lions," *Acid News*, special issue for International Air Pollution Week, May 27-June 5, 1989; Mike Leary, "Poisoned Environment Worries Eastern Europe," *Philadelphia Inquirer*, October 4, 1987.

33. MacKenzie and El-Ashry, *Ill Winds;* National Acid Precipitation Assessment Program (NAPAP), *Interim Assessment: The Causes and Effects of Acidic Deposition*, Volume IV (Washington, D.C.: U.S. Government Printing Office, 1987); economic loss estimated by Walter W. Heck, Chairman, Research Committee, National Crop Loss Assessment Network, cited in MacKenzie and El-Ashry, *Ill Winds.*

34. Heise, "Air Pollution Attacks China's Forests" ; see also James N. Galloway et al., "Acid Rain: China, United States, and a Remote Area," *Science*, June 19, 1987; "Acid Rain Eats Into Province," *China Daily*, May 31, 1989; Peter Muello, "Acid Rain Falling on Latin America," *Los Angeles Times*, September 30, 1984.

35. Henning Rodhe, "Acidification in a Global Perspective," *Ambio*, Vol. 18, No. 3, 1989; Marlise Simons, "High Ozone and Acid-Rain Levels Found Over African Rain Forests," *New York Times*, June 19, 1989; see also Rodhe and Herrera, *Acidification of Tropical Countries.*

36. See OTA, *Acid Rain*, for a description of materials at risk.

37. Economic Commission for Europe (ECE), *Air Pollution Across Boundaries* (New York: United Nations, 1985); Athens and Katowice estimated from John McCormick, *Acid Earth* (London: International Institute for Environment and Development, 1985); Paul Hofmann, "Italy's Endangered Treasures," *New York Times*, July 30, 1989.

38. Henry L. Magaziner, American Institute of Architects, Testimony at Hearings on Acid Rain Control Proposals, Subcommittee on Health and the Environment, Committee on Energy and Commerce, U.S. House of

Representatives, Washington, D.C., April 6, 1989; R. Drummond Ayres, Jr., "Gettysburg Journal," *New York Times*, June 20, 1989; Bernard Smith et al., "Elusive Solution to Monumental Decay," *New Scientist*, June 2, 1988.

39. McCormick, *Acid Earth*; John Noble Wilford, "New Threat to Maya Ruins: Acid Rain," *New York Times*, August 8, 1989; Merle Green Robertson, Pre-Columbian Art Research Institute, San Francisco, "Investigation of Color on Maya Sculpture and the Effects of Acid Precipitation," Mesoamerica Foundation, Mérida, Mexico, 1989.

40. Swedish and Dutch studies from McCormick, *Acid Earth*; for a discussion of various studies done in the United States see Magaziner, Testimony; for information on a draft U.S. study that was never published in final form see Philip Shabecoff, "Draft Study Puts Acid Rain Damage at $5 Billion for 17 States," *New York Times*, July 18, 1985.

41. International Energy Agency (IEA), *Emission Controls in Electricity Generation and Industry* (Paris: Organisation for Economic Co-operation and Development, 1988).

42. NAPAP, *Interim Assessment*, Vol. 1; IEA, *Emissions Controls.*

43. Czechoslovakia information from "Ambitious Program Under Way to Improve Environment by 2000," *International Environment Reporter*, January 1989; Wellington Chu, "Coal Pollution Plagues China's Modernization," *Journal of Commerce*, August 26, 1988.

44. IEA, *Emissions Controls.*

45. Ibid.; NAPAP, *Interim Assessment*, Vol. II.

46. IEA, *Emissions Controls.*

47. Howard S. Geller, Testimony at Hearings on Acid Rain Control Proposals, Subcommittee on Health and the Environment, Committee on Energy and Commerce, U.S. House of Representatives, Washington, D.C., April 6, 1989; for a more complete

analysis see Howard S. Geller et al., *Acid Rain and Electricity Conservation* (Washington, D.C.: American Council for an Energy-Efficient Economy/Energy Conservation Coalition, 1987).

48. Geller, Testimony; Geller et al., *Acid Rain and Electricity Conservation*.

49. David H. Moscovitz, "Cutting the Nation's Electric Bill," *Issues in Science and Technology*, Spring 1989; Larry DeWitt, New York Public Service Commission, private communication, October 13, 1989; New York State Department of Public Service, "Consideration of Environmental Externalities in Competitive Bidding Programs of Orange and Rockland Utilities," unpublished paper, undated; New York State Department of Public Service, "Consideration of Environmental Externalities in Competitive Bidding Programs to Acquire Future Electric Capacity Needs of Niagara Mohawk Power Corporation," unpublished paper, undated; Sury N. Putta, "Competition in Electric Generation—Environmental Externalities," New York State Department of Public Service, Albany, N.Y., unpublished paper, undated.

50. Roberta Forsell Stauffer, "Energy Savings From Recycling," *Resource Recycling,* January/February 1989; Cynthia Pollock, *Mining Urban Wastes: The Potential for Recycling,* Worldwatch Paper 76 (Washington, D.C.: Worldwatch Institute, April 1987); Postel, *Air Pollution.*

51. Renner, *Rethinking the Automobile.*

52. "Athens Smog Forces Another Auto Ban," *Multinational Environmental Outlook*, September 29, 1988; Clyde Haberman, "Is It Over Then? City Closes Its Heart to the Car," *New York Times*, October 10, 1988; "Central Budapest Car Ban Seeks to Curb Air Pollution," Reuters, March 30, 1988; "New Mexican President Promises Action to Control Air Pollution," *Multinational Environmental Outlook*, December 22, 1988; "Smog-Filled Santiago to Clean Up," *Multinational Environmental Outlook*, May 2, 1989; O'seun

Ogunseitan, "Wednesday is Odd Day in Lagos," *New Internationalist*, March 1989.

53. Formaldehyde number from Renner, *Rethinking the Automobile*; carbon implications from James MacKenzie, World Resources Institute, Testimony at Hearings on Alternative Fuels, Committee on Energy and Natural Resources, U.S. Senate, October 17, 1989; information on safety from ibid., and from James S. Cannon, *Drive for Clean Air: Natural Gas and Methanol Vehicles* (New York: INFORM, 1989). See also David E. Gushee, "Alternative Fuels from Motor Vehicles: Some Environmental Issues," U.S. Congressional Research Service, Washington, D.C., September 20, 1988, and David E. Gushee, "Carbon Dioxide Emissions from Methanol as a Vehicle Fuel," U.S. Congressional Research Service, Washington, D.C., June 3, 1988.

54. OTA, *Catching Our Breath: Next Steps for Reducing Urban Ozone* (Washington, D.C.: U.S. Government Printing Office, 1989). The numbers given are estimates for reduction costs in 1994 in areas not in attainment with Clean Air Act standards.

55. Richard D. Wilson, EPA, and Steve Plotkin, Office of Technology Assessment, Testimony at Hearings on Automotive Fuel Economy, Subcommittee on Energy and Power, Committee on Energy and Commerce, U.S. House of Representatives, Washington, D.C., July 13, 1989; Chris Calwell, "Fuel Economy/Emissions Interactions in Automobiles," Testimony before State of California Energy Resources Conservation and Development Commission Hearing on *1989 Fuels Report*, June 12, 1989. Calwell also found that the 50 most efficient models had 24 percent lower carbon monoxide emissions than the average car and 35 percent lower emissions than the 50 least efficient ones. He found no correlation one way or the other with nitrogen oxides, stating that "the only conclusion that can be safely drawn about NO_x is that raising or lowering fuel economy is not likely to have a statistically

significant impact on NO$_x$ emissions." For more detail, see Chris Calwell, "The Near-Term Potential for Simultaneous Improvements in the Fuel Efficiency and Emissions of U.S. Automobiles," Masters Thesis, Energy and Resources Group, University of California, Berkeley, June 1989.

56. Deborah Gordon and Leo Levenson, *Drive +: A Proposal for California to Use Consumer Fees and Rebates to Reduce New Motor Vehicle Emissions and Fuel Consumption* (Berkeley, Calif.: Lawrence Berkeley Laboratory, 1989); Calwell, "Near-Term Potential."

57. Joel S. Hirschhorn, "Cutting Production of Hazardous Waste," *Technology Review*, April 1988; OTA, *From Pollution to Prevention: A Progress Report on Waste Reduction* (Washington, D.C.: U.S. Government Printing Office, 1987); Sandra Postel, *Defusing the Toxics Threat: Controlling Pesticides and Industrial Waste*, Worldwatch Paper 79 (Washington, D.C.: Worldwatch Institute, September 1987).

58. Project '88, sponsored by Senators Timothy E. Wirth and John Heinz, *Harnessing Market Forces to Protect Our Environment: Initiatives for the New President* (Washington, D.C.: 1988).

59. David Sarokin, EPA, Washington, D.C., private communication, October 10, 1989.

60. Helmut Weidner, *A Survey of Clean Air Policy In Europe* (Berlin: Wissenschaftszentrum Berlin für Sozialforschung, 1989); "European Commission Considers Directive on Freedom of Information on the Environment," *International Environment Reporter*, October 1988.

61. For a detailed description of Bush's proposal to Congress, see "Section-by-Section Analysis of the 'Clean Air Act Amendments of 1989'," Executive Office of the President, Washington, D.C., July 20, 1989. At this stage, many different proposals are being considered by Congress. It is impossible to predict exactly what shape a final bill might take.

62. ECE, "Acid Rain: Protocol on Emissions Enters into Force," press release, Geneva, August 25, 1987; "25 ECE Members Sign Protocol To Limit Emissions of Nitrogen Oxides," *International Environment Reporter*, November 1988; see also Peter H. Sand, "Air Pollution in Europe: International Policy Responses," *Environment*, December 1987.

63. Sand, "Air Pollution in Europe"; "25 ECE Members Sign Protocol," *International Environment Reporter*; "Probability Seen as Highly Desirable for Volatile Organic Compound Protocol," *International Environment Reporter*, August 1989.

64. Commission of the European Communities, cited in Jim Skea, "Reducing Fossil Fuel Emissions: The Policy Framework," Science Policy Research Unit, University of Sussex, Brighton, U.K., unpublished paper, September 1988; see also Nigel Haigh, "New Tools for European Air Pollution Control," *International Environmental Affairs*, Winter 1989, and "Limits Set for Large Combustion Plants," *Acid News*, special issue for International Air Pollution Week, May 27-June 5, 1989.

65. "Statement," NGO Strategy Meeting on Air Pollution, Wageningen, Netherlands, April 3–4, 1989; Christer Agren, "Critical Loads Remain Most Important Factor," *Acid News*, Special Issue for International Air Pollution Week, May 27-June 5, 1989; see also Jan Nilsson, ed., *Critical Loads for Sulphur and Nitrogen: Report from a Nordic Working Group* (Solna: National Swedish Environmental Protection Board, 1986).

66. "European Community Environment Ministers Agree on New Emission Levels for Small Cars," *International Environment Reporter*, June 1989; "Common Market Finally Adopts Tight Vehicle Standards," *Car Lines* (Michael P. Walsh, Arlington, Va.), June 1989.

67. South Coast Air Quality Management District, *Air Quality Management Plan: South Coast Air Basin* (El Monte, Calif.: March 1989); Robert Reinhold, "Southern California Takes Steps to Curb Its Urban Air Pollution," *New York Times*, March 18, 1989.

68. Marjorie Sun, "Environmental Awakening in the Soviet Union," *Science*, August 26, 1988; "Report of the Round Table Subunit on Ecology," Warsaw, March 1989, submitted as supplementary material to Congressional Hearings on East-West Environmental Cooperation, Commission on Security and Cooperation in Europe, September 28, 1989.

69. Robert Watson, Natural Resources Defense Council, Speech to "Meeting on Soviet Energy Options and Global Environmental Issues," Georgetown University, June 29, 1989; "NRDC & Soviet Academy of Sciences Cooperation on Global Warming and Energy Efficiency," Status Report, unpublished memorandum, October 1989; "Strong Soviet Interest in RMI's Efficiency Message," *RMI Newsletter*, Rocky Mountain Institute, Old Snowmass, Colo., November 1987; "Back in the U.S.S.R.," *RMI Newsletter*, Rocky Mountain Institute, Old Snowmass, Colo., February 1989; "Swedish-Soviet Industry Committee Proposes Baltic Sea Environmental Plan," *International Environment Reporter*, August 1989.

70. George Ingram, Staff Consultant, Committee on Foreign Affairs, U.S. House of Representatives, Washington, D.C., private communication, November 21, 1989; "Swedish Government to Give Poland $45 Million for Environmental Activities," *International Environment Reporter*, November 1989; "Antipollution Technology Crosses from West Germany to the East," *New Scientist*, July 15, 1989; see also Helmut Schreiber, "East and West Germany Cooperate to Control Pollution," *European Environment Review*, December 1988.

71. Debbie Macklin, "Rebirth in Death Valley," *South*, March 1989; Tyler Bridges, "Brazilian City Battles Pollution," *New York Times*, May 23, 1988; "New Mexican President Promises Action to Control Pollution," *Multinational Environment Outlook*, December 22, 1988; William Branigin, "Japan Plans $1 Billion in Aid for Mexico to Combat Severe Air Pollution," *Washington Post*, August 30, 1989; "Mexico, U.S. Sign Accords for Joint Clean-Up Action," *Multinational Environmental Outlook*, October 17, 1989; Walsh, private communication.

72. John Elkington and Jonathan Shopley, *Cleaning Up: U.S. Waste Management Technology and Third World Development* (Washington, D.C.: World Resources Institute, 1989); Fiona Sebastian, World Bank, Washington, D.C., private communication, August 1, 1989; Barber B. Conable, Address to the Board of Governors of the World Bank Group, Washington, D.C., September 26, 1989; information on pending AID legislation from Carol Werner, Environmental and Energy Study Institute, U.S. Congress, Washington, D.C., private communication, October 11, 1989.

Chapter 7. Cycling into the Future

1. Number of bicycles worldwide is author's estimate based on International Trade Centre, *Bicycles and Components: A Pilot Survey of Opportunities for Trade Among Developing Countries* (Geneva: UNCTAD/GATT, 1985); number of automobiles (394 million in 1987) is from Motor Vehicle Manufacturers Association (MVMA), *Facts and Figures '89* (Detroit, Mich.: 1989).

2. R.H. ter Heide, "Recreational Use of the Bicycle, Organizational and Economic Aspects," *Velo City 87 International Congress: Planning for the Urban Cyclist*, proceedings of the Third International Velo City Congress, Groningen, the Netherlands, September 22–26, 1987 (hereinafter cited as Velo City 87 Conference). In 1986 the Netherlands had

some 120 passenger cars per square kilometer. Only Hong Kong and Singapore had more, with 175 and 400 cars per square kilometer, respectively; MVMA, *Facts and Figures '88* (Detroit, Mich.: 1988).

3. "Population Control Comes in the Kingdom of Bikes," *China Daily*, January 9, 1989; Jun-Men Yang, "Bicycle Traffic in China," *Transportation Quarterly*, January 1985; "See the PRC in Your Xiali," *Asiaweek*, September 2, 1988; Zhou You Ma, "Ode to the Bicycle," *China Reconstructs*, 1987; *Japan Cycle Press International*, October 1988.

4. V. Setty Pendakur, "Formal and Informal Urban Transport in Asia," *CUSO Journal*, December 1987; for a full description of different kinds of paratransit, see Peter J. Rimmer, *Rikisha to Rapid Transit: Urban Public Transport Systems and Policy in Southeast Asia* (Sydney: Pergamon Press, 1986); Ken Hughes and Michael A. Replogle, "Sustainable Transportation," *Not Man Apart*, July/August 1987.

5. David Mozer, Director, International Bicycle Fund, Bellevue, Wash., private communication, August 4, 1989.

6. Ibid.; Richard Barret, World Bank, Washington, D.C., private communication, March 24, 1989.

7. Ricardo A. Navarro et al., *Alternativas de Transporte en América Latina: La Bicicleta y los Triciclos* (St. Gallen: Swiss Center for Appropriate Technology, 1985); Ken Hughes, Executive Director, Institute for Transportation and Development Policy (ITDP), Washington, D.C., private communication, April 7, 1989.

8. Werner Brög et al., "Promotion and Planning for Bicycle Transportation: An International Overview," paper presented at annual meeting of the Transportation Research Board, National Research Council, Washington, D.C., January 1984.

9. Number of bicycles per person based on International Trade Centre, *Bicycles and Components*; Andy Clarke, "Pro-bike: A Cycling Policy for the 1990s" (London: Friends of the Earth, 1987); the Bicycle Institute of America, Inc., estimates that in 1988 there were 2.7 million bicycle commuters in the United States.

10. See, for example, John Pucher, "Urban Travel Behavior as the Outcome of Public Policy: The Example of Modal-split in Western Europe and North America," *APA Journal*, Autumn 1988; W.G.M. Huyink, "Cycling Policy in the City of Groningen," Velo City '87 Conference.

11. Michael A. Replogle, *Bicycles and Public Transportation: New Links to Suburban Transit Markets*, 2nd ed. (Washington, D.C.: The Bicycle Federation, 1988); Andy Clarke, Government Relations Director, League of American Wheelmen, Baltimore, Md., private communication, March 16, 1989; Clarke, "Pro-bike: A Cycling Policy for the 1990s."

12. Alethea Morison, Bicycle Institute of New South Wales, Sydney, Australia, private communication, June 1, 1989; James C. McCullagh, "10 Best Cycling Cities," *Bicycling*, November 1988.

13. Production figures based on *Japan Cycle Press International*, various editions; "NICs Update," *Japan Cycle Press International*, June 1988.

14. For a thorough discussion of the consequences of excessive reliance on cars, see Michael Renner, *Rethinking the Role of the Automobile*, Worldwatch Paper 84 (Washington, D.C.: Worldwatch Institute, June 1988).

15. Michael P. Walsh, "The Global Importance of Motor Vehicles in the Climate Modification Problem," *International Environment Reporter*, May 1989; Richard Gould, "The Exhausting Options of Modern Vehicles," *New Scientist*, May 13, 1989; Renner, *Rethinking the Role of the Automobile*; "New Mexican President Promises Action to Control Air Pollution," *Multinational Environmental Outlook*, De-

cember 22, 1988; "Air Pollution in Brazil," *Multinational Environmental Outlook*, July 21, 1988.

16. James J. MacKenzie, *Breathing Easier: Taking Action on Climate Change, Air Pollution, and Energy Insecurity* (Washington, D.C.: World Resources Institute, 1988); James J. MacKenzie and Mohamed T. El-Ashry, "Ill Winds: Air Pollution's Toll on Trees and Crops," *Technology Review*, April 1989.

17. Renner, *Rethinking the Role of the Automobile*; Gould, "The Exhausting Options of Modern Vehicles"; Walsh, "The Global Importance of Motor Vehicles in the Climate Modification Problem."

18. Jeffry J. Erickson et al., "An Analysis of Transportation Energy Conservation Projects in Developing Countries," *Transportation*, Vol. 1, No. 5, 1988.

19. Deborah Bleviss, *The New Oil Crisis and Fuel Economy Technologies: Preparing the Light Transportation Industry for the 1990's* (New York: Quorum Press, 1988).

20. "Traffic Jams: The City, the Commuter and the Car," *The Economist*, February 18, 1989; Melinda Beck, "Smart Cars, Smart Streets," *Newsweek*, December 5, 1988; "Taipei Residents Troubled by Traffic Tangle," *China Daily*, August 30, 1988; "Commuters' Nine to Five Favourites," *South*, November 1988.

21. MacKenzie, *Breathing Easier*; Andy Clarke, "How to Use Bicycling as a Cost-effective Means of Reducing Air Pollution," League of American Wheelmen, Baltimore, Md., 1989.

22. John Young, "Methanol Moonshine," *World Watch*, July/August 1988.

23. Jillian Beardwood and John Elliott, "Roads Generate Traffic," paper prepared for the Greater London Council, 1985; Farris quoted in Robert D. Ervin and Kan Chen, "Toward Motoring Smart," *Issues in Science and Technology*, Winter 1988–89.

24. MVMA, *Facts and Figures '88*; Clarke, "Pro-bike: A Cycling Policy for the 1990s."

25. Mary C. Holcomb et al., *Transportation Energy Data Book: Edition 9* (Oak Ridge, Tenn.: Oak Ridge National Laboratory, 1987).

26. U.N. Center for Human Settlements (Habitat), *Transportation Strategies for Human Settlements in Developing Countries* (Nairobi: 1984); World Bank, *Urban Transport: A World Bank Policy Study* (Washington, D.C.: 1986).

27. For a discussion of the health benefits of cycling, see Nelson Pena, "Survival of the Fittest," *Bicycling*, June 1988.

28. Charles Komanoff is president of Transportation Alternatives, a cycling advocacy group in New York City; quote from Charles Komanoff, "The Bike Ban is Bad Medicine," *The New York Observer*, January 25, 1988.

29. Automobile prices based on 1983 market price for an economy car in some 25 developing countries, from World Bank, *Urban Transport*; bicycle prices assumed to range from $80–250; V. Setty Pendakur, *Urban Growth, Urban Poor and Urban Transport in Asia* (Vancouver: University of British Columbia, 1986).

30. World Bank, *Urban Transport*, and other World Bank documents.

31. Val Curtis, *Women and the Transport of Water* (London: Intermediate Technology Publications, 1986); Charles K. Kaira, "Transportation Needs of the Rural Population in Developing Countries: An Approach to Improved Transportation Planning," doctoral dissertation, University of Karlsruhe, West Germany, 1983; I. Barwell et al., *Rural Transport in Developing Countries* (London: Intermediate Technology Publications, 1985).

32. Paul Zille, "Transport Research: An Investigation into Local Level Transport Characteristics and Requirements in Ismani

Division, Iringa, Tanzania," Project Report, Intermediate Technology Transport, Ltd., Oxon, U.K., 1988; Gordon Hathway, *Low-cost Vehicles: Options for Moving People and Goods* (London: Intermediate Technology Publications, 1985); Habitat, *Transportation Strategies for Human Settlements in Developing Countries*.

33. Thampil Pankaj, World Bank, Washington, D.C., private communication, February 22, 1989.

34. Hathway, *Low-cost Vehicles*; "Bicycle Adapted for New Uses for Indian Farmers," *New York Times*, October 16, 1980; Job S. Ebenezer, Associate Director, World Hunger Program of the Evangelical Lutheran Church in America, Chicago, private communication, August 1, 1989; "Rescue the Rickshaw," *New Internationalist*, May 1989.

35. S.V. Sethuraman, ed., *The Urban Informal Sector in Developing Countries: Employment, Poverty and Environment* (Geneva: International Labor Organization, 1981); "Bike—an Indispensable Tool," *China Daily*, February 1989; Pendakur, "Formal and Informal Urban Transport in Asia"; Robin Stallings, "The Present Role of the Bicycle in Iran, Afghanistan, Pakistan, and India," ITDP, Washington, D.C., 1981; Rebecca L. Reichmann and Ron Weber, "Solidarity in Development: The Tricicleros of Santo Domingo," *Grassroots Development*, Vol. 11, No. 2, 1987.

36. Pendakur, *Urban Growth, Urban Poor and Urban Transport in Asia*; Pendakur, "Formal and Informal Urban Transport in Asia."

37. S. Carapetis et al., *The Supply and Quality of Rural Transport Services in Developing Countries: A Comparative Review* (Washington, D.C.: World Bank, 1984); Wilfred Owen, *Transportation and World Development* (Baltimore, Md.: Johns Hopkins University Press, 1987).

38. Michael A. Replogle, "Transportation Strategies for Sustainable Development," paper prepared for Fifth World Conference on Transport Research, Yokohama, Japan, July 1989; Alain Boebion, "Indonesia Sinks

Symbol of Free Enterprise," *The Australian*, April 26, 1989; U.N. Department of International Economic and Social Affairs, *Population Growth and Policies in Mega-Cities: Dhaka* (New York: United Nations, 1987); Guy Dimond, "Wheels of Change in Bangladesh," *New Cyclist*, Winter 1988; Sethuraman, *The Urban Informal Sector in Developing Countries: Employment, Poverty and Environment*.

39. Replogle, "Transportation Strategies for Sustainable Development"; Pendakur, "Formal and Informal Urban Transport in Asia."

40. Michael A. Replogle, "Sustainable Transportation Strategies for Third World Development," paper presented at annual meeting of the Transportation Research Board, National Research Council, Washington, D.C., January 1988 (document he cited is "China Transport Sector Study," World Bank, Washington, D.C., 1985).

41. Hughes, numerous private communications; Barbara Francisco, "The Third World: Making it Mobile," *Washington Post*, January 30, 1989; Keith Oberg, "Beira: Low-cost Vehicle Demonstration Project," paper prepared for ITDP Bikes for Africa Program, Washington, D.C., 1989. Other groups active in research and technical support for bicycle promotion in developing countries include the Swiss Center for Appropriate Technology (SKAT), the Netherlands-based Center for International Cooperation and Appropriate Technology, and the German Appropriate Technology Center.

42. Pankaj, private communication.

43. Asian producers include China, Taiwan, Japan, India, Indonesia, Malaysia, Korea, and Thailand; Centro Salvadoreno de Tecnologia Apropiada, "Promotion of Bicycle and Tricycle Use in El Salvador," San Salvador, El Salvador, 1986; Navarro et al., *La Bicicleta y los Triciclos*.

44. Navarro et al., *La Bicicleta y los Triciclos*; International Trade Center, *Bicycles and Components*.

45. Navarro et al., *La Bicicleta y los Triciclos*.

46. Pucher, "Urban Travel Behavior."

47. Carapetis et al., *Supply and Quality of Rural Transport Services in Developing Countries*.

48. Wang Dazi, "Touring Beijing on a Rented Bike," *China Daily*, January 25, 1989; Li Jia Ying, "Management of Bicycling in Urban Areas," *Transportation Quarterly*, October 1987; Michael A. Powills and Chen Sheng-Hong, "Bicycles in Shanghai: A Major Transportation Issue," paper presented at annual meeting of the Transportation Research Board, National Research Council, Washington, D.C., January 1989; Wang Zhi Hao, "Bicycles in Large Cities in China," *Transport Reviews*, Vol. 9, No. 2, 1989.

49. Ryozo Tsutsumi, "Bicycle Safety and Parking Systems in Japan," *Pro Bike '88*, proceedings of the Fifth International Conference on Bicycle Programs and Promotions, Tucson, Ariz., October 8–12, 1988; Michael Replogle, Institute for Transportation and Development Policy, Washington, D.C., private communication, August 9, 1989.

50. Replogle, *Bicycles and Public Transportation*.

51. Tsutsumi, "Bicycle Safety and Parking Systems in Japan"; Replogle, private communication; Replogle, *Bicycles and Public Transportation*.

52. Transportation Environmental Studies of London (TEST), *Quality Streets: How Traditional Urban Centres Benefit from Traffic-calming* (London: 1988).

53. Ter Heide, "Recreational Use of the Bicycle."

54. James McGurn, *On Your Bicycle: An Illustrated History of Cycling* (London: John Murray Publishers, Ltd., 1987); A. Wilmink, "The Effects of State Subsidizing of Bicycle Facilities," Velo City '87 Conference; Michael A. Replogle, "Major Bikeway Construc-

tion Effort in the Netherlands," *Urban Transportation Abroad*, Winter 1982.

55. Ter Heide, "Recreational Use of the Bicycle"; M.J.P.F. Gommers, "The Bicycle Network of Delft, Influence on Trip-level, Network Use and Route Choice of Cyclists," Velo City '87 Conference.

56. Clarke, private communication; Dutch Ministry of Transport and Public Works, "Evaluation of the Delft Bicycle Network: Final Summary Report," The Hague, Netherlands, 1987; Dirk H. ten Grotenhuis, "The Delft Cycle Plan: Characteristics of the Concept," Velo City '87 Conference.

57. TEST, *Quality Streets*.

58. Debora MacKenzie, "Dutch Lead Drive to Banish Car Pollution," *New Scientist*, December 24, 1988; "Dutch Courage," *New Cyclist*, Spring 1989.

59. Gunna Starck, in Niels Jensen and Jens Erik Larsen, "Cycling in Denmark: From the Past into the Future," Road Directorate, Ministry of Transport, and the Municipality of Copenhagen, Copenhagen, 1989; Pucher, "Urban Travel Behavior"; Replogle, *Bicycles and Public Transportation*.

60. David Peltz, Department of Public Works, Davis, Calif., private communication, July 28, 1989.

61. Gail Likens, transportation planner, Palo Alto, Calif., private communication, July 28, 1989; Jane Gross, "Palo Alto Journal: Where Bicycle is King (And the Queen, Too)," *New York Times*, June 26, 1989; Lisa Lapin, "CBS News Sees P.A. as the Bicycle Capital," *San Jose Mercury News*, August 23, 1988; Ellen Fletcher, City Council member, Palo Alto, Calif., private communication, August 7, 1989.

62. Lapin, "CBS News Sees P.A. as the Bicycle Capital."

63. Beck, "Smart Cars, Smart Streets."

64. Renner, *Rethinking the Role of the Automobile*.

65. Pucher, "Urban Travel Behavior."

66. Gasoline tax revenues in particular can also support research in automotive fuel efficiency. Potentially regressive effects of increased fuel taxes can be mitigated by a number of direct and indirect measures: low-income drivers can be eligible, for example, for tax coupons presented at the gas pump. Tax revenues can also be used for home weatherization assistance to low-income households and aid in paying heating fuel bills; Nancy Hirsh, Environmental Action Foundation, Washington, D.C., private communication, July 31, 1989.

67. Pucher, "Urban Travel Behavior"; Replogle, private communication; Erickson et al., "Transportation Energy Conservation Projects in Developing Countries."

68. Replogle, *Bicycles and Public Transportation*; Walter Grabe and Joachim Utech, "The Importance of the Bicycle in Local Public Passenger Transport: Facts and Experience from Selected Countries," International Commission on Traffic and Urban Planning, *UITP Revue*, March 1984; Bróg et al., "Promotion and Planning for Bicycle Transportation."

69. Bicycle industry examples, including factory closings in Kenya and Tanzania, from Mozer, private communication.

70. World Bank, "Sub-Saharan Africa Transport Program: Rural Travel and Transport Project" (draft project proposal), unpublished, 1989; John D. N. Riverson and Bernard M. Chatelin, World Bank, Washington, D.C., private communication, February 22, 1989.

71. Peter Harnik, *The Bicycle Advocate's Handbook* (Baltimore, Md.: League of American Wheelmen, 1989).

72. Rails-to-Trails Conservancy, "Restoring Life to Abandoned Railroad Corridors," Washington, D.C., 1988; Rails-to-Trails Conservancy, "27 Million Americans Enjoyed over 200 Rail-Trails in '88," *Trailblazer*, January-March 1989.

73. For a thorough guide to safe cycling techniques and other aspects of bicycle transportation, see John Forester, *Effective Cycling* (Cambridge, Mass.: MIT Press, 1984); Carl Hultberg, "Bike/Ped Accidents Drop: Motor Mayhem Continues," *City Cyclist*, March/April 1989; Nadine Brozan, "Bicycle Riding Up; Accidents Decline," *New York Times*, June 10, 1989.

74. Among the many recent publications on urban form and growth management is Madis Pihlak, ed., *The City of the 21st Century*, proceedings of The City of the 21st Century Conference, Tempe, Ariz., April 7–9, 1988.

75. For a comprehensive manual on bicycle advocacy, see Harnik, *The Bicycle Advocate's Handbook*; Clarke, private communication, August 14, 1989.

76. In a subsequent meeting with U.S. Transportation Secretary Samuel Skinner, representatives of the League of American Wheelmen and other advocacy groups "secured a promise of full consultation in the current development of national transportation policy"; League of American Wheelmen, "Better Bicycling Sought for 120 Million Bicyclists by Year 2000," press release, Baltimore, Md., August 3, 1989.

77. "Argentina: Menem Faces Economic Disaster and Military Intimidation," *Latinamerica Press*, July 6, 1989.

78. McGurn, *On Your Bicycle*.

Chapter 8. Ending Poverty

1. Quoted in Ben Whitaker, *A Bridge of People: A Personal View of Oxfam's First Forty Years* (London: Heinemann, 1984).

2. Economy and industry from Jim MacNeill, "Strategies for Sustainable Economic

Development," *Scientific American*, September 1989; travel from Richard Critchfield, "Science and the Villager: The Last Sleeper Wakes," *Foreign Affairs*, Fall 1982.

3. Billionaires from Alan Farnham, "The Billionaires: Do They Pay Their Way?" *Fortune*, September 11, 1989; millionaires estimated from John Steele Gordon, "Why It Costs So Much to Be Rich," *Washington Post*, May 21, 1989; homeless from U.N. Centre for Human Settlements, New York, private communication, November 1, 1989; diet expenditures from Calorie Control Council, Atlanta, Ga., private communication, October 4, 1989; undernutrition estimated from World Bank, *Poverty and Hunger: Issues and Options for Food Security in Developing Countries* (Washington, D.C.: 1986); drinking water and sanitation based on United Nations Children's Fund (UNICEF), *State of the World's Children 1989* (New York: Oxford University Press, 1989); and on World Bank, *Social Indicators of Development 1988* (Baltimore, Md.: Johns Hopkins University Press, 1988).

4. Early eighties estimates from U.N. Food and Agriculture Organization (FAO), *Dynamics of Rural Poverty* (Rome: 1986); World Bank, *World Development Report 1980* (New York: Oxford University Press, 1980) and *World Development Report 1982* (New York: Oxford University Press, 1982).

5. In 1988, Robert Summers and Alan Heston of the University of Pennsylvania used the voluminous and complicated results of 20 years of United Nations, World Bank, and Organisation for Economic Co-operation and Development research to compile per capita gross domestic product figures adjusted to reflect purchasing power for 130 of the world's 168 nations; Robert Summers and Alan Heston, "A New Set of International Comparisons of Real Product and Price Level Estimates for 130 Countries, 1950–1985," *Review of Income and Wealth*, March 1988. A database on computer diskettes that accompanied the Summers and Heston article was used to develop many of the tables and figures in this chapter; updated figures for China from Alan Heston, University of Pennsylvania, Philadelphia, Pa., private communication, September 22, 1989. In citations hereafter, Summers and Heston refers to their article, database, and private communication. Groupings in Figure 1 are based on countries' 1985 per capita income: rich nations are those above $6,000, middle-income countries are between $2,500 and $6,000, poor countries are between $1,000 and $2,500, and poorest are below $1,000; data for 1986–88 are Worldwatch estimates.

6. Income distribution figures tend to understate disparities because they are calculated for households despite differences in household size: poor households are usually larger; share of population below average income estimated from data in sources for Table 8–1; Egypt-Peru comparison based on Summers and Heston and on World Bank, *World Development Report 1988* (New York: Oxford University Press, 1988).

7. Nations' records estimated from Summers and Heston; World Bank, *World Development Report 1988*; World Bank, *World Development Report 1989* (New York: Oxford University Press, 1989).

8. Most malnutrition records from UNICEF, *State of the World's Children 1989*; Nicaraguan malnutrition from Stephen Kinzer, "Nicaragua's Economic Crisis Is Seen as Worsening," *New York Times*, October 16, 1988; Salvadoran malnutrition from Lindsey Grusen, "Salvador's Poverty Is Called Worst in Century," *New York Times*, October 16, 1988; African life expectancy and hunger from World Bank and International Monetary Fund (IMF), *Strengthening Efforts to Reduce Poverty*, Development Committee Paper 19 (Washington, D.C.: World Bank, 1989).

9. Infant mortality and quote from UNICEF, *State of the World's Children 1989*.

10. Worldwatch examined available estimates of absolute poverty and income distribution on a country-by-country basis for

both rural and urban areas, and compared them with data on economic and social indicators and against estimates for similar countries. Worldwatch then made its own estimate for each country, trying to reconcile available data and apply a single standard of poverty. Poverty in industrial countries is not included in the global figures, because very little industrial-country poverty involves conditions of deprivation severe enough to be classified as absolute poverty. Industrial countries together contain fewer than 10 million individuals who live in absolute poverty—a quantity too small to alter the global estimate substantially.

11. World Bank figure based on *World Development Report 1980* and *World Development Report 1982*, which estimated absolute poverty at "nearly 1 billion" in 1980. In 1980, 1 billion people constituted 22.3 percent of the world population. In fact, figures in the *World Development Reports* cited add up to about 970 million (if allowance is made for countries the World Bank did not count, such as Vietnam, Laos, North Korea, and Kampuchea), or 21.7 percent of the global 1980 population.

12. Urban-rural division of poor estimated as above for overall poverty rates.

13. Extent of female poverty from Michael Lipton, *The Poor and the Poorest: Some Interim Findings*, Discussion Paper 25 (Washington, D.C.: World Bank, 1988); Irene Tinker, "Feminizing Development—for Growth and Equity," Briefs on Development Issues 6, CARE, New York, undated.

14. Lower earnings from FAO, *Women in Food Production* (Rome: 1983); additional work time from Ben White, "No Time to Spare," *New Internationalist*, March 1988; literacy lag from UNICEF, *State of the World's Children 1989*; violence and lack of rights from Lori Heise, "Crimes of Gender," *World Watch*, March/April 1989; Brazilian woman quoted in Maria Luiza de Melo Carvalho, "Earth Flies in All Directions: The Daily Lives of Women in Minas Gerais," *Grassroots Development*, Vol. 11, No. 2, 1987.

15. Lipton, *Poor and the Poorest*; Michael Lipton, *Poverty, Undernutrition, and Hunger*, Staff Working Paper 597 (Washington, D.C.: World Bank, 1983); percentage of poor under age of 15 and child mortality among the poor estimated based on these two sources and on UNICEF, *State of the World's Children 1989*.

16. Michael Lipton, *Land Assets and Rural Poverty*, Staff Working Paper 744 (Washington, D.C.: World Bank, 1985); Robert Chambers, *Rural Development: Putting the Last First* (Harlow, Essex, U.K.: Longman Scientific and Technical, 1983).

17. Esther B. Fein, "Glasnost Is Opening the Door on Poverty," *New York Times*, January 29, 1989.

18. U.S. poverty from Spencer Rich, "Poverty Level Stabilizes at 31 Million," *Washington Post*, October 19, 1989; equity trends from Stephen Rose and David Fasenfest, *Family Incomes in the 1980s: New Pressure on Wives, Husbands, and Young Adults* (Washington, D.C.: Economic Policy Institute, 1988); Robert D. Hamrin, "Sorry Americans—You're Still Not 'Better Off'," *Challenge*, September/October 1988.

19. Eduardo Galeano, *Open Veins of Latin America: Five Centuries of the Pillage of a Continent* (New York: Monthly Review Press, 1973).

20. William Thiesenhusen, *Economic Development in the Third World*, 2nd ed. (New York: Longmans, 1985); World Resources Institute, *World Resources 1988–89* (New York: Basic Books, 1988); Radha Sinha, *Landlessness: A Growing Problem*, Economic and Social Development Paper 28 (Rome: FAO, 1984); M. Riad El Ghonemy, ed., *Development Strategies for the Rural Poor*, Economic and Social Development Paper 44 (Rome: FAO, 1984)

21. Poor increasingly laborers from Michael Lipton with Richard Longhurst, *New*

Seeds and Poor People (Baltimore, Md.: Johns Hopkins University Press, 1989); landlessness estimates from Sinha, *Landlessness: A Growing Problem*; World Commission on Environment and Development (WCED), *Our Common Future* (New York: Oxford University Press, 1987).

22. Wage and price trends from Per Pinstrup-Andersen, "Food Security and Structural Adjustment," in Colleen Roberts, ed., *Trade, Aid, and Policy Reform: Proceedings from the Eighth Agricultural Symposium* (Washington, D.C.: World Bank, 1988); labor force growth from H. Jeffrey Leonard, "Overview: Environment and the Poor," in H. Jeffrey Leonard et al., *Environment and the Poor: Development Strategies for a Common Agenda* (New Brunswick, N.J.: Transaction Books, 1989); unemployment and underemployment from Lipton, *Poor and the Poorest*.

23. Vulnerability from Chambers, *Putting the Last First*; various articles in *IDS Bulletin* (Institute for Development Studies), April 1989.

24. Lipton, *Poverty, Undernutrition, and Hunger*.

25. Lipton, *Poor and the Poorest*; Chambers, *Putting the Last First*.

26. Use of laws against the poor from Robert Chambers et al., *To the Hands of the Poor: Water and Trees* (New Delhi: Oxford University Press and IBH, 1989); Amnesty International, *Brazil: Authorized Violence in Rural Areas* (London: 1988); violence in India from Chambers, *Putting the Last First*; Richard W. Franke and Barbara H. Chasin, *Radical Reform as Development: Kerala State, India* (San Francisco: Institute for Food and Development Policy, in press).

27. James Brooke, "Ivory Coast Church to Tower Over St. Peter's," *New York Times*, December 19, 1988.

28. World Bank, *World Development Report 1988*; "Social Sector Pricing Policy," *Research Brief* (World Bank), June 1989.

29. World Bank, *World Development Report 1988*; World Health Organization from Susan Okie, "Health Crisis Confronts 1.3 Billion," *Washington Post*, September 25, 1989.

30. World Bank, *World Debt Tables: External Debt of Developing Countries*, 1988–89 Edition, First Supplement and Vol. II (Washington, D.C.: 1989).

31. Net resource transfers from World Bank, *World Debt Tables*, First Supplement; environmental impact of debt from WCED, *Our Common Future*.

32. Primary commodity dependence from Walden Bello, *Brave New Third World? Strategies for Survival in the Global Economy*, Food First Development Report 5 (San Francisco: Institute for Food and Development Policy, 1989).

33. European tariff and cost of protectionism from World Bank, *World Debt Tables*, Vol. I.

34. Total capital flight from Matt Moffett, "Mexico's Capital Flight Still Racks Economy Despite the Brady Plan," *Wall Street Journal*, September 25, 1989; Venezuela's capital flight from David R. Francis, "A U.S. Siren Song for Capital Flight," *Christian Science Monitor*, March 27, 1989; Venezuela's debt from World Bank, *World Debt Tables*, First Supplement.

35. UNICEF, *State of the World's Children 1989*.

36. "Security Helps Productivity," *Research Brief* (World Bank), Spring 1987; Robert Chambers, "Poverty, Environment and the World Bank: The Opportunity for a New Professionalism," Institute for Development Studies (IDS), Brighton, Sussex, U.K., September 1987.

37. Shubh K. Kumar and David Hotchkiss, *Consequences of Deforestation for Women's Time Allocation, Agricultural Production, and Nutrition in Hill Areas of Nepal* (Washington, D.C.: In-

ternational Food Policy Research Institute (IFPRI), 1988).

38. Ibid.

39. Sheldon Annis, "Costa Rica's Dual Debt: A Story About a Little Country that Did Things Right" (unpublished), prepared for World Resources Institute, Washington, D.C., June 1987.

40. Ibid.

41. Leonard, "Overview: Environment and the Poor."

42. Ibid.

43. N.S. Jodha, "Common Property Resources and Rural Poor in Dry Regions of India," *Economic and Political Weekly*, July 5, 1986.

44. Ibid.; Botswana from Solomon Bekure and Neville Dyson-Hudson, "The Operation and Viability of the Second Livestock Development Project," International Livestock Center for Africa, cited by Rick Lomba, independent filmmaker, Testimony before Subcommittee on Foreign Operations, Committee on Appropriations, U.S. Senate, Washington, D.C., May 1, 1986.

45. Joshua Bishop, "Indigenous Social Structures, Formal Institutions, and the Management of Renewable Natural Resources in Mali" (draft), Sahelian Department, World Bank, Washington, D.C., August 1988.

46. Tim Campbell, "Urban Development in the Third World: Environmental Dilemmas and the Urban Poor," in Leonard et al., *Environment and the Poor*; air pollution from United Nations Environment Programme and World Health Organization, *Assessment of Urban Air Quality* (Nairobi: Global Environment Monitoring System, 1988).

47. Leonard, "Overview: Environment and the Poor"; Centre for Science and Environment, *The State of India's Environment 1984–85* (New Delhi: 1985).

48. Public Data Access, *Toxic Wastes and Race in the United States: A National Report on the Racial and Socio-Economic Characteristics of Communities with Hazardous Waste Sites* (New York: United Church of Christ Commission for Racial Justice, 1987); Jay M. Gould, *Quality of Life in American Neighborhoods: Levels of Affluence, Toxic Waste, and Cancer Mortality in Residential Zip Code Areas* (Boulder, Colo.: Westview Press, 1986); Dick Russell, "Environmental Racism," *Amicus Journal*, Spring 1989.

49. UNICEF, *State of the World's Children 1989*; impacts of hard times on poor from, for instance, Michael Mandel, "And If It Comes, the Poor Will Really Take a Hit," *Business Week*, April 10, 1989.

50. Quote from Robert Chambers, Hyderabad, India, private communication, September 7, 1989.

51. Quoted in Chambers, *Putting the Last First*.

52. Alan B. Durning, *Action at the Grassroots: Fighting Poverty and Environmental Decline*, Worldwatch Paper 88 (Washington, D.C.: Worldwatch Institute, January 1989).

53. Inter-American Foundation (IAF), "Project History: Centro de Educación Técnica Humanistica Agropecuaria (CETHA)," Arlington, Va., undated; Kevin Healy, IAF, Arlington, Va., private communication, October 4, 1989; female literacy benefits from Lipton, *Poor and the Poorest*; Jodi L. Jacobson, *Planning the Global Family*, Worldwatch Paper 80 (Washington, D.C.: Worldwatch Institute, December 1987).

54. Mahbub Hossain, *Credit for Alleviation of Rural Poverty: The Grameen Bank in Bangladesh*, Research Report 65 (Washington, D.C.: IFPRI, 1988); Kristin Helmore, "Banking on a Better Life," *Christian Science Monitor*, March 15, 1989.

55. Tushaar Shah, "Gains from Social Forestry: Lessons from West Bengal," IDS

Discussion Paper 243, Brighton, Sussex, U.K., April 1988.

56. Anil Agarwal and Sunita Narain, "The Greening of India," *Illustrated Weekly of India*, June 4, 1989.

57. UNICEF, *State of the World's Children 1989*.

58. Bangladesh Rural Advancement Committee, "Unraveling Networks of Corruption," in David C. Korten, ed., *Community Management: Asian Experience and Perspectives* (West Hartford, Conn.: Kumarian Press, 1986); Chambers, *Putting the Last First*.

59. Franke and Chasin, *Radical Reform as Development: Kerala State*.

60. Ibid.

61. Ibid.

62. Pinstrup-Andersen, "Food Security and Structural Adjustment."

63. Linda Feldman, "Bush Announces Plan to Forgive Sub-Saharan Africa's Debt," *Christian Science Monitor*, July 10, 1989; Michael Quint, "Brady's Debt Plan Forcing Big Banks to Increase Losses," *New York Times*, September 21, 1989.

Chapter 9. Converting to a Peaceful Economy

1. For a more detailed discussion, see Michael Renner, *National Security: The Economic and Environmental Dimensions*, Worldwatch Paper 89 (Washington, D.C.: Worldwatch Institute, May 1989).

2. For global military spending and arms exports, see Arms Control and Disarmament Agency (ACDA), *World Military Expenditures and Arms Transfers 1988* (Washington, D.C.: U.S. Government Printing Office, 1989); Michael R. Gordon, with Stephen Engelberg, "Military Ordered to Draft a 5% Cut in 1992–94 Spending," *New York Times*, November 18, 1989.

3. For military contractors' profitability, see Seymour Melman, "An Economic Alternative to the Arms Race: Conversion from Military to Civilian Economy," SANE Education Fund, Washington, D.C., 1987; Inga Thorsson, "Disarmament and Development: An Idea Whose Time Should Have Come," *IDS Bulletin* (Institute for Development Studies), October 1985; Lloyd J. Dumas and Suzanne Gordon, "Economic Conversion: An Exchange," *Bulletin of the Atomic Scientists*, June/July 1986. Even in the case of a military contract cancellation, contractors can look to indemnification; no comparable payments assist workers whose livelihoods are at stake; Jonathan Feldman, "An Introduction to Economic Conversion," Briefing Paper No. 1, National Commission for Economic Conversion and Disarmament, Washington, D.C., May 1988.

4. Because technical and scientific personnel in military industries make up a much larger portion of the work force than in civilian enterprises (in the United States, 15 percent compared with 3 percent), engineers and scientists may find it difficult to find employment in the civilian sphere; see "Labor and Conversion," *Economic Notes*, February 1986.

5. Jonathan Feldman et al., "Criteria for Economic Conversion Legislation," Briefing Paper No. 4, National Commission for Economic Conversion and Disarmament, Washington, D.C., December 1988; Dumas and Gordon, "Economic Conversion: An Exchange."

6. Employment Research Associates, *Bankrupting America* (Lansing, Mich.: 1989).

7. U.S. arms production employment, 1950–91, from U.S. Department of Defense, *National Defense Budget Estimates for FY 1990/91* (Washington, D.C.: Office of the Assistant Secretary of Defense (Comptroller), 1989).

8. Klaus Schomacker et al., *Alternative Produktion Statt Rüstung. Gewerkschaftliche Initia-*

tiven für Sinnvolle Arbeit und Sozial Nützliche Produkte (Bund Verlag: Köln, 1987).

9. Employment loss in 1963–78 from Steve Schofield, "Employment and Security—Alternatives to Trident. An Interim Report," Peace Research Reports No. 10, University of Bradford, U.K., July 1986; 1978–86 job losses from John Lovering, "Arms Conversion into the 1990s: The British Experience," paper presented at conversion conference in Cortona, Italy, April 22, 1989.

10. William Hartung, *The Economic Consequences of a Nuclear Freeze* (New York: Council on Economic Priorities, 1984); Employment Research Associates data from Peter Wilke and Herbert Wulf, "Manpower Conversion in Defense-Related Industry," Disarmament and Employment Program, Working Paper No. 4, International Labour Organisation, Geneva, 1986. In 1985 dollars, $1 billion spent on pollution control would create 19,-547 jobs; in 1981 dollars, the same nominal amount would generate 16,380 jobs; Worldwatch Institute, based on Roger H. Bezdek et al., "The Economic and Employment Effects of Investments in Pollution Abatement and Control Technologies," *Ambio*, Vol. 18, No. 5, 1989.

11. West German data from Wilke and Wulf, "Manpower Conversion in Defense-Related Industry"; Indian figures from Mary Kaldor, *The Baroque Arsenal* (New York: Hill and Wang, 1981).

12. For a more detailed discussion, see Michael Dee Oden, *A Military Dollar Really Is Different: The Economic Impacts of Military Spending Reconsidered* (Lansing, Mich.: Employment Research Associates, 1988).

13. For a discussion of corporate diversification attempts, see Feldman et al., "Criteria for Economic Conversion Legislation."

14. Renner, *National Security: The Economic and Environmental Dimensions*.

15. Employment total from ibid.; assumptions on productivity gains and inflation from Wilke and Wulf, "Manpower Conversion in Defense-Related Industry." These scenarios deal only with the direct employment effects; they take into account neither the considerable indirect job losses in supplier industries nor the impact on what economists call "induced employment": jobs sustained by purchases made by defense workers.

16. Because a START accord will not prescribe how to meet the numerical weapons limits it codifies, additional options are possible within that range; see Stephen Alexis Cain, *The START Agreement: Strategic Options and Budgetary Savings* (Washington, D.C.: Center on Budget and Policy Priorities, 1988). The Soviet Union might save as much as $11 billion per year. For START's impact on Soviet military spending and for the conventional and naval forces scenarios, see Barry Blechman with Ethan Gutmann, "A $100 Billion Understanding," *SAIS Review* (School of Advanced International Studies, Johns Hopkins University), Summer-Fall 1989.

17. Feldman et al., "Criteria for Economic Conversion Legislation."

18. Seymour Melman, "Problems of Conversion from Military to Civilian Economy—An Agenda of Topics, Questions, and Hypotheses," *Bulletin of Peace Proposals*, Vol. 16, No. 1, 1985.

19. Ibid.

20. Seymour Melman, *Profits Without Production* (New York: Alfred A. Knopf, 1983).

21. See Greg Bischak, "Facing the Second Generation of the Nuclear Weapons Complex: Renewal of the Nuclear Production Base or Economic Conversion?" in Lloyd Dumas and Marek Thee, eds., *Making Peace Possible* (Elmsford, N.Y.: Pergamon Press, 1989).

22. Dumas and Gordon, "Economic Conversion: An Exchange."

23. For a discussion of the implications of the question "conversion to what?" see Michael Renner, "Conversion to a Peaceful Economy. Criteria, Objectives, and Constituencies," *Bulletin of Peace Proposals*, Vol. 19, No. 1, 1988.

24. Wilke and Wulf, "Manpower Conversion in Defense-Related Industry."

25. The Conference of Mayors study found a net gain of 6,600 jobs, on average, for every $1 billion shifted from military spending to the examined urban programs. Over five years, the proposed transfer of $150 billion would generate close to 1 million jobs *more* than the same amount spent in the military sector would have created. See Employment Research Associates, *A Shift in Military Spending to America's Cities*, Report Prepared for the U.S. Conference of Mayors (Washington, D.C., and Lansing, Mich.: 1988).

26. Auto industry employment figures are for 1986; see Motor Vehicle Manufacturers Association, *Facts and Figures '89* (Detroit, Mich.: 1989). For an analysis of the automobile's contribution to air pollution and global warming, see Michael Renner, *Rethinking the Role of the Automobile*, Worldwatch Paper 84 (Washington, D.C.: Worldwatch Institute, June 1988).

27. U.S. pollution control employment is a Worldwatch estimate, based on Bezdek et al., "The Economic and Employment Effects of Investments in Pollution Abatement and Control Technologies." The authors calculate that business investment for pollution control (12 percent of total U.S. spending for environmental protection in 1985) created some 167,000 jobs. Assuming that the remaining 88 percent of expenditures has a similar employment effect, the total number of jobs is just under 1.4 million. For European Community, see Werner Schneider, "Arbeit und Umwelt," in Lothar Gündling and Beate Weber, eds., *Dicke Luft in Europa* (Heidelberg, West Germany: C.F. Müller, 1988).

28. Michael Renner, "Swords Into Consumer Goods," *World Watch*, July/August 1989.

29. ACDA, *World Military Expenditures and Arms Transfers 1988*.

30. "China Invents the Entrepreneurial Army," *Economist*, May 14, 1988; "Military Plant Goes Civilian," *China Daily*, November 23, 1987; Lu Yun, "War Industry Turns Out Civilian Goods," *Beijing Review*, August 3, 1987.

31. Renner, "Swords Into Consumer Goods."

32. Marta Dassu, "The Problem of Reconversion of the Military Industry: The Case of China," Occasional Papers No. 2 (Rome: Center for the Study of International Relations, 1988); SIPRI Yearbook 1988, *World Armaments and Disarmament* (New York: Oxford University Press, 1988).

33. Alexandri Remisov, "Disarmament Treaties and Conversion of Military Production in USSR," *IDOC Internazionale*, September/October 1988; Alexei I. Izyumov, "Soviet Economic Conversion," *The New Economy*, October/November 1989; William J. Broad, "A Potential New Soviet Export Product: Cold-War-Surplus Missiles," *New York Times*, October 22, 1989; "In Gorbachev's Words: 'To Preserve the Vitality of Civilization'," *New York Times*, December 8, 1988.

34. Number of Soviet factories to be converted from John Lovering, University of Bristol, U.K., private communication, October 28, 1989; Peter Gumbel, "Gorbachev Announces Spending on Soviet Military for First Time," *Wall Street Journal*, May 31, 1989; overall arms production cut and Ryzhkov announcement from Michael Dobbs, "Soviets Admit Massive Deficit, Call for Defense Spending Cuts," *Washington Post*, June 8, 1989; tank cuts from Michael R. Gordon, "Soviets Cite 40% Cut in Tank Output," *New*

York Times, July 22, 1989; trade-off between military spending and housing construction from L. Veselovsky, Deputy Head of the All-Union Central Council of Trade Unions in Moscow, in "Notes on the Discussions at the U.S.-U.S.S.R. Symposium on Conversion from Military to Civilian Economy" (co-chaired by Prof. Seymour Melman, Columbia University, New York, and Dr. Ivan Ivanov, Institute of World Economy and International Relations, Moscow), Moscow, June 14–16, 1984; environmental protection budget from Maria Welfens, "Umweltpolitik im Sozialismus: Diagnose, Analyse, Perspektive," in Kurt P. Tudyka, ed., *Umweltpolitik in Ost- und Westeuropa* (Opladen, West Germany: Leske & Budrich, 1988).

35. Civilian production in military plants from Paul Grenier with Eric Stubbs, "A Farewell to Arms? Soviets Rethink Defense Spending," *CEP Research Report*, Council on Economic Priorities, New York, 1989; washing machine example from "Notes on the Discussions at the U.S.-U.S.S.R. Symposium"; increased food equipment and consumer goods production by military plants from Central Intelligence Agency and the Defense Intelligence Agency (CIA/DIA), "The Soviet Economy in 1988: Gorbachev Changes Course," report presented to Subcommittee on National Security Economics, Joint Economic Committee, U.S. Congress, Washington, D.C., April 14, 1989, and from U.S. Department of Agriculture, Economic Research Service, *USSR Agriculture and Trade Report: Situation and Outlook Series*, Washington, D.C., May 1989.

36. Remisov, "Disarmament Treaties and Conversion of Military Production in USSR"; CIA/DIA, "The Soviet Economy in 1988."

37. Izyumov, "Soviet Economic Conversion."

38. U.S. arms production employment from U.S. Department of Defense, *National Defense Budget Estimates*; recent military industry job losses from "Overcoming Defense Dependency and Revitalizing America's Economy," Appendix 1: Selected Recent Defense Cutbacks and Layoffs, Center for Economic Conversion (CEC), Mountain View, Calif., May 1988, and from Seymour Melman, Chair of National Commission for Economic Conversion and Disarmament, "Economic Conversion: The Missing Link for Economic Security," Statement to Subcommittee on Economic Stabilization, Committee on Banking, Finance, and Urban Affairs, U.S. House of Representatives, Washington, D.C., June 13, 1989; job losses from base closures from U.S. Congressional Budget Office, "Past Base Closures and Realignments: Costs and Savings," Staff Working Paper, Washington, D.C., February 1989; Gordon, with Engelberg, "Military Ordered to Draft a 5% Cut."

39. Office of Economic Adjustment, *1961–1986: 25 Years of Civilian Reuse. Summary of Completed Military Base Economic Adjustment Projects* (Washington, D.C.: 1986); Robert Krinsky, "Community Programs Aid the Pentagon," *Bulletin of the Atomic Scientists*, October 1986.

40. CEC, "Conversion Organizer's Update," Mountain View, Calif., various issues; Louise McNeilly, CEC, Mountain View, Calif., private communication, August 26, 1989; Martin Melkonian and Russell M. Moore, "Cutbacks in Defense Spending: Outlook and Options for the Long Island Economy," Business Research Institute, Hofstra University, Hempstead, N.Y., February 1989; Wilbur Maki et al., "Military Production and the Minnesota Economy," Minnesota Task Force on Economic Conversion, St. Paul, Minn., May 1989.

41. For an overview of current conversion initiatives, see CEC, "Conversion Organizer's Update," various issues.

42. For International Association of Machinists, see Hartung, *The Economic Consequences of a Nuclear Freeze*. Even during 1982 and 1985, when military procurement spend-

ing rose by 25 percent, employment in military industry only increased by 8.6 percent; "Labor and Conversion," *Economic Notes*, February 1986.

43. "A Healthy Economy in a Peaceful World" (flyer), National Jobs With Peace Campaign, Boston, undated; Frank Clemente, Jobs With Peace Washington office, private communication, February 3, 1987; Lisa Peattie, "Economic Conversion as a Set of Organizing Ideas," *Bulletin of Peace Proposals*, Vol. 19, No. 1, 1988.

44. U.K. military spending (in fiscal year 1986/87 prices) grew from 15.6 billion pounds (1 pound, at 1987 exchange rates, equals $1.87) in 1978/79 to a peak of 18.8 billion in 1985/86, but then declined to 17.5 billion in 1988/89; Paul Quigley, "The Opportunity of Disarmament: Investigating Alternatives for Military-Supported Employment in Coventry," Coventry Alternative Employment Research, Coventry, U.K., October 1988. Lovering, "Arms Conversion into the 1990s: The British Experience." The Trident nuclear submarine program, estimated to absorb 6–8 percent of the total military budget, is likely to create no more than 2 percent of British military industry employment; Schofield, "Employment and Security—Alternatives to Trident."

45. Although no proposals were directly taken up, the company did eventually move into some of the areas identified in the plan; Schofield, "Employment and Security—Alternatives to Trident." Perhaps the most comprehensive account of the Lucas campaign can be found in Dave Elliott and Hilary Wainwright, *The Lucas Plan—A New Trade Unionism in the Making?* (London: Allison and Busby, 1982).

46. Lovering, "Arms Conversion into the 1990s: The British Experience"; Klaus Schomacker et al., *Sichere Arbeitsplätze und Nützliche Produkte*, *Militärpolitik Dokumentation*, Vol. 8, No. 39/40, 1984.

47. For Labour Party position, see Lovering, "Arms Conversion into the 1990s: The British Experience"; local conversion committees from Jim Barnes and John Chowcat, "Arms Conversion—from Words to Action," *Sanity*, March 1987.

48. Lovering, "Arms Conversion into the 1990s: The British Experience."

49. Ibid.; Paul Quigley, "Conversion in the United Kingdom: A Review of Recent Developments," *The New Economy*, October/November 1989.

50. West German adjustment conditions discussed in John E. Ullmann, "Conversion Studies in West Germany," *The New Economy*, October/November 1989; Schomacker et al., *Alternative Produktion Statt Rüstung*.

51. Alternative product proposals have focused on less-polluting equipment for power plants, water and waste treatment, railways, and wind energy; Schomacker et al., *Alternative Produktion Statt Rüstung*.

52. Ibid.

53. Klaus Mehrens, "Alternative Produktion und Gewerkschaftliche Politik," *Die Mitbestimmung*, April/May 1984; Schomacker et al., *Alternative Produktion Statt Rüstung*.

54. Mehrens, "Alternative Produktion und Gewerkschaftliche Politik"; Schomacker et al., *Alternative Produktion Statt Rüstung*.

55. Inga Thorsson, *In Pursuit of Disarmament. Conversion from Military to Civil Production in Sweden* (Stockholm: Liber Allmänna Förlaget, 1984).

56. Ibid.

57. Follow-up report information from Inga Thorsson, private communication, August 17, 1989; *Veckans Affärer* report from Bengt Ljung, "Under the Gun: How Much Defense Can Sweden Afford?" *Sweden Now*, No. 2, 1989; Björn Hagelin, "The Prospects for Conversion in Sweden," *The New Economy*, October/November 1989.

58. Pieter van Rossem, "Ontwapening en Ontwikkeling: Op Dood Spoor," *Derde Wereld*, July 1986; "An Act to Provide for the Conversion of Technologies and Skills Used in the Nuclear Weapons Industry to Civilian Uses," Bill 18, Introduced by Richard Johnson, MPP, Scarborough West, Canada, November 10, 1987; Mario Pianta, "Conversion in Italy," *The New Economy*, October/November 1989.

59. "Criteria for Conversion Legislation and Guide to Three Bills in the 101st Congress" (fact sheet), National Commission for Economic Conversion and Disarmament, Washington, D.C., July 1989; recent efforts to create a single conversion bill foundered; Jennifer Adibi, "Legislative Labor Pains," *Plowshare Press*, Summer 1989.

60. The Defense Economic Adjustment Council is to be composed of various cabinet members, including the Secretary of Defense, six management representatives from nondefense companies, and six union representatives; H.R. 101, "The Defense Economic Adjustment Act," 101st U.S. Congress.

61. "The Defense Economic Adjustment Act."

62. Ibid. As an alternative to "taxing" contractors' revenues, Representative Weiss is now considering providing the funding through a 1–2 percent set-aside within the defense budget; "Statement of Representative Ted Weiss," Hearings on Title II of the Defense Production Act Pertaining to Economic Conversion, Subcommittee on Economic Stabilization, Standing Committee on Banking, Finance, and Urban Affairs, U.S. House of Representatives, Washington, D.C., June 13, 1989.

63. "The Defense Economic Adjustment Act.," Title 1, Sec. 102, paragraph 2.

64. The U.N. Centre for Science and Technology for Development recently sponsored a gathering of conversion experts to make recommendations for policy initiatives by the U.N. Secretary General; Uppsala's Center for Military Conversion covered in CEC, "Conversion Organizers' Update," May 1989.

Chapter 10. Picturing a Sustainable Society

1. World Commission on Environment and Development, *Our Common Future* (New York: Oxford University Press, 1987).

2. U.S. Environmental Protection Agency, *Policy Options for Stabilizing Global Climate* (draft) (Washington, D.C.: 1989); Gregg Marland et al., *Estimates of CO_2 Emissions from Fossil Fuel Burning and Cement Manufacturing, Based on the United Nations Energy Statistics and the U.S. Bureau of Mines Cement Manufacturing Data* (Oak Ridge, Tenn.: Oak Ridge National Laboratory, 1989); United Nations Secretariat, "Long-Range Global Population Projections as Assessed in 1980," *Population Bulletin of the United Nations* (New York: 1983).

3. Worldwatch Institute estimate based on "World List of Nuclear Power Plants," *Nuclear News*, August 1989; Christopher Flavin, *Reassessing Nuclear Power: The Fallout from Chernobyl*, Worldwatch Paper 75 (Washington, D.C.: Worldwatch Institute, March 1987).

4. John J. Taylor, "Improved and Safer Nuclear Power," *Science*, April 21, 1989; "Outlook on Advanced Reactors," *Nucleonics Week*, March 30, 1989; Amory B. Lovins and L. Hunter Lovins, *Brittle Power: Energy Strategy for National Security* (Andover, Mass.: Brick House Publishing Co., 1982); R.H. Williams and H.A. Feiveson, "Diversion-Resistance Criteria for Future Nuclear Power," Center for Energy and Environmental Studies, Princeton University, Princeton, N.J., May 22, 1989; Klaus Michael Meyerabich and Bertram Schefold, *Die Grenzen der Atomwitschaft* (Munich: Verlag C.H. Beck, 1986).

5. Population figure is authors' estimate derived from United Nations (UN), Depart-

ment of International Economic and Social Affairs (DIESA), *World Population Prospects 1988* (New York: 1989); extrapolating from medium projection for 2025 yields 8.85 billion.

6. Population Information Program, "Population and Birth Planning in the People's Republic of China," *Population Reports*, January/February 1982; UN, DIESA, *World Population Prospects 1988;* Population Reference Bureau, *1989 World Population Data Sheet* (Washington, D.C.: 1989).

7. UN, *1987 Energy Statistics Yearbook* (New York: 1989); Meridian Corporation, "Characterization of U.S. Energy Resources and Reserves," prepared for Deputy Assistant Secretary for Renewable Energy, U.S. Department of Energy (DOE), Alexandria, Va., June 1989.

8. D. Groues and I. Segal, *Solar Energy in Israel* (Jerusalem: Ministry of Energy & Infrastructure, 1984); International Energy Agency (IEA), *Renewable Sources of Energy* (Paris: Organisation for Economic Co-operation and Development, 1987).

9. Paul Savoldelli, Luz International Ltd., private communication and printout, July 9, 1989; Charles Komanoff, Komanoff Energy Associates, New York, private communication and printout, February 10, 1989.

10. H.M. Hubbard, "Photovoltaics Today and Tomorrow," *Science*, April 21, 1989; Solar Energy Research Institute (SERI), "Photovoltaics: Electricity from Sunshine," unpublished, Golden, Colo., 1989.

11. Hubbard, "Photovoltaics"; Christopher Flavin, *Electricity From Sunlight: The Emergence of Photovoltaics* (Washington, D.C.: U.S. Government Printing Office, 1984).

12. Hubbard, "Photovoltaics"; V. Elaine Gilmore, "Solar Shingles," *Popular Science*, June 1989.

13. Robert R. Lynette, "Wind Energy Systems," Paper presented to the Forum on Renewable Energy and Climate Change, Washington, D.C., June 14–15, 1989; Worldwatch Institute estimates based on Shepard Buchanan, Bonneville Power Administration, Portland, Ore., private communication and printout, July 28, 1989.

14. Mark Newman, "West Germany to Build 150 MWe of Wind Farms by Mid-1990s," *International Solar Energy Intelligence Report*, November 22, 1988; Department of Non-Conventional Energy Sources, Ministry of Energy, *Annual Report 1987–88* (New Delhi, India: undated).

15. Christopher Flavin, *Wind Power: A Turning Point*, Worldwatch Paper 45 (Washington, D.C.: Worldwatch Institute, July 1981).

16. Worldwatch Institute estimate, based on Larry Langemeier, Kansas State University, Manhattan, Kans., private communication, November 3, 1989.

17. UN, *1987 Energy Statistics Yearbook*.

18. Worldwatch Institute estimate based on D.O. Hall et al., *Biomass for Energy in Developing Countries* (Elmsford, N.Y.: Pergamon Press, 1982); Daniel Deudney and Christopher Flavin, *Renewable Energy: The Power to Choose* (New York: W.W. Norton & Co., 1983); and British Petroleum, *BP Statistical Review of World Energy* (London: 1989).

19. Lynn Wright, Oak Ridge National Laboratory, Oak Ridge, Tenn., private communication, August 25, 1989; Norman Hinman, SERI, Boulder, Colo., private communication, August 25, 1989.

20. Worldwatch Institute estimate based on Brad Karmen, U.S. Department of Agriculture (USDA), Washington, D.C., private communication, October 25, 1989, on Energy Information Administration (EIA), *Monthly Energy Review*, March 1989 (Washington, D.C.: DOE, 1989), on Wright, private communication, and on Hinman, private communication.

21. UN, *1987 Energy Statistics Yearbook*.

22. Ronald DiPippo, "International Developments in Geothermal Power Development," *Geothermal Resources Council Bulletin*, May 1988.

23. Meridian Corporation, "Energy System Emissions and Matériel Requirements," prepared for U.S. DOE, Alexandria, Va., February 1989; Savoldelli, private communication.

24. Worldwatch Institute estimates based on Savoldelli, private communication and printout, and on EIA, *Monthly Energy Review*, March 1989, assuming a solar-only mode and a capacity factor of 26 percent for the solar thermal stations; Paul Gipe, Paul Gipe and Associates, Tehachapi, Calif., private communication and printout, October 22, 1989.

25. Christopher Flavin and Alan Durning, *Building on Success: The Age of Energy Efficiency*, Worldwatch Paper 82 (Washington, D.C.: Worldwatch Institute, March 1988).

26. Deborah Bleviss, *The New Oil Crisis and Fuel Economy Technologies: Preparing the Light Transportation Industry for the 1990's* (New York: Quorum Press, 1988).

27. Flavin and Durning, *Building on Success*; Arthur Rosenfeld and David Hafemeister, "Energy-Efficient Buildings," *Scientific American*, April 1988; Peter Weiss, "Lighting the Way Towards More Efficient Lighting," *Home Energy*, January/February 1989.

28. José Goldemberg et al., "An End-Use Oriented Global Energy Strategy" in Annual Review, Inc., *Annual Review of Energy*, Vol. 10 (Palo Alto, Calif.: 1985).

29. Jørgen Norgaard, Technical University of Denmark, Lyngby, Denmark, private communications, October 28–29, 1987; Howard Geller, "Energy-Efficient Residential Appliances: Performance Issues and Policy Options," *IEEE Technology and Society Magazine*, March 1986; David B. Goldstein and Peter Miller, "Developing Cost Curves for Conserved Energy in New Refrigerators and Freezers," American Council for an Energy-Efficient Economy, Washington, D.C., 1986.

30. Marc Ross, "Industrial Energy Conservation," *Natural Resource Journal*, August 1984; Marc Ross, "Industrial Energy Conservation and the Steel Industry," *Energy, The International Journal*, October/November 1987.

31. U.S. Congress, Office of Technology Assessment, *Industrial Energy Use* (Washington, D.C.: U.S. Government Printing Office, 1983).

32. Bundesministerium für Verkehr, *Verkehr in Zahlen 1987* (Bonn: 1987).

33. Marcia D. Lowe, *The Bicycle: Vehicle for a Small Planet*, Worldwatch Paper 90 (Washington, D.C.: Worldwatch Institute, September 1989); A. Wilmink, "The Effects of State Subsidizing of Bicycle Facilities," *Velo City '87 Congress: Planning for the Urban Cyclist*, proceedings of the Third International Velo City Congress, Groningen, the Netherlands, September 22–26, 1987; Michael A. Replogle, "Major Bikeway Construction Effort in the Netherlands," *Urban Transportation Abroad*, Winter 1982; David Peltz, Davis Department of Public Works, Davis, Calif., private communication, July 28, 1989; Michael A. Replogle, *Bicycles and Public Transportation: New Links to Suburban Transit Markets*, 2nd ed. (Washington, D.C.: The Bicycle Federation, 1988).

34. Worldwatch Institute estimate based on International Road Federation, *World Road Statistics 1981–1985* (Washington, D.C.: 1986), and on Motor Vehicle Manufacturers Association, *Facts and Figure '87* (Detroit, Mich.: 1987); "The Motorization of the 3rd World," *National Association of Railroad Passengers News*, July 1987.

35. Share discarded of various materials and energy savings of aluminum and glass recycling from Cynthia Pollock, *Mining Urban Wastes: The Potential for Recycling*, Worldwatch

Paper 76 (Washington, D.C.: Worldwatch Institute, April 1987); energy savings of steel recycling from William U. Chandler, *Materials Recycling: The Virtue of Necessity*, Worldwatch Paper 56 (Washington, D.C.: Worldwatch Institute, October 1983); energy savings of newsprint recycling from Roberta Forsell Stauffer, "Energy Savings From Recycling," *Resource Recycling*, January/February 1989.

36. Pollock, *Mining Urban Wastes*.

37. Chandler, *Materials Recycling*; Pollock, *Mining Urban Wastes*.

38. Donald F. Barnett and Robert W. Crandall, *Up From the Ashes: The Rise of the Steel Minimill in the United States* (Washington, D.C.: Brookings Institution, 1986).

39. Kirsten U. Oldenburg and Joel S. Hirschhorn, "Waste Reduction: A New Strategy to Avoid Pollution," *Environment*, March 1987.

40. "Cost of Packaging Food Could Exceed Farm Net," *Journal of Commerce*, August 12, 1986.

41. Lester R. Brown and Jodi L. Jacobson, *The Future of Urbanization: Facing the Ecological and Economic Constraints*, Worldwatch Paper 77 (Washington, D.C.: Worldwatch Institute, May 1987); Shanghai example from Yue-Man Yeung, "Urban Agriculture in Asia," The Food Energy Nexus Program of the United Nations University, Tokyo, September, 1985.

42. Peter Edwards, *Aquaculture: A Component of Low Cost Sanitation Technology* (Washington, D.C.: United Nations Development Programme and World Bank, 1985).

43. Assumes worldwide soil erosion rate of 24 billion tons per year and forest loss of 11 million hectares per year; see Lester R. Brown and Edward C. Wolf, *Soil Erosion: Quiet Crisis in the World Economy*, Worldwatch Paper 60 (Washington, D.C.: Worldwatch Institute, September 1984); U.N. Food and Agriculture Organization (FAO), *Tropical Forest Resources*, Forestry Paper 30 (Rome: 1982).

44. Per capita cropland area calculated from FAO, *Production Yearbook 1987* (Rome: 1988), and from Population Reference Bureau, *1989 World Population Data Sheet*; assumed cropland expansion from Francis Urban, "Agricultural Resources Availability," *World Agriculture Situation and Outlook Report*, USDA, Economic Research Service, Washington, D.C., June 1989.

45. Walter V. C. Reid, "Sustainable Development: Lessons from Success," *Environment*, May 1989; International Council for Research in Agroforestry (ICRAF)-Nitrogen Fixing Tree Association International Workshop, *Perennial Sesbania Species in Agroforestry Systems* (Nairobi: ICRAF, 1989).

46. Sandra Postel, "Halting Land Degradation," in Lester R. Brown et al., *State of the World 1989* (New York: W.W. Norton & Co., 1989).

47. Ibid.; grain fed to livestock from USDA, Foreign Agricultural Service (FAS), *World Wheat and Coarse Grains Reference Tables* (unpublished printout) (Washington, D.C.: August 1988), and Gerald Ostrowski, USDA, FAS, private communication, September 1989.

48. For discussion of the work of the Land Institute, see Evan Eisenberg, "Back to Eden," *Atlantic Monthly*, November 1989; Wes Jackson, *Altars of Unhewn Stone: Science and the Earth* (San Francisco: North Point Press, 1987).

49. Omar Sattaur, "The Shrinking Gene Pool," *New Scientist*, July 29, 1989; National Research Council, *Amaranth: Modern Prospects for an Ancient Crop* (Washington, D.C.: National Academy Press, 1984).

50. Sandra Postel and Lori Heise, *Reforesting the Earth*, Worldwatch Paper 83 (Washington, D.C.: Worldwatch Institute, April 1988).

51. Philip M. Fearnside, "A Prescription for Slowing Deforestation in Amazonia," *Environment*, May 1989.

52. Reid, "Sustainable Development"; Charles M. Peters et al., "Valuation of an Amazonian Rainforest," *Nature*, June 29, 1989.

53. Postel and Heise, *Reforesting the Earth*.

54. Lester R. Brown et al. "Outlining a Global Action Plan," in Brown et al., *State of the World 1989*.

55. Postel, "Halting Land Degradation."

56. William C. Thiesenhusen, ed., *Searching for Agrarian Reform in Latin America* (Boston: Unwin Hyman, 1989); for some interesting examples of village land management, see Anil Agarwal and Sunita Narain, "The Greening of India," *Illustrated Weekly of India*, June 10, 1989.

57. Peter M. Vitousek et al., "Human Appropriation of the Products of Photosynthesis," *BioScience*, June 1986; Paul R. Ehrlich et al., "Global Change and Carrying Capacity: Implications for Life on Earth," in Ruth S. DeFries and Thomas F. Malone, eds., *Global Change and Our Common Future: Papers From a Forum* (Washington, D.C.: National Academy Press, 1989).

58. For discussion of the shortcomings of traditional accounting measures, see Robert Repetto, *Wasting Assets: Natural Resources in the National Income Accounts* (Washington, D.C.: World Resources Institute, 1989); Herman E. Daly and John B. Cobb, Jr., *For the Common Good: Redirecting the Economy toward Community, the Environment, and a Sustainable Future* (Boston: Beacon Press, in press).

59. Herman E. Daly, "Sustainable Development: From Concept and Theory Towards Operational Principles," *Population and Development Review*, Spring 1990 (in press); Daly and Cobb, *For the Common Good*.

60. World military expenditures from U.S. Arms Control and Disarmament Agency, *World Military Expenditures and Arms Transfers 1988* (Washington, D.C.: U.S. Government Printing Office, 1989); abolition of Costa Rican army from Muni Figueres de Jimenez, acceptance speech, Better World Society Peace Advocacy and Arms Reduction Medal, New York, November 28, 1988.

61. Daly and Cobb, *For the Common Good*.

Index